Lecture Notes in Physics

The Lecture Notes in Physics

The series Lecture Notes in Physics (LNP), founded in 1969, reports new developments in physics research and teaching – quickly and informally, but with a high quality and the explicit aim to summarize and communicate current knowledge in an accessible way. Books published in this series are conceived as bridging material between advanced graduate textbooks and the forefront of research and to serve three purposes:

- to be a compact and modern up-to-date source of reference on a well-defined topic

- to serve as an accessible introduction to the field to postgraduate students and nonspecialist researchers from related areas

- to be a source of advanced teaching material for specialized seminars, courses and schools

Both monographs and multi-author volumes will be considered for publication. Edited volumes should, however, consist of a very limited number of contributions only. Proceedings will not be considered for LNP.

Volumes published in LNP are disseminated both in print and in electronic formats, the electronic archive being available at springerlink.com. The series content is indexed, abstracted and referenced by many abstracting and information services, bibliographic networks, subscription agencies, library networks, and consortia.

Proposals should be sent to a member of the Editorial Board, or directly to the managing editor at Springer:

Christian Caron
Springer Heidelberg
Physics Editorial Department I
Tiergartenstrasse 17
69121 Heidelberg / Germany
christian.caron@springer.com

S.N. Lyle

Self-Force and Inertia

Old Light on New Ideas

 Springer

Stephen N. Lyle
Andébu
09240 Alzen
France
steven.lyle@wanadoo.fr

Lyle S.N., *Self-Force and Inertia: Old Light on New Ideas*, Lect. Notes Phys. 796
(Springer, Berlin Heidelberg 2010), DOI 10.1007/978-3-642-04785-5

Lecture Notes in Physics ISSN 0075-8450   e-ISSN 1616-6361
ISBN 978-3-642-26225-8   e-ISBN 978-3-642-04785-5
DOI 10.1007/978-3-642-04785-5
Springer Heidelberg Dordrecht London New York

Cover design: Integra Software Services Pvt. Ltd., Pondicherry

Printed on acid-free paper

Springer is part of Springer Science+Business Media (www.springer.com)

To Samuel J. Lyle

Preface

Any student working with the celebrated Feynman Lectures will find a chapter in it with the intriguing title *Electromagnetic Mass* [2, Chap. 28]. In a way, it looks rather out of date, and it would be easy to skate over it, or even just skip it. And yet all bound state particles we know of today have electromagnetic mass. It is just that we approach the question differently. Today we have multiplets of mesons or baryons, and we have colour symmetry, and broken flavour symmetry, and we think about mass and energy through Hamiltonians. This book is an invitation to look at all these modern ideas with the help of an old light.

Everything here is quite standard theory, in fact, classical electromagnetism for the main part. The reader would be expected to have encountered the theory of electromagnetism before, but there is a review of all the necessary results, and nothing sophisticated about the calculations. The reader could be any student of physics, or any physicist, but someone who would like to know more about inertia, and the classical precursor of mass renormalisation in quantum field theory. In short, someone who feels it worthwhile to ask why $F = m\mathbf{a}$.

A spatially extended charge distribution will exert a force on itself if you try to accelerate it, and that force will lie along the direction of acceleration. This is not obvious. The classic case considered by the pioneers of relativity theory is a spherical shell of charge, but the calculations are not easy. In this book, I replace this by the simplest possible spatially extended charge distribution, viz., a dumbbell with an electrical charge at each end.

Some of the calculations may be original. Four cases are considered: velocity and acceleration parallel to one another and normal to the dumbbell axis, then parallel to the dumbbell axis; and then the two cases where the velocity and acceleration are normal to one another and the velocity is either normal to the system axis or parallel to it. Of course, the latter two cases involve rotation of the system about a center of rotation that is not located in the dumbbell. The calculations with standard classical electromagnetic theory are rather ugly and involve a certain level of approximation, so the contribution of this self-force to the inertia of the system would not appear to be a simple consequence of the theory.

Actually, any small charged object will exert this kind of inertial effect on itself whenever it is accelerated in any way relative to an inertial frame. Dirac showed this indirectly in 1938 by considering the electromagnetic energy–momentum tensor of such an object. He was on his way to devising the notorious Lorentz–Dirac equation of motion for accelerating charges, which takes into account their electromagnetic radiation. His calculations turned up a term proportional to the reciprocal of the spatial dimensions of the object. Or rather, since he was considering a point charge, the term in question was proportional to the reciprocal of the spatial dimensions of a worldtube he had constructed to contain the charge worldline, and whose radius he intended to set to zero at the end of the calculation. In this book, we note that, if the particle had had some spatial dimensions, he would have escaped the obvious problem that awaited him. But he was lucky. The thing that was about to go to infinity had exactly the right form to be absorbed into the mass times acceleration part of his equation, whereupon he could forget it. This is the miracle of renormalisation.

Like many other students of quantum field theory in the early 1980s, I found renormalisation mysterious. Disappointing by its messiness, but intriguing by its success. It seemed that one should be able to find a new quantum field theory that went straight to the right answers. Looking back at Dirac's problem, we find that the need for classical mass renormalisation can be avoided simply by denying point particles. But it is much simpler to say than to do! Once our particles have spatial extent, they become much more difficult to model. The intention here is not to hide that fact, nor to suggest that this hypothesis is a panacea for all ills. The point particle approximation has been extraordinarily successful. But one of the themes of this book is that we might understand physics better by knowing what can be done with spatially extended particles.

The electromagnetic bootstrap force described above would not be the only one affecting a spatially extended particle. If any of its components were sources for the strong force, for example, there would be a strong bootstrap force. Indeed, any interaction between its components would lead to a bootstrap effect. And what is more, this idea is actually a standard part of particle physics, although it appears in another guise, the one provided by relativistic quantum physics.

The reader should be quite clear that, although the basic ideas here are not new, and although there is not much about quantum theory in the book, this is not a denial of quantum theory, nor of any of the other wonderful hypotheses that make up modern physics. I include a chapter on elementary particle physics which reviews the state of the art with respect to inertial mass in a suitably simple way. I sketch the Higgs mechanism which is generally considered to cause the inertia of the truly elementary particles like quarks and leptons, but also the way we try to understand the inertial masses of the vast majority of particles, the mesons and baryons, today considered to be bound states of quarks and antiquarks, hence spatially extended.

Given the ubiquity of bootstrap contributions to inertia, there is an obvious possibility here, if the Higgs particle should continue to be elusive.

Acknowledgements

My first duty is to all those authors cited in this book, without whom it could not exist. Many thanks to Vesselin Petkov in Concordia, Montreal, for rekindling interest in this problem by pointing out the connection with the equivalence of inertial and passive gravitational mass, and Graham Nerlich in Adelaide for refusing to be bored stiff by the problem of rigidity. Also to Angela Lahee at Springer, who is still passionately concerned about books. And I must not forget the unerring, even increasing, support of Thérèse, Paul, and Martin.

Alzen, France   *Stephen N. Lyle*
November 2009

Contents

Acronyms

Here are some of the abbreviations used throughout the book:

EDM	energy-derived electromagnetic mass
EM	electromagnetic
FW	Fermi–Walker transport
GR	general relativity
GWS	Glashow–Weinberg–Salaam theory
HOS	hyperplane of simultaneity
ICIF	instantaneously comoving inertial frame
ICIO	instantaneously comoving inertial observer
LD	Lorentz–Dirac equation
MDM	momentum-derived electromagnetic mass
MEME	minimal extension of Maxwell's equations to curved spacetime
NG	Newtonian gravitation
QCD	quantum chromodynamics
QED	quantum electrodynamics
QM	quantum mechanics
SE	semi-Euclidean frame or coordinates
SEP	strong equivalence principle
SFDM	self-force-derived electromagnetic mass
SHGF	static homogeneous gravitational field
SO	Schwarzschild observer
SR	special relativity
WEP	weak equivalence principle

Chapter 1
Introduction and Guide

This book considers the dichotomy between point particle and spatially extended particle, and aims to show that interesting things can be discovered by treating the difficult problems posed by the latter. It is not an argument against the extraordinary successes of the point particle approach as an approximation, but a reminder that it is likely only to be an approximation.

Any spatially extended charge distribution will exert a force on itself when accelerated. In order to calculate this self-force, one needs to use Maxwell's theory of electromagnetism. The relevant results are reviewed in Chap. 2. But it would be difficult and unwise for someone with no knowledge of Maxwell's theory to take this as an introduction, even though it is fairly complete.

Chapter 3 goes straight to the heart of the matter. In fact it takes us back to the time when the pioneers of relativity theory were beginning to grasp the full implications of Maxwell's theory for our understanding of the world. They were trying to model the electron by a charged spherical shell, and discovered that it would oppose being accelerated, by exerting a bootstrap force. Its EM fields held energy and momentum, and this fact corroborated the idea that all its inertia might be due to the bootstrap effect. At the very least, part of its inertial mass could be attributed to its being a source of EM fields. And oddly, that contribution to its inertial mass had to increase as a function of its speed v according to a factor $\gamma(v) := (1 - v^2/c^2)^{-1/2}$, where c is the speed of light.

But the EM self-force of the charge shell does not only contribute to its inertia. There are other terms. If one expands the force as a power series in the radius a of the shell, the inertial term goes as $1/a$, but there are terms of order a^0, a, and higher. The term going as a^0, i.e., independent of the system dimensions, is what powers the EM radiation of any accelerated charge, while the higher order terms may be negligible if a is small. In Sect. 3.6, there is a brief discussion of Dirac's attempt to include the radiation reaction term in the Lorentz force law that describes the motion of a charged particle through EM fields.

Chapter 4 takes us momentarily into the higher realms of general relativity to discuss what this theory has to tell us about inertial mass. Very little knowledge of GR is actually needed, but some understanding of the founding experimental ob-

Lyle, S.N.: *Introduction and Guide*. Lect. Notes Phys. **796**, 1–3 (2010)
DOI 10.1007/978-3-642-04785-5_1

servation that the passive gravitational mass of a particle, measure of the extent to which it will be affected by gravity, is actually exactly equal to its inertial mass. The main result discussed is that any self-force contribution to a particle's inertial mass will automatically contribute equally to its passive gravitational mass. A simplification of Newton's second law is proposed to account for this. The idea that Einstein's equations for the curvature of spacetime might explain inertia is opposed, despite a well known result according to which, under certain other assumptions, some small particles may follow geodesics, because one of the assumptions is effectively identified as being the relativistic extension of Newton's second law.

This chapter ends with a discussion of the idea that particles somehow acquire inertial mass through the overall distribution of matter and energy throughout the Universe. It is argued that GR does not implement such an idea, and further that Brans and Dicke's adjustment of GR goes no further toward explaining inertia, because it still basically assumes a geodesic principle.

Having examined and rejected these alternative approaches to inertia, we return to the self-force. One of the problems with it is the difficulty in actually carrying out calculations. In this respect, the charge shell is not ideal for revealing the advantages of the idea. In Chaps. 5–9, we thus examine what must be the simplest possible extended charge distribution, namely a pair of point charges of the same sign separated by a distance d, using the standard results of Chap. 2. In a way, it may seem paradoxical to oppose the idea of point particles by replacing them with two point particles, but it does provide a clear way of demonstrating what can come out of a spatially extended charge distribution, and it makes certain calculations tractable that would be hopeless otherwise.

In Chap. 5, we consider the energy in the EM fields of this charge dumbbell when it is stationary in some inertial frame, and then the momentum in the EM fields when it moves with some constant velocity relative to that frame. In the latter case, we contrast motion along the system axis and normal to it. It has long been known that there is a discrepancy between the EM mass when it is derived from the energy and when it is derived from the momentum of the EM fields, and this feature is exposed in detail for later analysis. We contrast the cases where the charges at each end of the dumbbell have the same sign and different signs, the latter being used to model a neutral particle. Finally, we briefly discuss the way momentum enters the EM fields when such a system is accelerated.

Chapters 6–9 are concerned with the EM force the dumbbell exerts on itself when accelerated in various ways:

1. Linear acceleration perpendicular to the system axis.
2. Linear acceleration along the system axis.
3. Circular orbit with velocity perpendicular to the system axis.
4. Circular orbit with velocity along the system axis.

It is found that the self-force always opposes acceleration, even in cases where it is really not obvious that it will do so, e.g., when the dumbbell is orbiting a center of rotation in different ways. These calculations apply the Lienard–Wiechert solution to Maxwell's equations and are quite involved, but they are given in detail, since this may be the first time such results have been shown explicitly. The reader is encouraged to ask what feature of Maxwell's theory leads to this simple outcome.

The connection with classical mass renormalisation is spelt out in detail. In the case where the system is accelerated along a straight line normal to its axis, dealt with in Chap. 6, we also calculate the radiation reaction, and show the consistency with the standard formulas for radiated EM power. In Chap. 7, where the system accelerates along its axis, we discuss the problem of relativistic contraction, in preparation for later discussion. When any system has uniform motion relative to an inertial frame, we expect it to contract spatially in the direction of motion by a factor of γ^{-1}, but when it is accelerated, we need to think more carefully about dynamical effects within it. This subject is postponed to its own chapter, under the heading of rigidity. Chapter 10 takes stock of all the results regarding EM effects in charge shells and charge dumbbells.

Chapter 11 gives a detailed analysis of the discrepancy between energy-derived and momentum-derived EM masses, showing exactly why this should be expected to come about. The solution is simple enough, but there has been much debate about it, and the idea here is to reject a purely mathematical ploy for getting around the problem, showing that it is not well motivated physically, at least in the context where spatially extended particles are taken seriously. This chapter also contains a great deal of theory that should be useful in getting a better understanding of this kind of problem.

Chapter 12 faces one of the main complicating factors when dealing with spatially extended objects under acceleration, namely, the problem of their relativistic contraction. It is shown that the usual assumption made when handling charge shells is a rigidity assumption, and this is criticised. We also refer to a somewhat controversial paper by Bell [5], in which he tackles the question directly. The whole issue here is how best to approximate. Calculations with spatially extended objects are tough.

Chapter 13 reviews the way inertial mass is treated in particle physics today. Section 13.1 is particularly important. It describes a typical deduction of the rule of thumb $E = mc^2$, the boldest hypothesis of them all, showing that merely on the basis of Lorentz symmetry considerations, we then declare that all inertial mass is energy and all energy is inertial mass. The reader is encouraged to wonder why this works so well.

We then deal with bound states, from hydrogen, through positronium and quarkonium, to the mesons and baryons which are today treated as composed of quarks, antiquarks, and gluons. Several attempts are made to estimate the inertial masses of mesons and baryons without knowing the masses of their component quarks, or to estimate the masses of the quarks by measuring certain features of the composites they make up. The connection is made clear between self-forces and binding energy contributions to inertia. The penultimate section of this chapter then outlines the Higgs mechanism, generally expected to explain the masses of truly elementary particles, while the last section provides an example to show that the Higgs mechanism is not necessary to obtain electroweak unification through massive intermediate vector bosons.

Chapter 2
Some Notions of Electromagnetism

A lot of this book is about electromagnetism, carrying out calculations with rather standard bits of Maxwell's theory. This is why we begin with an overview of the main results used later. The aim is not to produce a textbook account, with complete proofs and explanations, only to exhibit the sequence of results a reader would need to know in order to follow the rest and fill in some of the more apposite details. To help things along, and make it easier for someone to pick up what is required, the overview given in this chapter is largely based on the account in *The Feynman Lectures on Physics*, by Feynman, Leighton, and Sands [1, 2]. The discussion in these books is remarkable anyway, and strongly recommended to anyone who does not know it.

2.1 Maxwell's Equations and Their Solution

We shall write Maxwell's equations in the form [2, Chap. 21]

$$\nabla \cdot \mathbf{E} = \frac{\rho}{\varepsilon_0}, \qquad \nabla \cdot \mathbf{B} = 0,$$

$$\nabla \times \mathbf{E} = -\frac{\partial \mathbf{B}}{\partial t}, \qquad c^2 \nabla \times \mathbf{B} = \frac{\mathbf{j}}{\varepsilon_0} + \frac{\partial \mathbf{E}}{\partial t}, \qquad (2.1)$$

where $\mathbf{E}$ is the electric field, $\mathbf{B}$ is the magnetic field, $\mathbf{j}$ is the current density, ρ is the charge density, c is the speed of light in vacuum, and ε_0 is a constant.

Because $\nabla \cdot \mathbf{B} = 0$ and $\nabla \times \mathbf{E} = -\partial \mathbf{B}/\partial t$, there is a 3-vector field $\mathbf{A}$ and a scalar field ϕ such that

$$\mathbf{E} = -\nabla \phi - \frac{\partial \mathbf{A}}{\partial t}, \qquad \mathbf{B} = \nabla \times \mathbf{A}. \qquad (2.2)$$

Lyle, S.N.: *Some Notions of Electromagnetism*. Lect. Notes Phys. **796**, 5–29 (2010)
DOI 10.1007/978-3-642-04785-5_2

ϕ and $\mathbf{A}$ are called the scalar and vector potential, respectively. Together they form the 4-potential $A^\mu = (\phi/c, \mathbf{A})$, a 4-vector under Lorentz transformations (see below) [2, Chap. 25].

When $\mathbf{E}$ and $\mathbf{B}$ have the form given in (2.2), the two Maxwell equations that do not mention the sources ρ and $\mathbf{j}$ are automatically satisfied. Then it turns out that, with one proviso, the other two Maxwell equations take the form

$$\Box^2 \phi = \frac{\rho}{\varepsilon_0}, \qquad \Box^2 \mathbf{A} = \frac{\mathbf{j}}{\varepsilon_0 c^2}, \tag{2.3}$$

where $\Box^2$ is the differential operator

$$\Box^2 := \frac{1}{c^2} \frac{\partial^2}{\partial t^2} - \nabla^2. \tag{2.4}$$

The proviso here is that one must put a constraint on ϕ and $\mathbf{A}$, viz.,

$$\nabla \cdot \mathbf{A} = -\frac{1}{c^2} \frac{\partial \phi}{\partial t}. \tag{2.5}$$

This is possible because several choices of ϕ and $\mathbf{A}$ actually lead to (2.2), something known as gauge freedom. Then (2.5) is called a gauge condition, in fact the Lorenz gauge condition. It constrains the potentials somewhat, but not totally. There remains some gauge freedom, i.e., one can simultaneously change $\mathbf{A}$ to $\mathbf{A}' := \mathbf{A} + \nabla \psi$ and ϕ to $\phi' := \phi - \partial \psi / \partial t$ and obtain the same $\mathbf{E}$ and $\mathbf{B}$ from (2.2), and even maintain the condition (2.5) applied now to $\mathbf{A}'$ and ϕ', provided only that ψ satisfies $\Box \psi' = 0$.

The great thing about the step from $\mathbf{E}$ and $\mathbf{B}$ to ϕ and $\mathbf{A}$ is that (2.3) and (2.5) can be solved by

$$\phi(\mathbf{r}_0, t_0) = \int \frac{\rho(\mathbf{r}_1, t_0 - r_{01}/c)}{4\pi \varepsilon_0 r_{01}} dV_1, \qquad \mathbf{A}(\mathbf{r}_0, t_0) = \int \frac{\mathbf{j}(\mathbf{r}_1, t_0 - r_{01}/c)}{4\pi \varepsilon_0 c^2 r_{01}} dV_1, \tag{2.6}$$

where $(\mathbf{r}_0, t_0)$ is the field point, i.e., the event of spacetime at which we are evaluating the fields, while the dummy variable $\mathbf{r}_1$ in the integral ranges over all values in $\mathbb{R}^3$, and $\mathbf{r}_{01} := \mathbf{r}_0 - \mathbf{r}_1$ is the vector from the volume element dV_1 to the field point, with length r_{01}.

The time $t_+ := t_0 - r_{01}/c$ is called the retarded time, somewhat vaguely. It is the time when there would have had to have been some charge at the integration point $\mathbf{r}_1$ in order for something at the field point to be affected by a field due to that charge at that integration point. The little diagram in Fig. 2.1 is designed to illustrate this trivial point, but anyone with any doubts about this is advised to go through the account in [2] in detail, or consult some other elementary textbook.

One also speaks of the retarded point $\mathbf{r}_+$ for a given field point and a given charge element. In terms appropriate to the special theory of relativity (SR), the retarded point is the intersection of the past light cone of the field point with the worldline of the charge element. It is the spatial position of the charge element when it produced

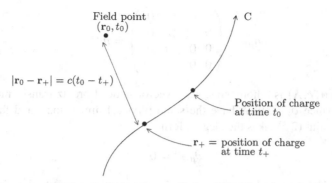

Fig. 2.1 A picture to show why the retarded time $t_+ := t_0 - r_{01}/c$ turns up in the solution to Maxwell's equations. The curve denotes the spatial trajectory of a charge element C. The latter would have had to have been at the retarded point $\mathbf{r}_+ = \mathbf{r}_1$ to affect the field point $(\mathbf{r}_0, t_0)$, because the distance $r_{01} := |\mathbf{r}_0 - \mathbf{r}_1|$ is precisely the distance that light would cover in the time $t_0 - t_+$

the fields that affect the field point. For given field point and given charge element, the retarded point is unique.

Note in passing that charge is conserved by any system satisfying Maxwell's equations, because they imply

$$\nabla \cdot \mathbf{j} = -\frac{\partial \rho}{\partial t} \,. \tag{2.7}$$

In integral form, this states that the flux of current out through any closed surface is equal to minus the rate of change of the amount of charge within the surface. Together the charge and current densities form the 4-current density $j^\mu = (c\rho, \mathbf{j})$, a 4-vector under Lorentz transformations (see below) [2, Chap. 18].

Note that (2.6) is a very general solution to Maxwell's equations (2.3), for any charge distribution with any motion. There are other solutions involving advanced times, rather than retarded times, and combinations of both, but in the present discussion, we shall stick with the retarded solutions in (2.6) and the old-fashioned notion of causality. One then obtains the electric and magnetic fields from (2.2).

2.2 Relativistic Notation

All this can be phrased in terms of four-vectors and other tensors. If we consider homogeneous coordinates, that is, each with the same physical dimensions,

$$x^\mu = (ct, x, y, z) \,, \qquad \partial_\mu = \frac{\partial}{\partial x^\mu} = \left(\frac{\partial}{\partial (ct)}, \frac{\partial}{\partial x}, \frac{\partial}{\partial y}, \frac{\partial}{\partial z} \right) \,,$$

and the Minkowski metric in the form

$$\eta_{\mu\nu} = \begin{pmatrix} 1 & 0 & 0 & 0 \\ 0 & -1 & 0 & 0 \\ 0 & 0 & -1 & 0 \\ 0 & 0 & 0 & -1 \end{pmatrix} = \eta^{\mu\nu} \, ,$$

then $A^\mu = (\phi/c, \mathbf{A})$ is a homogeneous 4-vector under Lorentz transformations [2, Chap. 25], since ϕ/c and $\mathbf{A}$ have the same physical dimensions, and the Lorenz gauge constraint (2.5) takes the elegant form

$$\partial_\mu A^\mu = 0 \, . \tag{2.8}$$

The charge density ρ and current density $\mathbf{j}$ form another homogeneous four-vector $j^\mu = (c\rho, \mathbf{j})$ [2, Chap. 18], and charge conservation (2.7) assumes the equally elegant form

$$\partial_\mu j^\mu = 0 \, . \tag{2.9}$$

The differential operator $\Box^2$ in (2.4) has the form $\Box^2 = \partial_\mu \partial^\mu$, and Maxwell's equations (2.3) for the potential take the form

$$\Box^2 A^\mu = \frac{j^\mu}{\varepsilon_0 c^2} \, . \tag{2.10}$$

Furthermore, if we define

$$F_{\mu\nu} = \partial_\mu A_\nu - \partial_\nu A_\mu \, , \tag{2.11}$$

where $A_\mu := \eta_{\mu\nu} A^\nu$, then (2.2) implies that

$$F_{\mu\nu} = \begin{pmatrix} 0 & E_x/c & E_y/c & E_z/c \\ -E_x/c & 0 & -B_z & B_y \\ -E_y/c & B_z & 0 & -B_x \\ -E_z/c & -B_y & B_x & 0 \end{pmatrix} \, . \tag{2.12}$$

This is also physically homogeneous in the sense that each component of the tensor has the same physical dimensions. The contravariant version is

$$F^{\mu\nu} := \eta^{\mu\sigma} \eta^{\nu\tau} F_{\sigma\tau} = \begin{pmatrix} 0 & -E_x/c & -E_y/c & -E_z/c \\ E_x/c & 0 & -B_z & B_y \\ E_y/c & B_z & 0 & -B_x \\ E_z/c & -B_y & B_x & 0 \end{pmatrix} \, . \tag{2.13}$$

The two Maxwell equations among (2.1) that do not mention the sources ρ and $\mathbf{j}$ take the form

$$F_{\mu\nu,\sigma} + F_{\sigma\mu,\nu} + F_{\nu\sigma,\mu} = 0 \, . \tag{2.14}$$

This can also be written

$$F_{[\mu\nu,\sigma]} = 0 \,, \tag{2.15}$$

where the square brackets on indices indicate antisymmetrisation, whence the latter says that

$$F_{\mu\nu,\sigma} + F_{\sigma\mu,\nu} + F_{\nu\sigma,\mu} - F_{\nu\mu,\sigma} - F_{\mu\sigma,\nu} - F_{\sigma\nu,\mu} = 0 \,,$$

and, since $F_{\mu\nu}$ is antisymmetric, this is just the same as (2.14). The two Maxwell equations mentioning ρ and $\mathbf{j}$ take the form

$$F^{\mu\nu}{}_{,\nu} = -\frac{j^{\mu}}{\varepsilon_0 c^2} \,. \tag{2.16}$$

Note. The reader should be warned that this is only one set of conventions for writing everything in relativistic notation. The Minkowski metric is sometimes the negative of the form shown here, and it contains a factor of c^2 if non-homogeneous coordinates (t, x, y, z) are chosen. Unfortunately, one encounters many variants, so it is better to just get used to that.

2.3 Lorentz Force Law

In a certain sense, the Lorentz force law tells us what the electric and magnetic fields actually do out in the real world when there is a charge there to probe them, because they give the force that those fields will exert on the charge, viz.,

$$\mathbf{F} = q(\mathbf{E} + \mathbf{u} \times \mathbf{B}) \,,$$

where q is the value of the charge in coulombs and $\mathbf{u}$ is its 3-velocity. This is a non-relativistic version. We would like to look here at the Lorentz force law in its special relativistic formulation. We need the 4-velocity

$$v^{\mu} := \frac{dx^{\mu}}{d\tau} = \frac{dt}{d\tau}(c, \mathbf{u}) \,, \tag{2.17}$$

where the coordinate 3-velocity $\mathbf{u}$ in the given inertial frame is defined by

$$\mathbf{u} := \left(\frac{dx}{dt}, \frac{dy}{dt}, \frac{dz}{dt} \right) \,, \tag{2.18}$$

and the proper time τ by

$$c^2 d\tau^2 = c^2 dt^2 - dx^2 - dy^2 - dz^2 \,, \tag{2.19}$$

whence

$$\left(\frac{d\tau}{dt}\right)^2 = 1 - \frac{u^2}{c^2}, \quad \frac{dt}{d\tau} = \gamma. \qquad (2.20)$$

defining $\gamma(u)$ in the usual way. Hence, for the 4-velocity,

$$v^\mu = \gamma(c, \mathbf{u}). \qquad (2.21)$$

In special relativity, the Lorentz force law takes the form

$$\boxed{m_0 \frac{d^2 x^\mu}{d\tau^2} = q F^\mu{}_\nu v^\nu} \qquad (2.22)$$

where m_0 and q are the particle mass and charge, respectively.

Let us see what these four equations tell us in terms of $\mathbf{E}$ and $\mathbf{B}$. The left-hand side is

$$m_0 \frac{d^2 x^\mu}{d\tau^2} = m_0 \frac{dt}{d\tau} \frac{dv^\mu}{dt} = m_0 \gamma \frac{dv^\mu}{dt}. \qquad (2.23)$$

On the right-hand side, we have

$$F^\mu{}_\nu v^\nu = \gamma \begin{pmatrix} \mathbf{E} \cdot \mathbf{u}/c \\ \mathbf{E} + \mathbf{u} \times \mathbf{B} \end{pmatrix}. \qquad (2.24)$$

Now the Lorentz force equation consists of one vector equation (components $i = 1, 2, 3$) and one scalar equation (the 0 component), but in fact the scalar equation follows from the vector equation as we shall see in a moment. The vector equation states that

$$\boxed{m_0 \frac{d}{dt}(\gamma \mathbf{u}) = q(\mathbf{E} + \mathbf{u} \times \mathbf{B})} \qquad (2.25)$$

This is the usual version of the relativistic Lorentz force law [2, Sect. 26.4]. It can be written more specifically in the form

$$\frac{d}{dt}\left[\frac{m_0}{(1 - u^2/c^2)^{1/2}} \mathbf{u}\right] = q(\mathbf{E} + \mathbf{u} \times \mathbf{B}). \qquad (2.26)$$

The left-hand side is the coordinate time rate of change of the relativistic 3-momentum (which has to multiplied by γ to give the spatial components of the 4-force).

Now in special relativity, the 4-force and 4-velocity are not independent. In fact, they are orthogonal in the Minkowski geometry. Let us see how this relationship comes out when the 4-force is equated with $q F^\mu{}_\nu v^\nu$. The scalar equation requires

$$m_0 c \frac{d\gamma}{dt} = q\mathbf{E} \cdot \mathbf{u}/c. \qquad (2.27)$$

But the vector equation already implies that

$$qE \cdot u = m_0 u \cdot \frac{d}{dt}(\gamma u) \,. \tag{2.28}$$

This is because the magnetic field does no work, i.e., the magnetic force $u \times B$ is orthogonal to the velocity u (in 3-space). This in turn means that

$$qE \cdot u = m_0 u^2 \frac{d\gamma}{dt} + \gamma m_0 u \cdot \dot{u} \,. \tag{2.29}$$

The right-hand side can be simplified here using

$$\frac{d\gamma}{dt} = \frac{u \cdot \dot{u}}{c^2} \gamma^3 \,, \tag{2.30}$$

whence the vector equation in the form (2.29) does indeed imply that

$$qE \cdot u = m_0 c^2 \frac{d\gamma}{dt} \,.$$

In fact, we have the following identity:

$$u^2 \frac{d\gamma}{dt} + \gamma u \cdot \dot{u} = c^2 \frac{d\gamma}{dt} \,, \tag{2.31}$$

which is precisely the relation which says that the 4-force and 4-velocity are orthogonal (a completely general result).

The point of the last short discussion is just to make it clear that the whole content of the relativistic relation (2.22) is expressed by the 3-vector relation (2.25).

2.4 Electromagnetic Energy–Momentum Tensor

The definition we shall use is

$$T^{\mu\nu} = -\varepsilon_0 c^2 \left(F^\mu{}_\sigma F^{\sigma\nu} + \frac{1}{4} F_{\sigma\tau} F^{\sigma\tau} \eta^{\mu\nu} \right) \,, \tag{2.32}$$

with the above definition for $F_{\mu\nu}$. If we work out the components of $T^{\mu\nu}$ in terms of the electric and magnetic fields, we obtain

$$T^{\mu\nu} = \varepsilon_0 \begin{pmatrix} -\frac{1}{2}(E^2 + c^2 B^2) & -cE \times B \\ -cE \times B & EE + c^2 BB - \frac{1}{2}(E^2 + c^2 B^2)\mathbb{I} \end{pmatrix} \,. \tag{2.33}$$

The 3×3 matrix in the bottom right is

$$\mathsf{T} := \varepsilon_0 \left[\mathbf{EE} + c^2 \mathbf{BB} - \frac{1}{2}(E^2 + c^2 B^2)\mathbb{I} \right] , \tag{2.34}$$

where

$$\mathbf{EE} := \begin{pmatrix} E_x \\ E_y \\ E_z \end{pmatrix} \begin{pmatrix} E_x & E_y & E_z \end{pmatrix} = \begin{pmatrix} E_x^2 & E_x E_y & E_x E_z \\ E_y E_x & E_y^2 & E_y E_z \\ E_z E_x & E_z E_y & E_z^2 \end{pmatrix} , \tag{2.35}$$

and similarly for $\mathbf{BB}$.

We can now bring in the energy density u and its flow $\mathbf{S}$ with the same conventions as Feynman in [2, Sect. 27.5]. Hence,

$$u = \frac{\varepsilon_0}{2} \mathbf{E} \cdot \mathbf{E} + \frac{\varepsilon_0 c^2}{2} \mathbf{B} \cdot \mathbf{B} = \frac{\varepsilon_0}{2}(E^2 + c^2 B^2) \tag{2.36}$$

and

$$\mathbf{S} := \varepsilon_0 c^2 \mathbf{E} \times \mathbf{B} . \tag{2.37}$$

We now have

$$T^{00} = -u , \qquad T^{0k} = -\frac{1}{c} S^k . \tag{2.38}$$

The vector quantity $\mathbf{S}$ is the energy flux of the field, i.e., the flow of energy per unit time across a unit area perpendicular to the flow. It is known as the Poynting vector. One can obtain the field momentum density (momentum per unit volume) from it in the form [2, Sect. 27.6]

$$\mathbf{g} := \frac{1}{c^2} \mathbf{S} = \varepsilon_0 \mathbf{E} \times \mathbf{B} , \tag{2.39}$$

a formula that will be put to use later.

The above quantities u and $\mathbf{S}$ are chosen because they satisfy the constraint of energy conservation, viz.,

$$\frac{\partial u}{\partial t} = -\nabla \cdot \mathbf{S} - \mathbf{E} \cdot \mathbf{j} ,$$

which expresses the idea that the total field energy in a given volume decreases either because field energy flows out of the volume (the term $\nabla \cdot \mathbf{S}$) or because it loses energy to matter by doing work on it (the term $\mathbf{E} \cdot \mathbf{j}$). In relativistic language, this becomes $T^{0\nu}{}_{,\nu} = F^{0\nu} j_\nu$.

It is important to see that the field does work on each unit volume of matter at the rate $\mathbf{E} \cdot \mathbf{j}$. The force on a particle is $\mathbf{F} = q(\mathbf{E} + \mathbf{v} \times \mathbf{B})$, and the rate of doing work is $\mathbf{F} \cdot \mathbf{v} = q\mathbf{E} \cdot \mathbf{v}$. If there are N particles per unit volume, the rate of doing work per unit volume is $Nq\mathbf{E} \cdot \mathbf{v}$, and $Nq\mathbf{v}$ is of course what we have called $\mathbf{j}$. The quantity $\mathbf{E} \cdot \mathbf{j}$ is also called the Lorentz force density.

With the definition (2.32) for $T^{\mu\nu}$, Maxwell's equations in the relativistic form (2.14) and (2.16) in fact imply

$$\boxed{T^{\mu\nu}{}_{,\nu} = F^{\mu\nu} j_\nu} \tag{2.40}$$

It is instructive to see this derivation. We begin with Maxwell's equations in the form

$$F^{\mu\nu}{}_{,\nu} = -\frac{j^\mu}{\varepsilon_0 c^2}, \qquad F_{[\mu\nu,\sigma]} = 0. \tag{2.41}$$

Now, from

$$T^{\mu\nu} = -\varepsilon_0 c^2 \left(F^\mu{}_\sigma F^{\sigma\nu} + \frac{1}{4} F_{\sigma\tau} F^{\sigma\tau} \eta^{\mu\nu} \right), \tag{2.42}$$

we have

$$T^{\mu\nu}{}_{,\nu} = -\varepsilon_0 c^2 \left(F^\mu{}_{\sigma,\nu} F^{\sigma\nu} + F^\mu{}_\sigma F^{\sigma\nu}{}_{,\nu} + \frac{1}{4} F_{\sigma\tau,\nu} F^{\sigma\tau} \eta^{\mu\nu} + \frac{1}{4} F_{\sigma\tau} F^{\sigma\tau}{}_{,\nu} \eta^{\mu\nu} \right). \tag{2.43}$$

Now note that, using the Maxwell equation in (2.41) that refers to the sources, the second term on the right-hand side is

$$\text{second term} = -\varepsilon_0 c^2 F^\mu{}_\sigma F^{\sigma\nu}{}_{,\nu} = F^\mu{}_\sigma j^\sigma, \tag{2.44}$$

and this is the 4-force density. One can then show that the other terms in (2.43) sum to zero, i.e.,

$$F^\mu{}_{\sigma,\nu} F^{\sigma\nu} + \frac{1}{4} F_{\sigma\tau,\nu} F^{\sigma\tau} \eta^{\mu\nu} + \frac{1}{4} F_{\sigma\tau} F^{\sigma\tau}{}_{,\nu} \eta^{\mu\nu} = 0,$$

using the antisymmetry of $F^{\sigma\tau}$ and the source-free Maxwell equations in (2.41). Our conclusion here is therefore

$$T^{\mu\nu}{}_{,\nu} = F^{\mu\nu} j_\nu, \tag{2.45}$$

as claimed. In words, Maxwell's equations ensure that the divergence $T^{\mu\nu}{}_{,\nu}$ of the electromagnetic energy–momentum tensor is equal to the electromagnetic force density on the charge distribution according to the Lorentz force law. We conclude that the electromagnetic energy–momentum tensor is conserved at a point if and only if the electromagnetic force density on the charges there is zero. One situation where this happens is if there are no charges!

Consider for a moment how this analysis would fit in with a charged dust model [8, pp. 104–118], where one would have a total energy–momentum tensor of the form

$$T_{\text{charged dust}} = \frac{m}{q} \rho u \otimes u + T_{\text{em}} ,$$

where u is the 4-velocity field of the dust, m and q are the mass and charge of each dust particle, respectively, and ρ is the proper charge density distribution. When the mass part of the tensor (proportional to $u \otimes u$ here) is included, conservation of the total tensor says two separate things. The component of the resulting equation parallel to the 4-velocity field u gives conservation of mass, whilst the component orthogonal to the 4-velocity field says that charges follow non-geodesic curves as given by the Lorentz force law.

Of course, in most situations, other forces will be involved, i.e., the distribution could not be treated as a charged dust. However, considering this model for a moment, let us just ask what has changed by adding the mass term. In fact, when we obtained (2.45) above, we concluded that the electromagnetic energy–momentum tensor alone is conserved at a point if and only if the electromagnetic force density on the charges there is zero. Of course, this never happens unless there are no charges there! In that case, we can define a field energy density and momentum flow, and interpret the resulting equation. When there are charges, we have to equate the right-hand side of (2.45) with the missing bit of the global conservation equation, including the mass terms, and this delivers the Lorentz force law dictating how the charged dust must flow in the given fields.

As a final point regarding the quantities u and $\mathbf{S}$, it is important to note that there is some ambiguity here, because there are other definitions that lead to the right relations. However, they are more complex [2, Sect. 27.4]. There is also some strangeness in the way energy is conserved. A good discussion and examples can be found in [2, Chap. 27].

2.5 Solution for Point Charge with Arbitrary Motion

One case that interests us in this book is a point charge with arbitrary motion. The notion of point charge is clearly idealistic and one of the themes here is that there are no point particles. However, this does not mean that the point charge is not a useful mathematical approximation. It is in this sense that it is presented here.

2.5.1 Fields Due to a Single Point Charge

Four-Current Vector

The first step is to obtain the 4-vector current describing such a source. As usual in the relativistic context, we begin by arbitrarily choosing some inertial frame to describe things in. The trajectory of the source is described in Minkowski spacetime by $x(\tau)$, where τ is the proper time. The trajectory can also be parametrised by

$t = x^0(\tau)/c$. Then the 4-vector current density is

$$j^{\mu}(\mathbf{y}, t) = q \frac{dx^{\mu}}{dt} \delta^3 (\mathbf{y} - \mathbf{x}(\tau)) \Big|_{ct=x^0(\tau)} , \qquad (2.46)$$

where δ^3 is the 3D Dirac delta distribution.

It is worth being quite clear about this formula. In a simplistic model of a charge distribution,

$$j(y) = \rho_0(y)V(y) , \qquad (2.47)$$

where

$$V(y) := \frac{dx}{d\tau} = \gamma(v)(c, \mathbf{v}) , \quad v := |\mathbf{v}| , \quad \gamma(v) := (1 - v^2/c^2)^{-1/2} ,$$

is the 4-velocity field of the charge, with τ the proper time of the charge, and $\rho_0(y)$ is the local charge density as measured at each point in an instantaneous rest frame of the charge. Now $dt = \gamma(v)d\tau$, so in the simplistic model (2.47),

$$j^{\mu}(y) = \rho_0 \gamma(v) \frac{dx^{\mu}}{dt} .$$

But $\rho_0 \gamma(v) = \rho$, the local charge density as measured from our chosen inertial frame, so our formula (2.46) will be right if the term

$$q \delta^3 (\mathbf{y} - \mathbf{x}(\tau)) \Big|_{ct=x^0(\tau)} \qquad (2.48)$$

corresponds to the charge density distribution as measured in that inertial frame. Notice that the condition $ct = x^0(\tau)$ says what τ is supposed to be, given t, and not the other way round. Then the delta function is a point distribution located at the space point $\mathbf{x}(\tau)$, the point of the trajectory corresponding to the given t, so it is intuitively reasonable to suggest that (2.48) is good model for the charge density distribution of our point charge with worldline $x(\tau)$.

Note that the distribution (2.46) can be rewritten in covariant form:

$$j^{\mu}(\mathbf{y}, t) = q \int d\tau \frac{dx^{\mu}}{d\tau} \delta^4 (y - x(\tau)) , \qquad (2.49)$$

as can be seen by changing the variable to t and doing the integration over this variable.

Since the charge density distribution in space is supposed to be

$$\rho = \frac{j^0(\mathbf{y}, t)}{c} = q \delta^3 (\mathbf{y} - \mathbf{x}(\tau)) ,$$

at any instant of time t, the total charge at that time is

$$\int d^3y \, j^0(\mathbf{y}, t) = q \, ,$$

which is constant in time. We ought to check that $\partial_\mu j^\mu = 0$, as required by (2.9). We need to see why the flux of current out of any closed surface is equal to the rate of change of charge contained within it. The current density is

$$q\mathbf{v}\big(\mathbf{x}(\tau)\big) \, \delta^3\big(\mathbf{y} - \mathbf{x}(\tau)\big)\Big|_{ct=x^0(\tau)} = \mathbf{v}\rho \, ,$$

and this is intuitively sufficient.

Potential Equation

The equation for the 4-vector potential is (2.10), viz.,

$$\Box A^\mu = \frac{j^\mu}{\varepsilon_0 c^2} \, ,$$

and we will maintain the Lorenz gauge, so that

$$A^\mu{}_{,\mu} = 0 \, .$$

We use the retarded Green function

$$G_{\text{ret}}(x) = \frac{c}{2\pi} \theta(+x^0) \delta(x^2)$$

to solve for A^μ, where θ is the step function, equal to zero for negative values of the argument and $+1$ for positive values. This has the property that

$$\Box^2 G_{\text{ret}} = \delta^4(x) \, .$$

The solution for the 4-potential is thus

$$A^\mu(y) = \frac{1}{2\pi\varepsilon_0 c} \int d^4z \, \theta(y^0 - z^0) \delta\big((y-z)^2\big) j^\mu(z) \, , \tag{2.50}$$

by the standard property of Green functions. We should check that this satisfies the Lorenz gauge condition. Taking the relativistic divergence, the operator ∂_μ can be transferred to the 4-current inside the integral, and this time charge conservation implies the Lorenz gauge condition.

Then, inserting our 4-current density (2.49) into (2.50) and carrying out the integral over z using the 4D delta function from (2.49), we obtain

$$A^\mu(y) = \frac{q}{2\pi\varepsilon_0 c} \int_{-\infty}^{\infty} d\tau \, \theta\big(y^0 - x^0(\tau)\big) \delta\Big(\big[y - x(\tau)\big]^2\Big) \dot{x}^\mu(\tau) \, , \tag{2.51}$$

where the dot on x^μ indicates the proper time derivative. This formula clearly exhibits the fact that our A^μ depends only on the worldline at earlier times.

Retarded Points

As mentioned earlier, for all y, there exists a unique $x_+ = x(\tau_+)$ on the trajectory of
the charged particle, such that

$$(y - x_+)^2 = 0, \qquad x_+^0 < y^0 .$$

This point is referred to as the retarded point. In relativistic jargon, it is the intersection of the worldline with the past null cone through y (see Fig. 2.2).

Let us understand the significance of these points via the delta function approach.
Consider first a standard piece of distribution theory. Suppose we have

$$I = \int d\tau h(\tau)\, \delta\big(g(\tau)\big) ,$$

where $g(\tau) = \big[y - x(\tau)\big]^2$ only has the one zero for $y^0 > x^0(\tau)$, namely at $\tau = \tau_+$.
Expanding out

$$g(\tau) = g(\tau_+ + \tau - \tau_+) \approx g(\tau_+) + (\tau - \tau_+)g'(\tau_+) ,$$

the first term is zero and the second is supposed small (for τ close to τ_+). Consequently,

$$I = \int d\tau h(\tau)\, \delta\big((\tau - \tau_+)g'(\tau_+)\big) ,$$

and we can now use

$$g'(\tau) = 2\big[y - x(\tau)\big] \cdot \dot{x}(\tau)$$

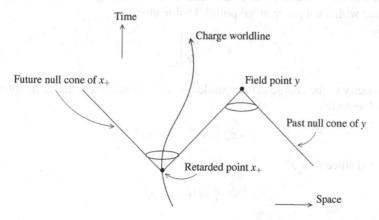

Fig. 2.2 A spacetime picture of the retarded point, to be contrasted with the space picture in
Fig. 2.1

to deduce

$$I = \frac{h(\tau_+)}{|g'(\tau_+)|} \, .$$

The modulus comes in from the variable change

$$\tau' := g'(\tau_+)\tau \, .$$

However, it is unnecessary, because it can be shown that

$$\dot{x}_+ \cdot (y - x_+) > 0 \, ,$$

by observing that $\dot{x}_+$ is timelike, with $\dot{x}_+^0 > 0$, and $y - x_+$ is null, with $y^0 - x_+^0 > 0$.

Lienard–Wiechert Retarded Potential

The above study of retarded points can be applied to the potential (2.51) we obtained earlier. The result is the Lienard–Wiechert retarded potential

$$A_{\text{ret}}^{\mu}(y) = \frac{q}{4\pi\varepsilon_0 c} \frac{\dot{x}_+^{\mu}}{\dot{x}_+ \cdot (y - x_+)} \, . \tag{2.52}$$

This is the relativistic generalisation of the Coulomb potential. The above derivation displays the power of the distribution approach using the step function and Dirac delta. A longer but much more physical derivation can be found in [2, Chap. 21], and is strongly recommended.

We would like to write this in more familiar terms. Define $\mathbf{r}_+ := \mathbf{y} - \mathbf{x}_+$, the vector from the retarded point to the field point (recalling that the field point is associated with a unique retarded point). Define also

$$\mathbf{v}_+ = \left. \frac{d\mathbf{x}}{dt} \right|_{\tau_+} ,$$

the 3-velocity of the charge at the retarded point associated with the field point.

We observe that

$$y - x_+ = (y^0 - x_+^0, \, \mathbf{r}_+)$$

is null, and since $y^0 > x_+^0$,

$$y^0 - x_+^0 = |\mathbf{r}_+| = r_+ \, .$$

Furthermore,

$$\dot{x}_+ = \left.\frac{dx}{d\tau}\right|_{\tau_+} = \left.\frac{dx}{dt}\right|_{\tau_+} \frac{dt}{d\tau} \,,$$

and the derivative $dt/d\tau$ will cancel in the ratio of terms.

Finally, we obtain the equations

$$A_{\text{ret}}^0(y) = \frac{q}{4\pi\varepsilon_0 c(r_+ - \mathbf{r}_+\cdot\mathbf{v}_+/c)} \,, \qquad \mathbf{A}_{\text{ret}}(y) = \frac{q\mathbf{v}_+}{4\pi\varepsilon_0 c^2(r_+ - \mathbf{r}_+\cdot\mathbf{v}_+/c)} \qquad (2.53)$$

Retrieving the Coulomb Potential

Choose the frame in which

$$\dot{x}_+^\mu = (1, 0, 0, 0) \,,$$

namely, the instantaneous rest frame of the source charge when it was at the retarded point. Then $\mathbf{v}_+ = 0$. In this frame,

$$\phi_{\text{ret}} = cA_{\text{ret}}^0(y) = \frac{q}{4\pi\varepsilon_0 r_+} \,, \qquad \mathbf{A}_{\text{ret}}(y) = 0 \,,$$

where r_+ is the distance between the test point y and the retarded point related to y. Notice that the velocity of our frame depends on the field point, because it is equal to the velocity of the charge at the retarded point for that field point. We cannot say that all effects due to acceleration of the source charge disappear everywhere in space in this frame, not even instantaneously, but we can say that the scalar potential ϕ goes as $1/r_+$, and not as the reciprocal of the distance to the point where the charge is now. (The word 'now' refers to simultaneity in whatever inertial frame we have selected at the outset.)

Electromagnetic Fields

The electromagnetic fields due to the charge can now be calculated, using the relations (2.2) on p. 5, viz.,

$$\mathbf{E} = -c\nabla A^0 - \frac{\partial \mathbf{A}}{\partial t} \,, \qquad \mathbf{B} = \nabla \times \mathbf{A} \,.$$

This deduction will be sketched here because heavy use will be made of the resulting formulas for $\mathbf{E}$ and $\mathbf{B}$ later on, and because the kind of calculation arising here illustrates something about the physics.

First for the notation, we drop the subscript indicating that we are talking about the retarded solutions, and stick with the notation $y^\mu = (y^0, \mathbf{y})$ for the (homogeneous) coordinates of the field point and $x^\mu(t_+)$ for the event coinciding with the charge at the appropriate retarded time t_+ for the given field point. The retarded displacement vector from the charge at the appropriate retarded time to the field point is then denoted by $\mathbf{r}_+ := \mathbf{y} - \mathbf{x}(t_+)$, as in the discussion of the Lienard–Wiechert potential above.

This immediately reminds us that the key thing about t_+, or indeed $\mathbf{x}_+ := \mathbf{x}(t_+)$, $\mathbf{r}_+$, and $\mathbf{v}_+$, is that they are functions of the field point. If we are to carry out the derivatives in

$$\frac{4\pi\varepsilon_0}{q} \mathbf{E} = -\nabla \frac{1}{r_+ - \mathbf{r}_+\cdot\mathbf{v}_+/c} - \frac{\partial}{\partial y^0} \frac{\mathbf{v}_+}{c(r_+ - \mathbf{r}_+\cdot\mathbf{v}_+/c)} , \qquad (2.54)$$

then first of all, we need to find the partial derivatives of t_+ with respect to y^μ. This is the point about the physics just mentioned: everything here hinges on the retarded time.

So the key relation here is the one defining the retarded time for the given field point, viz.,

$$r_+ = |\mathbf{y} - \mathbf{x}(t_+)| = y^0 - ct_+ ,$$

which can also be written explicitly in the form

$$\left[y^1 - x^1(t_+)\right]^2 + \left[y^2 - x^2(t_+)\right]^2 + \left[y^3 - x^3(t_+)\right]^2 = (y^0 - ct_+)^2 . \qquad (2.55)$$

Taking partial derivatives of this relation with respect to y^μ, we soon arrive at

$$\frac{\partial t_+}{\partial y^0} = \frac{r_+}{c(r_+ - \mathbf{r}_+\cdot\mathbf{v}_+/c)} , \qquad \nabla t_+ = -\frac{\mathbf{r}_+}{c(r_+ - \mathbf{r}_+\cdot\mathbf{v}_+/c)} . \qquad (2.56)$$

It is then a simple matter to obtain

$$\frac{\partial \mathbf{r}_+}{\partial y^0} = -\frac{r_+\mathbf{v}_+}{c(r_+ - \mathbf{r}_+\cdot\mathbf{v}_+/c)} , \qquad \frac{\partial r_+}{\partial y^0} = -\frac{\mathbf{r}_+\cdot\mathbf{v}_+}{c(r_+ - \mathbf{r}_+\cdot\mathbf{v}_+/c)} , \qquad (2.57)$$

$$\frac{\partial r_+^j}{\partial y^i} = \delta_i^j + \frac{r_+^i v_+^j}{c(r_+ - \mathbf{r}_+\cdot\mathbf{v}_+/c)} , \qquad i,j \in \{1,2,3\} , \qquad (2.58)$$

$$\nabla r_+ = \frac{\mathbf{r}_+}{r_+ - \mathbf{r}_+\cdot\mathbf{v}_+/c} , \qquad (2.59)$$

and

$$\frac{\partial \mathbf{v}_+}{\partial y^0} = \frac{r_+\mathbf{a}_+}{c(r_+ - \mathbf{r}_+\cdot\mathbf{v}_+/c)} , \qquad \frac{\partial v_+^j}{\partial y^i} = -\frac{r_+^i a_+^j}{c(r_+ - \mathbf{r}_+\cdot\mathbf{v}_+/c)} . \qquad (2.60)$$

Now with a few applications of the Leibniz and chain rules and a certain amount of book-keeping, (2.54) leads to the result

$$E = \frac{q}{4\pi\varepsilon_0} \frac{\left(\mathbf{r}_+ - \frac{r_+ \mathbf{v}_+}{c}\right)\left(1 - \frac{v_+^2}{c^2}\right) + \frac{\mathbf{r}_+}{c^2} \times \left[\left(\mathbf{r}_+ - \frac{r_+ \mathbf{v}_+}{c}\right) \times \frac{d\mathbf{v}}{dt}\bigg|_{t=t_+}\right]}{\left(r_+ - \frac{\mathbf{r}_+ \cdot \mathbf{v}_+}{c}\right)^3}$$ (2.61)

while the relation

$$\mathbf{B} = \nabla \times \mathbf{A} = \nabla \times \frac{q\mathbf{v}_+}{4\pi\varepsilon_0 c^2 (r_+ - \mathbf{r}_+ \cdot \mathbf{v}_+/c)}$$

implies

$$\mathbf{B} = \frac{\mathbf{r}_+ \times \mathbf{E}}{c r_+}$$ (2.62)

2.5.2 Larmor Formula for Radiated Power

We said in Sect. 2.4 that the Poynting vector $\varepsilon_0 c^2 \mathbf{E} \times \mathbf{B}$ determines the flux of energy in the EM fields, i.e., the flow of energy per unit time across a unit area perpendicular to the flow [2, Chap. 27]. In the case of a point charge source, we can find the flux of energy across a sphere S centred on the retarded point, all points on such a sphere having the same retarded point, namely the centre of S. Taking $\mathbf{r}$ as the vector from the common retarded point, and using $d\mathbf{S} = r r d\Omega$,

$$\frac{d\mathcal{E}}{dt} = \varepsilon_0 c^2 \int_S d\mathbf{S} \cdot \mathbf{E} \times \mathbf{B} = \varepsilon_0 c \int_S d\Omega (\mathbf{r} \times \mathbf{E})^2 > 0 .$$

An approximation can be made in the case where the source charge has a small velocity. The result will be the well known Larmor formula for radiation by an accelerating charge.

Both $\mathbf{E}$ and $\mathbf{B}$ contain terms in $1/r^2$ contributing only at short distances, and terms in $1/r$ which are called radiative terms. We shall keep only the latter here. Then, by (2.61) and (2.62), respectively, dropping terms in v/c and dropping the subscript $+$ that indicates that all quantities are evaluated relative to the retarded time,

$$\mathbf{E}_{rad} = \frac{q}{4\pi\varepsilon_0 c^2 r^3}\left[\mathbf{r} \times \left(\mathbf{r} \times \frac{d\mathbf{v}}{dt}\right)\right] ,$$

and

$$\mathbf{B}_{rad} = -\frac{q}{4\pi\varepsilon_0 c^3 r^2}\mathbf{r} \times \frac{d\mathbf{v}}{dt} .$$

Making this small velocity approximation, the radiated power is

$$\frac{d\mathcal{E}}{dt} = \frac{q^2}{(4\pi)^2 \varepsilon_0 c^3} \int \frac{d\Omega}{r^2} \left(\mathbf{r} \times \frac{d\mathbf{v}}{dt} \right)^2 .$$

Choosing a polar axis instantaneously parallel to the 3-acceleration of the source charge, and denoting by θ the angle between this axis and $\mathbf{r}$,

$$\frac{d\mathcal{E}}{dt} = \frac{q^2}{8\pi\varepsilon_0 c^3} \left(\frac{d\mathbf{v}}{dt} \right)^2 \int d\theta \, \sin^3 \theta = \frac{4}{3} \frac{q^2}{8\pi\varepsilon_0 c^3} \left(\frac{d\mathbf{v}}{dt} \right)^2 .$$

This is the Larmor formula for the power radiated by an accelerating source charge.

The formula can be made to look considerably neater by making the replacement $e^2 := q^2/4\pi\varepsilon_0$, whence

$$\frac{d\mathcal{E}}{dt} = \frac{2}{3} \frac{e^2 a^2}{c^3} , \tag{2.63}$$

where a is the magnitude of the acceleration. If $q = q_e = 1.60206 \times 10^{-19}$ C is the electron charge in coulombs, and since $1/4\pi\varepsilon_0 = 8.98748 \times 10^9$ in the mks system of units, it turns out that e is numerically equal to 1.5188×10^{-14} [1, Chap. 32].

2.5.3 Alternative Formula for Fields Due to a Point Charge

This section is really a digression to advertise an elegant version of (2.61) which Feynman apparently devised himself as a way of explaining synchrotron radiation [2, Sect. 21.4]. Indeed, the formula (2.61) for the electric fields due to a point source charge can be rewritten in a revealing way:

$$\mathbf{E} = -\frac{q}{4\pi\varepsilon_0} \left[\frac{\mathbf{e}_+}{r_+^2} + \frac{r_+}{c} \frac{d}{dt} \left(\frac{\mathbf{e}_+}{r_+^2} \right) + \frac{1}{c^2} \frac{d^2}{dt^2} \mathbf{e}_+ \right] , \tag{2.64}$$

reverting to the form q for the charge and introducing the unit vector $\mathbf{e}_+$ from the field point to the corresponding retarded point, and the retarded distance r_+.

This can be interpreted as follows. Firstly, there is a term which looks just like the Coulomb field, but relating to the retarded point. Then there is a term in which nature appears to allow for the fact that the effect is retarded, by means of a correction equal to the rate of change of the main term multiplied by the retarded time r_+/c. In other words, we add something like the change which has taken place in the Coulomb term whilst the information is being transferred to the field point. Then there is yet another term, which turns out to be the one describing radiation, and which Feynman uses in [1] to derive all the physical effects related to EM radiation, viz., interference, diffraction, refraction, light scattering, polarisation, and so on.

The magnetic field is just

$$\mathbf{B} = -\frac{\mathbf{e}_+ \wedge \mathbf{E}}{c} ,$$

as we found in (2.62), noting that

$$\mathbf{e}_+ := -\frac{\mathbf{r}_+}{r_+} . \tag{2.65}$$

Although this section is frankly a digression, the above formulas for $\mathbf{E}$ and $\mathbf{B}$ illustrate something about the classical theory of electromagnetism which supports one of the main themes in this book. Hopefully, the reader will agree that there is something thoroughly remarkable about the EM fields produced by a point charge, which is fully expressed by (2.64), in particular the second term which adjusts for the retardation in the sense explained, and the third term which clearly brings out the role of the acceleration in producing radiation effects, among other things (see below for a further note on that).

It is one of the themes here that Maxwell's theory, although pre-quantum, is still extremely rich. Just as it told us to move on to the special theory of relativity, which in a sense could be said to explain some of the mysteries of uniform velocity motion, maybe it can also tell us something about accelerating motions and inertia. This is not to say that one could then ignore what quantum theory has done to improve on things in QED. The idea put forward here is not to try to do away with that and promote a rebirth of classical theory, but to look back at the classical theory and ask whether it cannot still teach us something that could then be recognised also within a quantum theoretical version of that.

Relating the Two Formulas

This is not done explicitly in [2], but it is not difficult. The best approach is to start with the new formula and convert it to the original one. This will involve converting the time derivatives of the direction vector $\mathbf{e}_+$ into the notation $\mathbf{r}_+ := \mathbf{y} - \mathbf{x}_+$ and $t = y^0/c$ used in Sect. 2.5.1, although preferably with some simplifications such as dropping the $+$ subscript and adopting the common trick of setting $c = 1$. Most of the exercise is straightforward book-keeping of terms generated by the derivatives and it would not be useful to display all that here. The following is therefore just a pointer.

From (2.65), we have

$$\frac{d\mathbf{e}_+}{dt} = -\frac{1}{r_+}\frac{d\mathbf{r}_+}{dt} + \frac{\mathbf{r}_+}{r_+^2}\frac{dr_+}{dt} .$$

Of course, d/dt corresponds to $c\partial/\partial y^0$ in the notation of Sect. 2.5.1, and we have the results (2.57), viz.,

$$\frac{\partial \mathbf{r}_+}{\partial y^0} = -\frac{r_+ \mathbf{v}_+}{c(r_+ - \mathbf{r}_+ \cdot \mathbf{v}_+/c)} , \qquad \frac{\partial r_+}{\partial y^0} = -\frac{\mathbf{r}_+ \cdot \mathbf{v}_+}{c(r_+ - \mathbf{r}_+ \cdot \mathbf{v}_+/c)} ,$$

which imply easily that

$$\frac{d\mathbf{e}_+}{dt} = \frac{\mathbf{v}_+ - (\mathbf{r}_+ \cdot \mathbf{v}_+)\mathbf{r}_+/r_+^2}{r_+ - \mathbf{r}_+ \cdot \mathbf{v}_+/c} = \frac{\mathbf{v}_+ - (\mathbf{e}_+ \cdot \mathbf{v}_+)\mathbf{e}_+}{r_+ - \mathbf{r}_+ \cdot \mathbf{v}_+/c} .$$

A lot of terms are generated in taking the next derivative to obtain $d^2\mathbf{e}_+/dt^2$, where one also requires the expression (2.60) for $\partial \mathbf{v}_+/\partial y^0$, viz.,

$$\frac{\partial \mathbf{v}_+}{\partial y^0} = \frac{r_+ \mathbf{a}_+}{c(r_+ - \mathbf{r}_+ \cdot \mathbf{v}_+/c)} .$$

The book-keeping is left to the reader. The point of doing this calculation is to illustrate once again the great care needed to manipulate time derivatives in this context.

Using the New Formula

As mentioned, the remarkable relation (2.64) can be used to derive all the basic results concerning EM radiation by accelerating charges [1, 2]. The idea is to select the piece of $\mathbf{E}$ which varies inversely as the distance, and neglect the terms varying inversely as the square of the distance. Indeed, this could be taken as a definition of radiative terms. The point is that the energy density of this part of the field, proportional to $\mathbf{E}^2$, will go as $1/r^2$, whence its flux through a series of spheres centered on some retarded point will not diminish with distance from the retarded point (radius of the spheres).

It turns out after a little analysis that the radiative part of the electric field is

$$\mathbf{E}_{\mathrm{rad}} = -\frac{q}{4\pi\varepsilon_0 c^2} \frac{d^2}{dt^2} \mathbf{e}_+ . \tag{2.66}$$

The picture we thus get is as follows. We look at the charge, in its apparent position, and note the direction of the unit vector (projecting the direction vector onto the unit sphere centred on ourselves). As the charge moves around, the unit vector wiggles, and the acceleration of that unit vector is what gives the radiative field. Now this unit vector will have both a transverse and a radial component of acceleration. The latter is due to the fact that the end point must stay on the surface of a sphere. One can argue that this radial component of acceleration is inversely proportional to the square of the distance of the source charge, and hence does not contribute to the radiation. So finally, one only need consider the transverse component of the field in (2.66), because this is the only component of the fields that escapes to infinity in the usual sense that physicists understand it.

2.5.4 Point Charge with Constant Velocity

It is very instructive indeed to examine the potentials and fields due to a point charge with constant velocity, and they will be used in what follows, so here are the details [2, Sect. 21.6]. We choose the x axis along the trajectory of the charge, so that this trajectory is given by $x = vt$, $y = 0$, and $z = 0$ (see Fig. 2.3).

We choose a field point (t, x, y, z) at which to evaluate the Lienard–Wiechert potentials (2.53) on p. 19, viz.,

$$A_{\text{ret}}^0(y) = \frac{q}{4\pi\varepsilon_0 c(r_+ - \mathbf{r}_+ \cdot \mathbf{v}_+/c)}, \quad \mathbf{A}_{\text{ret}}(y) = \frac{q\mathbf{v}_+}{4\pi\varepsilon_0 c^2(r_+ - \mathbf{r}_+ \cdot \mathbf{v}_+/c)}. \quad (2.67)$$

In the present case, $\mathbf{v}_+ = \mathbf{v} = (v, 0, 0)$. We need to find the position of the charge at the retarded time

$$t_+ = t - \frac{r_+}{c}, \quad (2.68)$$

where r_+ is the distance from the field point to the charge at the retarded time. Now because the charge motion is so simple, we know that it was at $x_+ = vt_+$ at the retarded time, so we know that

$$r_+^2 = (x - vt_+)^2 + y^2 + z^2,$$

which combines with (2.68) to give the usual condition

$$c^2(t - t_+)^2 = (x - vt_+)^2 + y^2 + z^2.$$

Solving this quadratic equation in t_+, we find

$$\gamma^{-2}t_+ = t - \frac{vx}{c^2} - \frac{1}{c}\left[(x - vt)^2 + \gamma^{-2}(y^2 + z^2)\right]^{1/2}, \quad \gamma^{-2} := \left(1 - \frac{v^2}{c^2}\right). \quad (2.69)$$

Then r_+ is obtained from $r_+ = c(t - t_+)$.

The scalar potential $\phi := cA^0$ is now found from

$$\phi(t, x, y, z) = \frac{q}{4\pi\varepsilon_0} \frac{1}{r_+ - \mathbf{r}_+ \cdot \mathbf{v}/c} = \frac{q}{4\pi\varepsilon_0} \frac{1}{r_+ - (x - vt_+)v/c}.$$

The denominator here is

$$c(t - t_+) - \frac{v}{c}(x - vt_+) = c\left[t - \frac{vx}{c^2} - \gamma^{-2}t_+\right],$$

and substituting in the formula (2.69) for $\gamma^{-2}t_+$, this gives

$$c(t - t_+) - \frac{v}{c}(x - vt_+) = \left[(x - vt)^2 + \gamma^{-2}(y^2 + z^2)\right]^{1/2}.$$

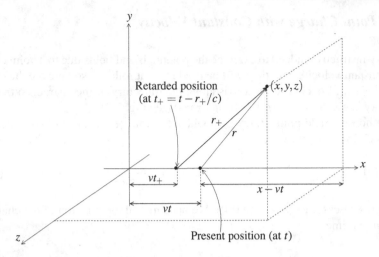

Fig. 2.3 Charge moving with constant velocity along the x axis. The field at the chosen point and chosen time can be neatly expressed in terms of the present position of the charge, whereas it is the behaviour of the charge at the appropriate retarded position that determines its value

Finally then, the scalar potential due to a charge moving with constant speed v along the x axis is

$$\phi = cA^0 = \frac{q}{4\pi\varepsilon_0} \frac{1}{\left[(x-vt)^2 + \gamma^{-2}(y^2+z^2)\right]^{1/2}} \tag{2.70}$$

Another way to put this is

$$\phi = \frac{q}{4\pi\varepsilon_0} \frac{\gamma}{\left[\gamma^2(x-vt)^2 + y^2 + z^2\right]^{1/2}} \tag{2.71}$$

From (2.67), $\mathbf{A} = \mathbf{v}\phi/c^2$, whence

$$\mathbf{A} = \frac{q}{4\pi\varepsilon_0 c^2} \frac{\gamma\mathbf{v}}{\left[\gamma^2(x-vt)^2 + y^2 + z^2\right]^{1/2}} \tag{2.72}$$

Note that these can be obtained from the Coulomb potential, with $\mathbf{A} = 0$, by a Lorentz transformation, because in the rest frame of the charge, that would be the potential. However, Lorentz actually obtained the form of the Lorentz transformation by looking at the way $(\phi/c, \mathbf{A})$ changes in going from one inertial frame to another [2, Sect. 21.6].

It is interesting to note that the potentials given here at x, y, z and at time t, relative to some inertial frame, for a charge whose present position in this frame is $(vt, 0, 0)$, are neatly expressed in terms of the coordinates $(x - vt, y, z)$ of the field point as measured from the current position of the moving charge, despite the fact that it is the behaviour of the charge back at the appropriate retarded position that really counts (see Fig. 2.3). This has a consequence when we come to look at the electric and magnetic fields.

It also allows one to get back the general formulas (2.67) in a remarkable way. This account is adapted from [2, Sect. 26.1]. In Fig. 2.4, the charge is moving in an arbitrary way. We seek the potentials at (x,y,z) at time t. We first identify the retarded point P_+ and retarded time t_+, because we know that it is the doings of the charge at the retarded time that determine what happens now at our field point. If v_+ is the velocity of the charge at the retarded time, then we construct what Feynman calls the projected position P_{proj}, where the charge would be now (at time t) if it had continued on without acceleration since time t_+. In fact, its real position now is P on the diagram. Then the potentials at (x,y,z) at time t are just what (2.71) and (2.72) would give for the imaginary charge at P_{proj}, because the potentials depend solely on what the charge was doing at the retarded time, so they will be the same at (x,y,z) now whether the charge continued moving at constant velocity v_+, or however else it may actually move thereafter.

The whole of electromagnetism as expressed by Maxwell's equations thus follows from the three axioms:

- A^μ is a four-vector.
- The potential for a stationary charge in an inertial frame is the Coulomb potential $\phi = q/4\pi\varepsilon_0 r$, $A = 0$.

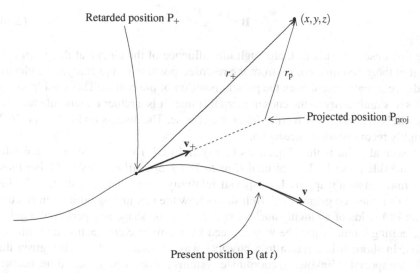

Fig. 2.4 Charge with arbitrary motion. The potentials at the chosen field point and time are determined by the position P_+ and velocity v_+ at the retarded time $t_+ = t - r_+/c$, but they are neatly expressed in terms of the coordinates relative to what Feynman calls the projected position P_{proj}. Note that the distance from P_+ to P_{proj} is just the length of the vector $v_+(t - t_+)$, which is just $r_+ v_+/c$

- The potentials produced by a charge with arbitrary motion depend only on the velocity and position of the charge at the retarded time.

Knowing that A^μ is a four-vector, we transform the Coulomb potential to get the potentials (2.71) and (2.72) for a charge with constant velocity. We then use the rule that, even for an arbitrary motion of the charge, the potentials depend only on the position and velocity at the retarded time, together with the argument above, to find the potentials in that case.

As so clearly explained in [2, Sect. 26.1], this does not mean that the whole of electrodynamics can be deduced solely from the Lorentz transformation and Coulomb's law. We do need to know that the scalar and vector potentials form a four-vector, and we do need to know that the potentials for the arbitrarily moving charge depend only on the position and velocity, and not for example the acceleration, at the retarded time. We see from (2.61) and (2.62) on p. 21 that the electric and magnetic fields do in fact depend on the acceleration of the charge at the retarded time, as well as the position and velocity then.

But let us return to the electric and magnetic fields for a charge moving with constant velocity. Naturally, we apply the relations (2.2) on p. 5, and this leads to

$$\mathbf{E} = \frac{q\gamma}{4\pi\varepsilon_0 \left[\gamma^2(x-vt)^2 + y^2 + z^2\right]^{3/2}} \begin{pmatrix} x-vt \\ y \\ z \end{pmatrix} \tag{2.73}$$

and

$$\mathbf{B} = \frac{\mathbf{v} \times \mathbf{E}}{c^2}. \tag{2.74}$$

The first observation is that, although the influence of the charge at the given field point at the given time comes from the retarded position of the charge, the electric fields are actually radial from the present position of the charge. The word 'present' refers to simultaneity in the chosen inertial frame. It is a rather remarkable fact that this will be true whatever inertial frame we choose. The discussion in [2, Sect. 26.2] is highly recommended once again.

A second point is this. Equations (2.73) and (2.74) were the starting point for a remarkable paper by Bell entitled *How to Teach Special Relativity* [5]. For those who find Einstein's approach to special relativity somewhat aphysical, and Minkowski's rather too geometrical, Bell shows how the readjusting orbit of an electron in the EM fields of a moving nucleus will change the shape and period of a gently accelerating atom, in just the way decreed by relativistic contraction and time dilation. In short, it is a return to a physical way of understanding why, given that there are special (Minkowski) coordinate systems for describing spacetime, adapted to certain (inertially moving) observers, these coordinate systems should be related by Lorentz transformations. This is not just deduced in a pseudo-axiomatic way as is often the case in sophisticated textbooks, nor imputed without further ado to the Minkowskian geometry of spacetime.

This reading is once again highly recommended and more will be said about the subject later on. It is relevant here, where one of the themes is that our theories of the fundamental forces (mainly the theory of electromagnetism in this book) are telling us things that we may not have heard. It was Maxwell's electromagnetism that told us about the special theory of relativity, although there is a clear tendency to turn things around and start with relativity in a dry and mathematical way. We should not forget where the theory of relativity came from. The view in this book is that Maxwell's electromagnetism is telling us more, if only we would listen.

The reader should be warned regarding Bell's paper that some scientists, and in particular philosophers of science it seems, are radically opposed to Bell's approach in [5]. This may just be because their epistemological concerns are radically different, but the debate is interesting. See for example [9].

Chapter 3
Electromagnetic Mass

The title of this chapter is a deliberate plagiarism of a chapter in *The Feynman Lectures on Physics* [2, Chap. 28]. We rapidly review some of the points made there, which clearly intrigued Feynman, Dirac, and many other scientists before classical electrodynamics was swept away by quantum electrodynamics with all its glorious successes. This section is a manifesto for the ideas presented in this book, and at the same time a demonstration that the subject here is not some weird offshoot of standard physics, but a straight application of the latter to a context that may have been largely forgotten simply because it is not very tractable from a mathematical standpoint.

The heart of the matter is this. In both classical and quantum physics, the electron is treated as a point particle. This means that it occupies a mathematical point in the space of any spacelike hypersurface in spacetime, or that it occupies a worldline rather than a worldtube in spacetime as a whole. It may seem surprising to see the claim that the electron is treated as a point particle in quantum physics, where it is of course related in some way to a wave function, but there is a very real sense in which it is still modelled by a pointlike object in quantum physics. This shows up in the need to renormalise quantum field theories.

Even classical theories need to be renormalised. Dirac was one of the first to work on this back in 1938 [3]. The paper is beautifully written and, like all the classics, well worth the detour. Renormalisation is a physical and mathematical fix for something that goes drastically wrong when one comes to squeeze numerical predictions out of a theory, the problem being that the numbers are actually infinite. It is certainly the most disappointing thing about great theories like QED, and one feels that there ought to be a way to avoid it. But any student who studies either classical or quantum theoretical renormalisation of electromagnetism, for example, will be struck by the remarkable fact that renormalisation is actually possible at all. This is undoubtedly telling us something about the theory that we have missed.

Lyle, S.N.: *Electromagnetic Mass*. Lect. Notes Phys. **796**, 31–46 (2010)
DOI 10.1007/978-3-642-04785-5_3   © Springer-Verlag Berlin Heidelberg 2010

3.1 Energy in the EM Fields of a Charged Particle

In Maxwell's electromagnetism, the first problem arises when we consider the energy in the EM fields around a stationary point charge. Since the electric field has magnitude $E = q_e/4\pi\varepsilon_0 r^2$, and since the energy density of the field is supposed to be given by u in (2.36) on p. 12, we find that

$$u = \frac{\varepsilon_0}{2} E^2 = \frac{q_e^2}{32\pi^2 \varepsilon_0 r^4} \; .$$

But this means that the total energy in the field, given by the integral of this over all of space, is going to be the divergent integral

$$U_{\text{electron}} = \int_{\mathbb{R}^3} \frac{q_e^2}{32\pi^2 \varepsilon_0 r^4} 4\pi r^2 dr \; .$$

Now the problem obviously occurs for small values of r, so an easy way to avoid this disaster is to cut off the integral for small values of r. But this is also straightforwardly justified from a physical point of view. We only need to say that the electron is not a point particle after all, but a little sphere. Its charge might be considered as distributed in some way through the sphere. Let us say that the charge is distributed uniformly over a spherical surface of radius a, since that simplifies calculations enormously. The fields inside the surface will be zero, and the total energy in the field around this electron will now be

$$U_{\text{electron}} = \int_{r=a}^{\infty} \frac{q_e^2}{32\pi^2 \varepsilon_0 r^4} 4\pi r^2 dr = \frac{1}{2} \frac{q_e^2}{4\pi\varepsilon_0} \frac{1}{a} = \frac{e^2}{2a} \; , \tag{3.1}$$

with the usual definition of e.

It is clear from (3.1) that $U_{\text{electron}} \to \infty$ when $a \to 0$. Of course, one could just about live with an infinite energy if there were no physical manifestations. This is the way one is supposed to view the renormalisation process in quantum field theory and it is the suggestion in [2]. An alternative, however, would be to say that the electron is not a point particle, but has some spatial extent.

3.2 Momentum in the EM Fields of a Charged Particle

Now let us consider the momentum of the fields of a moving charge q_e. We shall consider once again the spherical shell of charge with positive uniform charge density (strictly, a positron then), moving at a constant velocity $\mathbf{v}$. To begin with we shall assume that $\mathbf{v}$ is small compared with c. For a point P at distance r from the present position of the charge center C, such that the line CP makes an angle θ with the velocity $\mathbf{v}$, the electric field lies radially outward from the present position of C. In fact, we know from (2.73) on p. 28 that

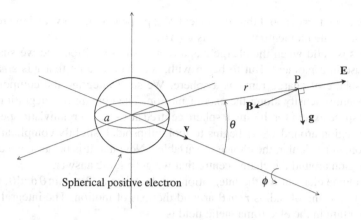

Fig. 3.1 Momentum density **g** in the fields **E** and **B** of a spherical charged shell with uniform velocity **v** along the x axis

$$E_x = \frac{q_e \gamma(v)}{4\pi\varepsilon_0} \frac{x - vt}{[\gamma^2(x - vt)^2 + y^2 + z^2]^{3/2}}, \tag{3.2}$$

$$E_y = \frac{q_e \gamma(v)}{4\pi\varepsilon_0} \frac{y}{[\gamma^2(x - vt)^2 + y^2 + z^2]^{3/2}}, \tag{3.3}$$

$$E_z = \frac{q_e \gamma(v)}{4\pi\varepsilon_0} \frac{z}{[\gamma^2(x - vt)^2 + y^2 + z^2]^{3/2}}. \tag{3.4}$$

Further, according to (2.74), the magnetic field is $\mathbf{B} = \mathbf{v} \times \mathbf{E}/c^2$. The fields are shown in Fig. 3.1. Note that, if the sign of the charge were in fact negative, both **E** and **B** would be reversed, but the momentum density **g** as given by (2.39) on p. 12, viz.,

$$\mathbf{g} = \varepsilon_0 \mathbf{E} \times \mathbf{B},$$

would remain the same. Because the magnetic field has magnitude $vE\sin\theta/c^2$, the momentum density has magnitude

$$g = \frac{\varepsilon_0 v}{c^2} E^2 \sin\theta,$$

and points down toward the path of the charge, making an angle θ with the vertical (see Fig. 3.1).

Note that it is because there is motion that there is a magnetic field, and it is because there is a magnetic field that there is a momentum carried by the electromagnetic fields. The corresponding momentum density is not a radiation field in the present case, because it drops off as quickly as the Coulomb field of a static charge.

If we integrate the momentum density over the whole space outside the sphere of radius a in order to work out the total momentum **p** in the fields, only a component along the axis of motion will remain. This is a straightforward symmetry consideration. So we can forget about the component of **g** perpendicular to **v**, and the relevant component of the momentum density is $g \sin \theta$.

All this is valid when the charge has a relativistic speed too and we shall turn to that case in a moment. But to begin with, we shall assume that v is small and that the sphere of charge remains a sphere. We shall integrate the component of the momentum density along the axis of motion. The domain of integration is the whole of space except for the small sphere centred at $x - vt$. At relativistic speeds the excluded region around $x - vt$ begins to look ellipsoidal, slightly complicating the integration, in particular, the choice of variables. Note that it is because we exclude a small region around the charge centre that we get a finite answer.

The volume element for the integration can be taken as $2\pi r^2 \sin \theta \, dr \, d\theta$, the volume of a thin ring of radius $r \sin \theta$ around the axis of motion. The integral giving the momentum in the electromagnetic field is

$$\mathbf{p} = \int \frac{\varepsilon_0 \mathbf{v}}{c^2} E^2 \sin^2 \theta \, 2\pi r^2 \sin \theta \, dr \, d\theta \,, \tag{3.5}$$

where $0 \le \theta \le \pi$ and $a \le r$. We soon obtain

$$\mathbf{p} = \frac{8\pi}{3} \frac{\varepsilon_0 \mathbf{v}}{c^2} \int_{r=a}^{\infty} E^2 r^2 dr \,. \tag{3.6}$$

since E^2 calculated from (3.2–3.4) is independent of θ when $\gamma \approx 1$.

The first thing to note is that this happens to be parallel to the motion of the charge! Of course, the situation is so symmetric, where else could it have been? Well, it could have been antiparallel, and it is not. We shall return to this point later. The second thing to note is the sense in which we approximate here for small velocity. We have already used the fact that $\gamma \approx 1$ in order to carry out the θ integration, but there was another more subtle assumption in the last relation. The question here is this: what shape does our electron have when in motion? Of course, the motion is a uniform velocity here, so one might expect it to FitzGerald contract in the direction of motion. It would then be ellipsoidal in shape when viewed from the particular inertial frame we have chosen, relative to which it has this motion. However, we have taken γ to be close to unity, hence the simple cutoff for r in (3.6).

Later we shall consider accelerating charge distributions, and the question of their shape will become a more pressing issue. But even here, the changed shape of the charge distribution raises the question of what holds it together. After all, the charge elements making up the spherical shell are all like charges and will therefore repel one another. Something is required to hold the system together, which we shall just refer to as binding forces (also known in the literature as Poincaré stresses). If the system is accelerated from rest into the state of uniform velocity, something must be happening inside it so that the various forces readjust to give an ellipsoid. This will be discussed in more detail later. The reader will spot the connection with Bell's ideas, mentioned in Sect. 2.5.4 [5].

But for the moment, let us return to (3.6). In fact, in this approximation, we may take E to have the nonrelativistic value

$$E = \frac{q_e}{4\pi\varepsilon_0 r^2} , \tag{3.7}$$

whence

$$\mathbf{p} = \frac{2}{3} \frac{q_e^2}{4\pi\varepsilon_0} \frac{\mathbf{v}}{ac^2} . \tag{3.8}$$

Making the replacement $e^2 = q_e^2/4\pi\varepsilon_0$, this becomes

$$\mathbf{p} = \frac{2}{3} \frac{e^2}{ac^2} \mathbf{v} . \tag{3.9}$$

So the momentum in the EM field is proportional to $\mathbf{v}$ and in the same direction. It is the momentum of a particle with mass $2e^2/3ac^2$, which Feynman calls the electromagnetic mass:

$$\boxed{m_{EM} = \frac{2}{3} \frac{e^2}{ac^2}} \tag{3.10}$$

We observe immediately that it would be infinite if the charge distribution occupied only a mathematical point in space ($a = 0$). What is significant here, from a qualitative standpoint, is that the momentum carried by the fields is indeed proportional to the electron velocity. The exact value of the coefficient, e.g., the appearance of a factor of 2/3, depends on the spatial structure attributed to the electron, as we shall see later.

Let us now carry out the relativistic calculation. There will be a bonus in doing this. The main point will be to show that the electromagnetic mass then contains a factor $\gamma(v)$, just as any good inertial mass should do according to special relativity theory.

Take the x axis along the motion. To begin with, we shall be integrating over all space outside an ellipsoid given by

$$\gamma^2(x - vt)^2 + y^2 + z^2 = a^2 . \tag{3.11}$$

Note that this is just the spherical shell when its dimensions in the x direction have been contracted by the FitzGerald factor. Better coordinates for dealing with such a domain of integration would be (X, y, z), where

$$X = \gamma(x - vt) . \tag{3.12}$$

In (X, y, z) space we can define the 'distance' from the origin as

$$R := (X^2 + y^2 + z^2)^{1/2} , \tag{3.13}$$

so that the ellipsoid has equation $R = a$ and we integrate over all of (X, y, z) such that $R \geq a$. We can define an angle θ' in this space by

$$X = R\cos\theta', \quad (y^2 + z^2)^{1/2} = R\sin\theta'. \tag{3.14}$$

The volume element in (X, y, z) space is then

$$dX\,dy\,dz = 2\pi R^2 \sin\theta'\,d\theta'\,dR. \tag{3.15}$$

Now as argued above, the momentum in the field is

$$\mathbf{p} = \int \frac{\varepsilon_0 \mathbf{v}}{c^2} E^2 \sin^2\theta\,dV, \tag{3.16}$$

where $dV = dx\,dy\,dz$ and E is the magnitude of $\mathbf{E}$ given by (3.2–3.4). We find

$$E^2 = \frac{q_e^2}{16\pi^2\varepsilon_0^2} \frac{\gamma^2\left[(x - vt)^2 + y^2 + z^2\right]}{\left[\gamma^2(x - vt)^2 + y^2 + z^2\right]^3}. \tag{3.17}$$

Everything in the integrand of (3.16) has to be expressed in terms of the new variables R and θ'. To begin with

$$dV = dx\,dy\,dz = \frac{dX\,dy\,dz}{\gamma}. \tag{3.18}$$

We also have

$$E^2 = \frac{q_e^2}{16\pi^2\varepsilon_0^2} \frac{\gamma^2\left(X^2/\gamma^2 + y^2 + z^2\right)}{\left(X^2 + y^2 + z^2\right)^3}, \tag{3.19}$$

where

$$\frac{X^2}{\gamma^2} + y^2 + z^2 = R^2\left(\frac{\cos^2\theta'}{\gamma^2} + \sin^2\theta'\right), \tag{3.20}$$

and

$$\sin^2\theta = \frac{y^2 + z^2}{(x - vt)^2 + y^2 + z^2} = \frac{\sin^2\theta'}{\dfrac{\cos^2\theta'}{\gamma^2} + \sin^2\theta'}. \tag{3.21}$$

Putting all this together, we now find that

$$\begin{aligned}
\mathbf{p} &= \frac{q_e^2\gamma^2\mathbf{v}}{16\pi^2\varepsilon_0 c^2} \int_{X^2+y^2+z^2\geq a^2} \frac{X^2/\gamma^2 + y^2 + z^2}{(X^2 + y^2 + z^2)^3} \sin^2\theta\,\frac{dX\,dy\,dz}{\gamma} \\
&= \frac{q_e^2\gamma\mathbf{v}}{16\pi^2\varepsilon_0 c^2} \int_{R\geq a,\,0\leq\theta'\leq\pi} \frac{2\pi\sin^3\theta'}{R^2}\,d\theta'\,dR.
\end{aligned}$$

Referring back to (3.5), we see that this is exactly γ times the nonrelativistic result. In other words, in the relativistic case

$$\mathbf{p} = \frac{2}{3} \frac{q_e^2}{4\pi\varepsilon_0} \frac{\gamma(v)\mathbf{v}}{ac^2} . \qquad (3.22)$$

Making the replacement $e^2 = q_e^2/4\pi\varepsilon_0$, this becomes

$$\mathbf{p} = \frac{2}{3} \frac{e^2\gamma(v)}{ac^2}\mathbf{v} . \qquad (3.23)$$

We thus attribute an electromagnetic mass

$$\boxed{m_{EM}(v) = \frac{2}{3} \frac{e^2}{ac^2}\gamma(v)} \qquad (3.24)$$

to the electron with spherical shell structure.

3.3 Inertial Mass

We do not know why things have mass. That is, we do not know how to predict the resistance something will show to being accelerated, which is quantified by its inertial mass.

In elementary particle physics, some particles are indeed considered to be elementary, in the sense of not being composed of anything smaller, e.g., all the leptons (electron, muon, tau lepton, their corresponding neutrinos, and all the associated antineutrinos), but also all the quarks (three generations, three colour charges for each, and all their associated antiparticles). The masses of all these particles must simply be fed into the Standard Model of particle physics as parameters whose values we attempt to provide experimentally.

Then there are an enormous range of bound state particles, built up from the elementary particles, e.g., the proton, comprising two up quarks and one down quark (whose colour charges combine to white). It is interesting to look at the way one tries to obtain the mass of a particle like the proton from the masses of the constituent quarks and other features, and we shall do that in detail in Chap. 13. For the moment, suffice it to say that there are three ingredients for the proton mass:

- The rest masses of the constituent quarks.
- The kinetic energies of the constituent quarks, divided by c^2 to get the right units.
- The binding energy of the system, divided by c^2 to get the right units.

The result is a long way from being the simple sum of the three rest masses of the constituent quarks. But this is in fact a good thing, because it is telling us something about what causes inertia.

It is interesting to ask students of physics why one should add in kinetic energies of constituents, or binding energies, divided by c^2. Most answer without hesitation that it is just an application of $E = mc^2$. Here we have something like a principle from special relativistic dynamics, and as a principle, it requires no further explanation. One of the aims of this book will be precisely to explain this feature of the inertial mass of a bound state.

But what about the masses of the elementary particles, like the electron? If it really were a point particle, it would be difficult to make any model for its inertia that were intrinsic to its structure, in order to predict its mass, or at least understand the origin of its resistance to acceleration, simply because one is effectively denying it any internal structure. But the very fact that we need to renormalise QED is presumably telling us that it should not really have been treated as a point particle. This is certainly what happens in the case of classical renormalisation, where one understands very clearly why the point particle idea must just be an 'approximation'. (This is not really the right word, because the error turns out to be infinite. It is rather a 'simplification', allowing one to do calculations and just feed in the measured value of the inertial mass. More about all that later.)

So how does the Standard Model explain the inertia of the truly elementary particles? The answer is the Higgs mechanism. The Standard Model predicts the existence of the Higgs boson, although it has not yet been detected. In popular accounts, moving through the Higgs field is rather like trying to move through honey. The particle gets its inertia from the outside, as it were, rather than from any intrinsic structure. The analogy is not perfect, however, because viscosity also opposes uniform velocities, while the Higgs field presumably does not.

One problem with the Higgs particle is that the theory cannot tell us what energy will be required of the particle accelerator that is to generate it, so the only solution is to keep ramping up the energy and hope that it will eventually be found. Since increasing the energy is something that we would have done anyway, out of pure curiosity, there is nothing to be lost by this strategy. But it does raise the question as to whether one should not have alternative theories up one's sleeve, just in case the Higgs particle never shows up.

In this book, we shall revive the old bootstrap idea that there are in fact no elementary particles, and that it is the very structure of each particle that causes it to resist its acceleration. We shall also mention a purely mathematical alternative to the Higgs mechanism which unifies the electromagnetic and weak interactions with the same predictions for the $W^\pm$ and Z bosons (see Sect. 13.5). Like the Higgs mechanism, these are merely hypotheses. In the end, experiments will hopefully guide us to the best hypothesis.

The reader may be wondering what the connection is with the last section. Well, as Feynman says [2, Sect.28.3], an inertial mass is associated with the electron, and that means that it in some sense carries a momentum proportional to its velocity. But, in the last section, we have seen how we can understand that a charged particle should carry a momentum proportional to its velocity. Could it be that all the mass of the electron comes from this electrodynamic effect?

Perhaps not. Perhaps there is also what physicists like to call a mechanical mass m_{mech}, with an associated momentum $m_{mech}\mathbf{v}$ when the object moves with velocity $\mathbf{v}$. When we measure the momentum of the particle in order to determine its mass, what we would then find is the total mass $m_{mech} + m_{EM}$. The momentum we measure would be this times the velocity of the charge. In this picture, the observed mass would comprise two parts, or maybe more if other fields were included. One thinks of the weak force, strong force, and even the gravitational force. Each force for which the object, or part of the object, serves as a source might contribute some further part to the mass.

But for the moment, all we know is that there is definitely an electromagnetic piece in the particle's mass, and there is the exciting possibility that there might actually be no mechanical piece m_{mech}, i.e., that all the mass might be electromagnetic. On this hypothesis, one can determine the radius the little spherical shell of charge must have if it is to model the electron, because one knows the rest mass m_e of the electron. Setting (3.10) equal to m_e, one obtains

$$a = \frac{2}{3} \frac{e^2}{m_e c^2} = \frac{2}{3} r_{classical} , \tag{3.25}$$

where

$$r_{classical} := \frac{e^2}{m_e c^2} = 2.82 \times 10^{-15} \text{ m}$$

is called the classical electron radius. This should be compared with the diameter of an atom, which is of the order of 10^{-10} m.

Note that one can carry out the calculations of the last section for other charge distributions and the results are very similar. Indeed, they differ only in the value of the constant in front of $r_{classical}$ in the last relation, and this is why $r_{classical}$ is singled out as perhaps being representative of the electron radius, rather than the actual value obtained in (3.25) with the factor of 2/3. For example, if the charge is uniformly distributed throughout the sphere of radius a, the prefactor changes to 4/5. Part of this book will be devoted to doing calculations with another charge distribution, a much simpler one, making certain calculations more tractable, in order to illustrate various features of this rather striking situation, and hopefully convince some readers that there are interesting things to be discovered.

But for now, let us reflect a moment on the relativistic calculation, leading to (3.24). The electromagnetic contribution to the inertial mass increases by a factor $\gamma(v)$ as v increases. This was discovered before the advent of the special theory of relativity, which decrees as one of its principles that the 3-momentum of a particle with rest mass m and velocity $\mathbf{v}$ should be taken as $m\gamma(v)\mathbf{v}$ in order to build up a consistent Lorentz symmetric theory of dynamics (see Sect. 13.1). The point we would like to make here is that special relativistic dynamics simply decrees that inertia should increase with speed, without other explanation than the idea that this leads to a consistent picture. But here we find that we have a little more understanding than that. At least the EM contribution to the inertial mass has to increase like this because this is how the momentum of the EM fields of the charge will increase.

The reader would be quite right to point out that this is not much of a mechanism, at least, not as it stands. After all, we have found the momentum of the EM fields, not the momentum of the charge itself! What brings about this momentum? Or for that matter, what brings about the energy in the fields, discussed in Sect. 3.1. And since we are on the subject of relativity theory, surely the electromagnetic mass we should be talking about is U_{electron}/c^2, where U_{electron} is as found in (3.1) on p. 32, which would give

$$
m_{\text{EM}}^{\text{EDM}} := \frac{U_{\text{electron}}}{c^2} = \frac{1}{2}\frac{e^2}{ac^2} \tag{3.26}
$$

where the superscript EDM indicates that this is the energy-derived mass. This differs from the momentum-derived mass $m_{\text{EM}}^{\text{MDM}}$ by a numerical factor. Put another way, combining (3.1) and (3.10), we obtain

$$
U_{\text{electron}} = \frac{3}{4}m_{\text{EM}}^{\text{MDM}}c^2 , \tag{3.27}
$$

whereas relativity theory would have preferred to do without the factor of 3/4.

That particular discrepancy has led to a long debate in the literature, and we shall be considering some of the issues raised in detail in Chap. 11. From the angle taken in this book, the appropriate solution will not be difficult to find. But first, what is the mechanism behind EM mass, either momentum-derived or energy-derived? When we try to accelerate a charged sphere, for example, what is it that we are having to push against?

3.4 Self-Force

Feynman is very clear [2, Sect. 28.4] about the problem raised at the end of the last section: we get a discrepancy because the binding forces in the system have been left out. This will indeed be our conclusion in Chap. 11, but the reader will also discover there exactly why there was such a debate about this issue, and hopefully obtain a better insight into this approach to explaining inertia.

So the point is that there are unbalanced forces in our spherical shell of like charges, which will naturally repel one another. As Poincaré was among the first to discuss this, the binding forces are often referred to as Poincaré stresses. And of course, these binding forces must be included in any energy and momentum calculations if we are to be able to use the conservation laws to make deductions. When we do this, the discrepancy disappears, i.e., we get an answer that is consistent with relativity, without the 3/4 factor in (3.27). Both $m_{\text{EM}}^{\text{MDM}}$ and $m_{\text{EM}}^{\text{EDM}}$ are changed, and each contains a contribution from EM effects and from the effects of the binding forces.

For Feynman, this need to include some other ingredient apart from electromagnetism spoilt the beauty of the idea and he complains about the complexity of the resulting theory [2, Sect. 28.4]. And there is no denying the complexity. How strong are the binding forces? What causes them? Can the system oscillate? What are its internal properties? If there are oscillation modes, why have we not observed them? On the other hand, as he points out, we may have observed such modes and just not recognised their origin. Today one immediately thinks of the muon and the tau lepton, for example, particles with very similar properties to the electron, but higher mass. More about that later.

But let us turn now to the electromagnetic forces within our system. So far we have discussed the momentum-derived mass, saying that we had a mass because the momentum in the fields was proportional to the velocity of the charge. But mass is also supposed to be a measure of inertia, resistance to being accelerated. A particle has a mass because a force has to be exerted on it to change its velocity, to accelerate it.

With regard to the EM contribution to the mass, we know there has to be a force here because we have a law of conservation of momentum for the fields in the form (2.40). When we push the charged particle for a while, there is some momentum in the EM field, and something must have got this momentum into the field. We conclude that there must have been a force in addition to the one required by the mechanical inertia of the electron (i.e., any inertia due to other contributions to its mass), in order to get it moving faster. But then there must have been a corresponding extra force back on the accelerating agent, in addition to the one due to the mechanical mass.

Now if our spherical shell of charge were just sitting there motionless in our inertial frame, we could consider the EM force of each charge element on each other charge element. The force of a surface element dS_1 on another surface element dS_2 would be equal and opposite to the force of dS_2 on dS_1. This is just the Coulomb force. The sum of the two forces, considered as acting not on the individual elements but as contributing to the force of the whole system on itself, would be zero (and this even if the charge distribution is non-uniform). The total EM self-force on the stationary system is trivially zero.

But when the system is moving, there are retardation effects. We know this from relations like (2.61) and (2.62) on p. 21, which give the fields somewhere due to a charge element in motion. The fields depend on what the charge was doing at the retarded time. The reader will see that there is a potentially very complicated calculation to be done here in order to find out how each charge element acts on each other charge element when the whole system has some arbitrary motion. Having a model in the form of a spherical shell of charge does not help things along here. In this book, we shall consider a much simpler system, which thus displays the relevant effects more clearly, although at first sight it may look like a step back from the ideal of doing away with point charges. More about that in a moment.

But first, here is the self-force on the uniform spherical shell of charge when it is being accelerated in some arbitrary way along the x axis, as quoted by Feynman without proof [2, Sect. 28.4]:

$$F_{\text{self}} = -\alpha \frac{e^2}{ac^2}\ddot{x} + \frac{2}{3}\frac{e^2}{c^3}\dddot{x} + \gamma\frac{e^2 a}{c^4}\ddddot{x} + O(a^2) \tag{3.28}$$

where $x(t)$ gives the position of the shell center at time t and α and γ are numerical coefficients of order unity. (The expression here has the opposite sign to the one given by Feynman, because he in fact quotes the force required to overcome the self-force.) This has been obtained by expanding everything in powers of a, treated as a small quantity. For the spherical shell, it turns out that $\alpha = 2/3$. So we have a term proportional to the instantaneous value of the acceleration $\ddot{x}$ which goes as $1/a$ and hence diverges as $a \rightarrow 0$. For a point particle, the self-force would be infinite.

Another thing that is immediately obvious from (3.28) is that the self-force will be zero when the spherical shell has a uniform velocity, because then $\ddot{x} = 0$, and all the higher time derivatives are zero. Thinking back to the above calculation of the self-force that takes retardations into account, it is actually quite remarkable that the self-force turns out to be zero in this case. We shall not demonstrate this here, although it is much easier than proving (3.28) for the general motion. The reason is that we shall soon introduce a simpler spatially extended charge distribution in which the calculations are generally easier anyway, so that basic issues like this become more transparent.

This is a basic issue. The self-force effect, at least the electromagnetic one, makes a clear distinction between uniform motion and accelerated motion. No extra force is required to keep electromagnetic mass in uniform motion. One only requires an extra force when there is acceleration. The reader should be thinking of Newton's first and second laws here. Newton's first law would have been in trouble if we had self-forces for uniform motions, because bound state particles like protons comprise a charge distribution, and the EM mass for such particles is indeed included in their inertial mass in the form of the EM binding energy (over c^2).

The idea here is that the EM mass contribution to inertia arises precisely because of the need to overcome the EM self-force. The coefficient of the first term in (3.28) is precisely what we denoted by $m_{\text{EM}}^{\text{MDM}}$. We express this by writing

$$\boxed{m_{\text{EM}}^{\text{MDM}} = m_{\text{EM}}^{\text{SFDM}}} \tag{3.29}$$

with the obvious notation $m_{\text{EM}}^{\text{SFDM}}$ for the self-force-derived mass, which is just defined to be the coefficient of $\ddot{x}$ in the self-force expression (3.28). In the proton example, it makes no difference to this conclusion whether the EM binding energy turns out to be positive or negative. When a binding energy is negative, it transpires that the EM self-force helps the acceleration along, i.e., it reduces the inertia. This will be demonstrated very clearly with the very simple spatially extended charge distribution to be introduced shortly.

One important and rather interesting problem with (3.28) is not mentioned by Feynman. If the reader attempts to carry out the self-force calculation for uniform motion, she/he will naturally take the shell to be ellipsoidal in shape as viewed from the chosen inertial frame. But what shape should it have when it has an arbitrary acceleration? One must assume something in order to get the result in (3.28). We shall go into this question in some detail, but for our simple charge distribution. A strong connection will be made with Bell's ideas in [5].

Suffice it to say for the moment that the usual assumption is as follows: at each instant of time the sphere is assumed to look exactly spherical in its instantaneous rest frame. This assumption amounts to a notion of rigidity in the relativistic context (see Chap. 12). It is something that the binding forces would have to enforce, and we shall see that there are problems with it, although it looks like a reasonable approximation in some sense. A good modern reference for the self-force calculations with a spherical shell of charge is [7], but it is not an easy subject.

3.5 Radiation Reaction

So much for the first term in (3.28). But what about the other terms? The second term is particularly interesting, because it turns out to be independent, not only of the dimension a of the charge distribution, but even of its shape! We shall see a detailed example of that in Sect. 6.6.2.

One reaction to the fact that the self-force is infinite for a point electron is to throw the whole thing away, decreeing that electrons are only affected by the fields due to other charged particles, not by their own fields. The problem with this approach is that one really needs the second term in (3.28) because it explains how accelerating electrons manage to radiate electromagnetic energy. Let us examine this claim.

When a charged particle like an electron is accelerated, it radiates EM waves and therefore loses energy (see Sect. 2.5.2). This suggests that more force would be needed to accelerate a charged particle than to accelerate a neutral one with the same mechanical mass, because otherwise energy would not be conserved. The extra bit of force is called the radiation reaction or radiation resistance. Indeed, the rate at which this extra bit of force does work on an accelerating charge must be equal to the rate of energy loss per second by radiation. But where does the extra bit of force come from, against which we must do this work?

Feynman cites the case of an antenna. When it radiates, the forces required come from the influence of one part of the current in the antenna on another part. So for a single accelerating electron radiating into otherwise empty space, there is only one place the force could come from, namely the action of one part of the electron on another part. Now that explanation is obviously in trouble if the electron really does just occupy a mathematical point. But when it has a spatial extent, like the spherical charged shell, or indeed any other shape, we always find a term

$$F_{\text{self}}^{\text{rad}} := \frac{2}{3} \frac{e^2}{c^3} \dddot{x} \qquad (3.30)$$

in the self-force, and we shall now show that this is precisely the right force to explain EM radiation by a charged particle.

We consider the rate of doing work on an electron against the bootstrap force in (3.28). The rate of doing work with a force F along the x axis is just $F\dot{x}$, so the first term in (3.28) requires a rate

$$\frac{dW_1}{dt} = \alpha \frac{e^2}{ac^2} \ddot{x}\dot{x} = \frac{1}{2}\alpha \frac{e^2}{ac^2}\frac{d}{dt}\left(\dot{x}^2\right) ,$$

and this is precisely the rate of change of the kinetic energy

$$\frac{1}{2}m_{\mathrm{EM}}^{\mathrm{SFDM}}v^2$$

associated with the (self-force-derived) EM mass. The second term in (3.28), picked out in (3.30) above, requires a rate

$$\frac{dW_2}{dt} = -\frac{2}{3}\frac{e^2}{c^3}\dddot{x}\dot{x} = -\frac{2}{3}\frac{e^2}{c^3}\frac{d}{dt}\left(\ddot{x}\dot{x}\right) + \frac{2}{3}\frac{e^2}{c^3}\ddot{x}^2 .$$

Now for a periodic motion $x \propto \cos \omega t$,

$$\ddot{x}\dot{x} \propto \sin 2\omega t ,$$

so the first term in dW/dt will average to zero. But the second term is always positive. It is precisely the Larmor radiation formula (2.63) on p. 22.

So the term (3.30) that depends on $\dddot{x}$ in the self-force is required to guarantee energy conservation in radiating systems, and cannot be simply thrown away. This discovery was made by Lorentz. As Feynman puts it, we must believe in the idea of the action of the electron on itself, and we have to keep the term (3.30). However, for Feynman, this created a difficulty, because he clearly considered the first term in (3.28) to be a problem, so for him, the issue was: how do we justify dropping the first term in the self-force but keeping the second? On the other hand, from our point of view, we may say that the first term is only really a problem if the electron is in fact spatially pointlike.

3.6 Lorentz–Dirac Equation

Feynman mentions what he describes as a peculiar possibility due to Dirac [3]. It amounted to saying that the electron would act on itself only through the second term (3.30) of (3.28) and not through the first. Actually, although it may appear peculiar, this is precisely the idea of mass renormalisation, carried over today into our most sophisticated theory of electromagnetic phenomena, viz., QED. The coefficient of $\ddot{x}$ in the first term of (3.28) is absorbed into the total inertial mass of the electron. One can then let the system size tend to zero, getting rid of all the higher order terms in (3.28), and not worrying about the fact that one has made an infinite adjustment to the inertial mass, since in the end, one just says that the inertial mass is the finite quantity we observe in practice.

Naively one has something like this:

$$F_{\mathrm{ext}} + F_{\mathrm{self}} = m_{\mathrm{bare}}\ddot{x} ,$$

where F_{ext} is the total external force on the system, and m_{bare} is the true (mechanical) inertial mass. By (3.28), this implies the following version of Newton's second law:

$$F_{\text{ext}} = -\frac{2}{3}\frac{e^2}{c^3}\dddot{x} + \left(m_{\text{bare}} + \alpha\frac{e^2}{ac^2}\right)\ddot{x}.$$

One defines the renormalised inertial mass to be

$$m_{\text{ren}} := m_{\text{bare}} + \alpha\frac{e^2}{ac^2}.$$

If we let $a \to 0$, this is infinite, unless m_{bare} was itself already infinite. But we do not worry about that, because we can just say that m_{ren} is what we actually measure to be the inertial mass. Then the equation of motion for the charge becomes

$$F_{\text{ext}} = m_{\text{ren}}\ddot{x} - \frac{2}{3}\frac{e^2}{c^3}\dddot{x}, \tag{3.31}$$

and when F_{ext} is a force due to the presence of EM fields, as described in Sect. 2.3, we have something like the Lorentz–Dirac equation for the motion of a charged particle in the presence of external EM fields. The only vestige of the self-force is the radiation reaction term. Note, however, that (3.31) is *not* the Lorentz–Dirac equation. The above manipulations are only intended to provide an idea of what is going on.

Indeed, this is not the place to go into all the details of Dirac's equation. It is not easy to obtain a rigorous derivation, and if it is used outside its domain of validity, it leads to some very strange conclusions, as attested by the vast literature on the subject over the past few decades. A good reference with a rigorous derivation that treats the domain of validity of the equation very carefully and reveals some of its limitations is the book by Parrott [8]. The original, elegantly written paper [3] by Dirac is also highly recommended, and a review of both in a slightly different context can be found in [4, Chap. 11].

3.7 A Toy Electron

The calculation leading to the self-force in (3.28) is not an easy one. A modern discussion of this case can be found in the brief and elegant book by Yaghjian [7]. A perfectly symmetrical charged sphere may well be the simplest continuous charge distribution one could think of and hence a good place to start when trying to carry out self-force calculations, or when trying to work out the energy and momentum in its fields. However, there is a much simpler way of obtaining a toy electron which satisfies the most basic requirement here of having some spatial extent.

Indeed, undoubtedly the simplest way to get some spatial extent is just to consider a system comprising two point charges of value $q_e/2$, separated by some fixed distance d. We have sacrificed the key idea of having a continuous charge distri-

bution, but kept the idea of a particle with spatial extent. We shall carry out the
following investigations with this system:

- We calculate the momentum in the fields of the system when it is moving with
  constant velocity along its axis in some inertial frame, assuming the appropriate
  FitzGerald contraction of the length and assuming that each point charge is actually a tiny spherical shell of charge of radius a (see Sect. 5.2).
- We calculate the momentum in the fields of the system when it is moving with
  constant velocity perpendicularly to its axis in some inertial frame, assuming
  once again that each point charge is actually a tiny spherical shell of charge of
  radius a (see Sect. 5.3).
- We calculate the self-force to leading order when the system is accelerating in a
  straight line perpendicularly to its axis in some inertial frame (see Chap. 6).
- We calculate the self-force to leading order when the system is accelerating along
  its axis in some inertial frame (see Chap. 7), discussing in detail what assumptions one ought to make about the system length (see Chap. 12).
- We calculate the self-force to leading order when the system is rotating about a
  fixed center, perpendicular to its axis, in such a way that it always lies along a
  radial line from the center of rotation (see Chap. 8).
- We calculate the self-force to leading order when the system is rotating about a
  fixed center, almost parallel to its axis, in such a way that it always lies perpendicular to a radial line from the center of rotation (see Chap. 9), discussing once
  again the assumptions one ought to make about the system length.

We shall then draw conclusions from the results (see Chap. 10), and also solve
the problem of the discrepancy between energy-derived EM mass and momentum-
derived EM mass for both the spherical charge shell and this dumbbell charge system (see Chap. 11).

The calculations in Chaps. 6–9 are given in some detail, partly because it is
important to understand the complexity of this kind of calculation, and partly because, to the author's knowledge, the self-force calculations for the rotating extended charge system may well be original. The complexity is important. The motions
in the four self-force scenarios are each radically different from the geometrical
standpoint, but Maxwell's equations deliver simple results with a pattern. Either
there is some very trivial explanation for this, or it reveals some deep message from
the classical theory of electromagnetism.

Chapter 4
A Brief Excursion into General Relativity

The reader is assumed to know a little about general relativity in this chapter, although only the most basic knowledge would suffice for this rather trivial application. An original and alternative discussion of some of the issues discussed here can be found in [9].

4.1 Static Homogeneous Gravitational Field

We consider the charged sphere in a static homogeneous gravitational field (SHGF), i.e., a spacetime with coordinates (y^0, y^1, y^2, y^3) in which the metric assumes the form

$$g_{\mu\nu} = \begin{pmatrix} \left(1 + \dfrac{gy^3}{c^2}\right)^2 & 0 & 0 & 0 \\ 0 & -1 & 0 & 0 \\ 0 & 0 & -1 & 0 \\ 0 & 0 & 0 & -1 \end{pmatrix} . \qquad (4.1)$$

Here we have to specify that $y^3 \neq -c^2/g$ to avoid degeneracy of the metric. These are supposed to be the coordinates that would be set up in a laboratory that is held fixed relative to some distant gravitational source, at least to within some approximation, with a parallel gravitational field in the negative y^3 direction. It should be a good approximation in a small laboratory on the surface of the Earth over some short time span. (Actually, the reader is invited to be more critical of this easy interpretation [4].)

It is natural to seek a locally inertial frame at the spatiotemporal origin of these coordinates. The existence of such a frame is guaranteed by the weak equivalence principle (WEP) when one uses the usual Levi-Civita connection. Consider the coordinates (t, x, y, z) defined by

Lyle, S.N.: *A Brief Excursion into General Relativity*. Lect. Notes Phys. **796**, 47–72 (2010)
DOI 10.1007/978-3-642-04785-5_4   © Springer-Verlag Berlin Heidelberg 2010

$$t = \frac{c}{g} \sinh \frac{gy^0}{c^2} + \frac{y^3}{c} \sinh \frac{gy^0}{c^2} , \tag{4.2}$$

$$x = y^1 , \qquad y = y^2 , \tag{4.3}$$

$$z = \frac{c^2}{g} \left(\cosh \frac{gy^0}{c^2} - 1 \right) + y^3 \cosh \frac{gy^0}{c^2} . \tag{4.4}$$

Note that the spatiotemporal origins of the two systems coincide. By simple algebraic manipulations, the inverse transformation is

$$y^0 = \frac{c^2}{g} \tanh^{-1} \frac{gt/c}{1 + gz/c^2} , \tag{4.5}$$

$$y^1 = x , \qquad y^2 = y , \tag{4.6}$$

$$y^3 = \left[\left(z + \frac{c^2}{g} \right)^2 - c^2 t^2 \right]^{1/2} - \frac{c^2}{g} . \tag{4.7}$$

Let us evaluate the metric components relative to these coordinates. The quickest approach is to consider the invariant

$$ds^2 = \left(1 + \frac{gy^3}{c^2} \right)^2 (dy^0)^2 - (dy^1)^2 - (dy^2)^2 - (dy^3)^2 , \tag{4.8}$$

using

$$dy^0 = \frac{\partial y^0}{\partial t} dt + \frac{\partial y^0}{\partial z} dz , \quad dy^3 = \frac{\partial y^3}{\partial t} dt + \frac{\partial y^3}{\partial z} dz . \tag{4.9}$$

The partial derivatives are

$$\frac{\partial y^0}{\partial t} = \frac{c \left(1 + \frac{gz}{c^2} \right)}{\left(1 + \frac{gz}{c^2} \right)^2 - \left(\frac{gt}{c^2} \right)^2} , \quad \frac{\partial y^0}{\partial z} = \frac{-gt/c}{\left(1 + \frac{gz}{c^2} \right)^2 - \left(\frac{gt}{c^2} \right)^2} , \tag{4.10}$$

$$\frac{\partial y^3}{\partial t} = \frac{-gt}{\left[\left(1 + \frac{gz}{c^2} \right)^2 - \left(\frac{gt}{c^2} \right)^2 \right]^{1/2}} , \quad \frac{\partial y^3}{\partial z} = \frac{1 + \frac{gz}{c^2}}{\left[\left(1 + \frac{gz}{c^2} \right)^2 - \left(\frac{gt}{c^2} \right)^2 \right]^{1/2}} . \tag{4.11}$$

We now work out

$$\left(1+\frac{gy^3}{c^2}\right)^2 (dy^0)^2 - (dy^3)^2 = c^2 dt^2 - dz^2 , \tag{4.12}$$

after a short calculation.

The upshot of this is that the metric takes the Minkowski form in the new coordinates. These coordinates are not therefore merely locally inertial, with Minkowski form at the event in question and zero connection coefficients at that same event, but globally inertial. The spacetime we are considering has zero curvature everywhere and is therefore identical to the Minkowski spacetime. That is a point in favour of its representing an SHGF, because there would be no tidal effects in such a gravitational field, and so there should be no curvature in the general relativistic model.

We can find the geodesics of our spacetime very simply by considering the geodesics in the inertial coordinates and transforming them to the coordinates (y^0, y^1, y^2, y^3). One obvious geodesic is the spatial origin of the inertial coordinate system as time goes by. If we take s as the inertial time parameter of this worldline, it has the form

$$t = s , \quad x = y = z = 0 . \tag{4.13}$$

In the original coordinates, this becomes

$$y^0 = \frac{c^2}{g}\tanh^{-1}\frac{gs}{c} , \quad y^1 = 0 = y^2 , \quad y^3 = \frac{c^2}{g}\left[\left(1-\frac{g^2 s^2}{c^2}\right)^{1/2} - 1\right] . \tag{4.14}$$

Eliminating the parameter s, this becomes

$$y^3 = \frac{c^2}{g}\left[\frac{1}{\cosh(gy^0/c^2)} - 1\right] , \quad y^1 = 0 = y^2 . \tag{4.15}$$

We are saying that this is a timelike geodesic and hence a worldline corresponding to free fall in this model.

We may also seek the equation for the worldline of the spatial origin of the coordinates (y^0, y^1, y^2, y^3) when described in the inertial coordinates. This time our worldline is given, with parameter σ, by

$$y^0 = \sigma , \quad y^1 = y^2 = y^3 = 0 . \tag{4.16}$$

In the inertial coordinates, we have

$$t = \frac{c}{g}\sinh\frac{g\sigma}{c^2} , \quad x = 0 = y , \quad z = \frac{c^2}{g}\left[\cosh\frac{g\sigma}{c^2} - 1\right] . \tag{4.17}$$

Eliminating the parameter, this gives the worldline

$$x = 0 = y , \quad z = \frac{c^2}{g}\left[\left(1+\frac{g^2 t^2}{c^2}\right)^{1/2} - 1\right] . \tag{4.18}$$

This is the worldline of a particle whose 4-acceleration a^μ has constant squared magnitude $a^2 = \eta_{\mu\nu}a^\mu a^\nu = -g^2$.

The best way to show this is to work out the 4-acceleration components in the coordinates (y^0, y^1, y^2, y^3), because the worldline then has the simple form (4.16). The proper time of the worldline as a function of the coordinate y^0 can be read straight from the metric as $\tau = y^0/c$, since $y^3 = 0$ along the worldline. The components of the 4-acceleration are thus

$$a^\mu := \frac{d^2 y^\mu}{d\tau^2} + \Gamma^\mu_{\alpha\beta} \frac{dy^\alpha}{d\tau} \frac{dy^\beta}{d\tau} = c^2 \Gamma^\mu_{00} , \tag{4.19}$$

where $\Gamma^\mu_{\alpha\beta}$ are the Levi-Civita connection coefficients, calculated from the metric by the usual formula

$$\Gamma^\mu_{00} = \frac{1}{2} g^{\mu\nu} \left[2 \frac{\partial g_{\nu 0}}{\partial y^0} - \frac{\partial g_{00}}{\partial y^\nu} \right] , \tag{4.20}$$

whence

$$\Gamma^0_{00} = \frac{1}{2} g^{00} g_{00,0} = 0 , \qquad \Gamma^1_{00} = -\frac{1}{2} g^{11} g_{00,1} = 0 , \tag{4.21}$$

$$\Gamma^2_{00} = -\frac{1}{2} g^{22} g_{00,2} = 0 , \qquad \Gamma^3_{00} = -\frac{1}{2} g^{33} g_{00,3} = \frac{g}{c^2} \left(1 + \frac{g y^3}{c^2} \right) . \tag{4.22}$$

On the worldline, we have $y^3 = 0$ and hence,

$$a^\mu = (0, 0, 0, g) . \tag{4.23}$$

The invariant $a^2 := g_{\mu\nu} a^\mu a^\nu = -g^2$ is therefore constant along the worldline. If we work this out in the inertial frame, we obtain the same result. Even the 4-acceleration components themselves are constant when viewed in the SHGF lab frame.

The above change of coordinates (4.2)–(4.4) seems to be somewhat miraculous. However, one could find it in this way. Starting with a Minkowski spacetime and inertial coordinates like (t, x, y, z), one can envisage an accelerating observer AO, following the worldline (4.18). One can set up what are known as semi-Euclidean coordinates $\{y^\mu\}$ for the accelerating observer, with the following properties (where the Latin index runs over $\{1, 2, 3\}$):

- Any curve with all three y^i constant is timelike and any curve with y^0 constant is spacelike.
- At any point along the worldline of AO, the zero coordinate y^0 equals the proper time along that worldline.
- At each point of the worldline of AO, curves with constant y^0 which intersect it are orthogonal to it where they intersect it.
- The metric has the Minkowski form along the worldline of AO.
- The coordinates y^i are Cartesian on every hypersurface of constant y^0.
- The equation for the worldline of AO has the form $y^i = 0$ for $i = 1, 2, 3$.

These semi-Euclidean coordinates are precisely the coordinates (y^0, y^1, y^2, y^3) that we started out with here. For example, we noted above that the time coordinate y^0 is precisely the proper time along the worldline of the spatial origin of the coordinates (y^0, y^1, y^2, y^3). There is more about this in Sect. 12.1.4.

4.2 Equality of Inertial and Passive Gravitational Mass

We are concerned here with an experimental result that already features in Newtonian physics, namely the equality of passive gravitational mass, the measure of how much a particle is supposed to feel the Newtonian force of gravity, and inertial mass, the measure of how much a particle is supposed to resist being accelerated. In Newtonian gravitational theory, this means that an observer could not tell whether the acceleration of a test particle relative to her Euclidean coordinate system was due to some gravitational field or due to her own acceleration and the consequent acceleration of her comoving Euclidean coordinate system. (The acceleration of that Euclidean system must not be rotational here.) In pre-relativistic theory, one had no difficulty setting up Euclidean coordinate systems in spatial hypersurfaces and time was the same for everyone.

Let us consider what we can say about the charged sphere of the last chapter, or a charged particle represented by such a sphere, if it is held fixed at the origin of the $\{y^\mu\}$ coordinate system in Sect. 4.1. We interpret this as meaning that the sphere is supported against the uniform gravitational field, e.g., resting on a table in our laboratory. Of course, this requires a force, as attested by the fact that the four-acceleration a^μ is non-zero. The sphere is being pushed off its geodesic. So as viewed from the globally inertial freely falling frame, the sphere is accelerating.

Now one can import the whole theory of electromagnetism into GR by applying the strong principle of equivalence (SEP). We said above that, when the connection is the symmetric metric connection known as the Levi-Civita connection, then for any event in spacetime, there is always a neighbourhood of that event in which the metric has approximately the Minkowski form and the connection coefficients are approximately zero. This statement of the existence of such locally inertial frames is often called the weak equivalence principle [4]. The strong equivalence principle then states that all other bits of physics, i.e., non-gravitational physics, will appear relative to the locally inertial frames roughly as they do in the flat Minkowski spacetime. This is formulated by taking all the equations of whatever non-gravitational physics one is considering and replacing all ordinary partial coordinate derivatives by covariant derivatives relative to the Levi-Civita connection.

So Maxwell's equations can be extended to any GR spacetime. In fact, this gives a minimal extension of Maxwell's equations (MEME). There are other extensions, in which curvature terms are introduced, but we shall only consider MEME. Now the reason for describing the SHGF in such detail is that there are in fact global inertial frames in that case, so we can view our charged sphere as accelerating uniformly in a flat spacetime without gravity. But we know in that context that the sphere will exert an EM self-force on itself. The strong equivalence principle then tells us that, since the electromagnetic effects will be exactly the same for the charged sphere when it is sitting on a table in our laboratory, according to this model using GR and SEP, the sphere will exert an EM self-force on itself in that case too. This self-force will oppose the supporting force of the table which pushes it off its geodesic.

To see the import of this deduction, imagine for a moment that all the inertial mass of the charge sphere is EM mass. Then we are saying that the supporting force of the table is precisely what is required to cancel the EM self-force, no more and no less. In this view of things, we understand why a force is required (from the table) to keep the sphere off its geodesic in this way, i.e., to stop it falling. Better than that, the passive gravitational mass of the sphere is precisely equal to its inertial mass. Let us try to formulate these ideas.

Suppose the sphere does have a mechanical mass m_{mech}, so that its equation of motion in the GR formulation is

$$m_{\text{mech}} \left(\frac{\mathrm{d}^2 y^\mu}{\mathrm{d}\tau^2} + \Gamma^\mu_{\alpha\beta} \frac{\mathrm{d}y^\alpha}{\mathrm{d}\tau} \frac{\mathrm{d}y^\beta}{\mathrm{d}\tau} \right) = F^\mu_{\text{self}} + F^\mu_{\text{supp}} ,$$

where F_{supp} is the supporting force. We saw above that, when the sphere is held fixed relative to the semi-Euclidean coordinates $\{y^\mu\}$, it has four-acceleration $a = (0, 0, 0, g)$, and by applying MEME, we know that

$$F_{\text{self}} = -m_{\text{EM}}^{\text{SFDM}}(0, 0, 0, g) ,$$

so we can rewrite the equation of motion in the form

$$\left(m_{\text{mech}} + m_{\text{EM}}^{\text{SFDM}} \right) (0, 0, 0, g) = F_{\text{supp}} .$$

The passive gravitational mass m_{PG} would be defined by the proportionality between the four-acceleration and F_{supp}, whence

$$\boxed{m_{\text{PG}} = m_{\text{mech}} + m_{\text{EM}}^{\text{SFDM}}} \qquad (4.24)$$

If there is no mechanical mass, then we have

$$m_{\text{PG}} = m_{\text{EM}}^{\text{SFDM}} = m_{\text{inertial}} .$$

If the mechanical mass is there, but derives from self-forces due to other fields (weak and strong) sourced by components of the particle, then the law of motion of the particle is no longer given fundamentally by a law like Newton's second law (the famous $\mathbf{F} = m\mathbf{a}$), but rather by a law of the form

$$\boxed{\sum_{\text{fields}} F_{\text{self}} + F_{\text{supp}} = 0} \qquad (4.25)$$

The typical form of Newton's second law then follows by analysing the self-forces into some multiple of the four-acceleration, and the whole problem of the research program suggested in this book is to show that this is always possible, not just for EM forces, but for the other forces too, and then to show that there is no other mechanical mass. It is a big program, but what we obtain automatically here is the equality of the inertial mass and the passive gravitational mass.

It is important to note that SEP really is necessary for this deduction. Even though the spacetime is flat, we are interpreting it as a spacetime in which there is a gravitational field, and we are applying GR to do so, and this means that we do need SEP to tell us how to do electromagnetism in this context. Apart from that, none of the algebra displayed at the beginning of this chapter is really necessary. It was only put in for concreteness, to define the context. Note also that we have ignored the higher order terms in the self-force (3.28), such as the radiation reaction, but these could be taken into account and that would not affect the conclusion here.

Physically, the spacetime metric plays a key role here. In the $\{y^\mu\}$ frame, the electromagnetic field and any other field sourced by the particle itself will be distorted in relation to their form in an inertial frame in which the particle is permanently at rest. The distortion will mean that the particle exerts a force on itself via its distorted fields, so it will need to be supported by some F_{supp} to remain at rest in the $\{y^\mu\}$ coordinate system. What we are saying then is that we need a metric that is not in the Minkowskian form in the relevant frame in order to distort the electromagnetic fields from the Coulomb form, in such a way that there is a self-force.

We have considered a spacetime that is not actually curved, so that the extension of Maxwell's theory is in fact identical to the flat spacetime theory, but the same arguments apply to any curved spacetime in GR. The difference is only in the details of the approximation because, in a generally curved spacetime, one can only find *locally* inertial frames. Put another way, the SHGF discussed above is only an approximation to the spacetime in an Earth-based laboratory, for example, due to variations in the gravitational effects across the laboratory or as time goes by. This means that the appropriate version MEME of Maxwell's theory in a locally inertial frame in the curved spacetime will not look exactly like the usual version of Maxwell's theory in an inertial frame in flat spacetime. But we will still deduce that, when a charged sphere (or particle) is prevented from free fall, this will require a supporting force to overcome self-force effects, and that any self-force contributions to the inertial mass will be exactly matched by equal self-force contributions to the passive gravitational mass, as described for the SHGF case.

Note also that, in this view, if the particle is at rest in a globally (locally) inertial frame in an SHGF (in a generally curved spacetime), i.e., following a geodesic like (4.15) for the SHGF case, its fields will not be distorted (will be only slightly distorted) from the Coulomb form in that frame, where Maxwell's equations assume exactly (approximately) the usual flat spacetime form, so they will not give a net force on it, and F_{supp} must then be zero if (4.25) is to be satisfied. We call this free fall.

It is worth mentioning that the theory of gravity looks very different in special relativity [4]. Here spacetime is always flat, with frames in which the metric has the

Minkowski form, and gravity is a force like any other. This means that electromagnetism can be described by Maxwell's theory in an inertial frame, even if there is a source of gravity sitting at the origin of that frame. The charged sphere may be supported in such a way that it does not move relative to the source of gravity, and yet be stationary in the inertial frame, and hence not exert any self-forces on itself.

4.3 Status of this Result

One might say that it is not really surprising to deduce that the inertial and passive gravitational masses are equal on the basis of GR plus SEP, since the whole theory of GR was inspired by this very observation. There are already two ways in which GR is considered as an attempt to explain this equality according to certain accounts:

- One could not even formulate the geodesic principle for the motion of massive particles if this were not the case, so the very possibility of a theory like GR implies the equality.
- The geodesic principle can be deduced from Einstein's equations, along with certain other assumptions, and this is even taken by some as an explanation of inertia.

So what is the geodesic principle?

4.3.1 Geodesic Principle

This states that, when point particles are not acted upon by forces (apart from gravitational effects), their trajectories take the form

$$\frac{d^2 x^i}{ds^2} + \Gamma^i_{jk} \frac{dx^j}{ds} \frac{dx^k}{ds} = 0 \,, \tag{4.26}$$

where $x^i(s)$ gives the worldline as a function of the proper time s of the particle and Γ^i_{jk} are the connection coefficients in the given coordinate system. In the literature, this is often derived from an action principle. One writes the worldline as a function $x^i(\lambda)$ of some arbitrary parameter λ, whence the appropriate action for the worldline between two points $P_1 = x(\lambda_1)$ and $P_2 = x(\lambda_2)$ of spacetime is

$$s(P_1, P_2) := \int_{\lambda_1}^{\lambda_2} \left(g_{ij} \frac{dx^i}{d\lambda} \frac{dx^j}{d\lambda} \right)^{1/2} d\lambda = \int_{\lambda_1}^{\lambda_2} L d\lambda = \int_{\lambda_1}^{\lambda_2} ds \,, \tag{4.27}$$

with Lagrangian

$$L := \left(g_{ij} \frac{\mathrm{d}x^i}{\mathrm{d}\lambda} \frac{\mathrm{d}x^j}{\mathrm{d}\lambda} \right)^{1/2}. \tag{4.28}$$

The Euler–Lagrange equations extremising the action under variation of the worldline are

$$\frac{\mathrm{d}}{\mathrm{d}\lambda} \left(\frac{\partial L}{\partial \dot{x}^i} \right) - \frac{\partial L}{\partial x^i} = 0, \tag{4.29}$$

with $\dot{x}^i := \mathrm{d}x^i/\mathrm{d}\lambda$, and these lead to the above geodesic equation (4.26).

The action for some particles labelled by a is

$$\mathscr{A} = -\sum_a cm_a \int \mathrm{d}s_a, \tag{4.30}$$

where m_a is the mass of particle a and s_a is its proper time. This is the action *because* variation of the worldline of particle a gives its equation of motion as

$$\frac{\mathrm{d}^2 a^i}{\mathrm{d}s_a^2} + \Gamma^i_{jk} \frac{\mathrm{d}a^j}{\mathrm{d}s_a} \frac{\mathrm{d}a^k}{\mathrm{d}s_a} = 0. \tag{4.31}$$

Likewise, if some of the particles are charged with charge e_a for particle a, and there are some EM fields F_{ik}, one declares the action to be

$$\mathscr{A} = -\sum_a cm_a \int \mathrm{d}s_a - \frac{1}{16\pi c} \int F_{ik}F^{ik}(-g)^{1/2}\mathrm{d}^4x - \sum_a \frac{e_a}{c} \int A_i \mathrm{d}a^i, \tag{4.32}$$

where A_i is a 4-vector potential from which F_{ij} derives, simply *because* variation of the worldline of particle a gives its equation of motion as

$$\frac{\mathrm{d}^2 a^i}{\mathrm{d}s_a^2} + \Gamma^i_{jk} \frac{\mathrm{d}a^j}{\mathrm{d}s_a} \frac{\mathrm{d}a^k}{\mathrm{d}s_a} = \frac{e_a}{m_a} F^i{}_j \frac{\mathrm{d}a^j}{\mathrm{d}s_a}, \tag{4.33}$$

the minimal generalisation of the Lorentz force law to a curved manifold, while variation of A_i gives the EM field equations as the minimal extension of Maxwell's equations to the curved spacetime. Of course, these actions are designed to give appropriate field equations, and we are just decreeing here that the appropriate field equation for the particle labelled by a is a geodesic equation, or an equation like (4.33).

So what is the physical motivation for the geodesic principle? It is claimed here that the appropriate physical argument supporting (4.26) is an application of the strong principle of equivalence. This is where we discover exactly how we are to link what happens mathematically in a curved manifold with measurements in our own world. We start from an action, but at some point we must say what the point of contact would be with physical reality. Let us suppose we impose a strong principle of equivalence, that is, we say roughly speaking that any physical interaction other than gravitation behaves in a locally inertial frame as though gravity were absent. Relative to such a frame, any particle that is not subject to (non-gravitational) forces will then move in a straight line with uniform velocity, i.e., it will follow the trajectory described by

$$\frac{dv^i}{ds} = 0 \,, \tag{4.34}$$

where v^i is its 4-velocity. This is expressed covariantly through (4.26). If there are non-gravitational forces, we start with $\mathbf{F} = m\mathbf{a}$ in the locally inertial frame and we find (4.26) with a force term on the right-hand side. It is quite clear that we still have a version of Newton's second law $\mathbf{F} = m\mathbf{a}$, so the present view is that we have not explained inertia and inertial effects by this ploy, but merely extended this equation of motion to the new theory.

It is worth looking more closely at the claim that (4.34) is expressed covariantly through (4.26). A cheap way is to set the connection coefficients equal to zero in (4.26). This is basically the observation that the two equations are the same relative to Cartesian coordinates in a flat spacetime. Such a claim misses out some of the machinery of the connection construction that lies at the heart of non-Euclidean geometry, but this is not the place to expose all that. A justification of sorts can be found in [4, Sect. 2.5].

It is not totally obvious from what has just been said that SEP is absolutely necessary here and some authors would claim that it is not. This will be discussed further in the following (see in particular Sect. 4.3.3).

4.3.2 Equality of Inertial and Passive Gravitational Mass Revisited

Equation (4.26) is thus taken as the equation of motion of a point particle upon which no forces are acting, unless one counts gravity as a force. We observe that there is no mention of any parameters characterising the point particle. In particular there is no mention of its inertial mass. Of course there is no mention of parameters describing its inner make-up. After all, it is supposed to be a point particle. In this book, we are considering what would happen to a slightly spatially extended particle, i.e., with a world tube that intersects spatial hypersurfaces in a small region rather than a single mathematical point. This object might be spinning in some sense, or contain a charge distribution, for example. We shall return to this point in a moment, but let us begin with the disappearance of the inertial mass since this is directly relevant to the first point mentioned at the beginning of this section.

So where did the inertial mass of the particle go? If we look back to (4.33), viz.,

$$m_a^{\text{inertial}} \left(\frac{d^2 a^i}{ds_a^2} + \Gamma_{jk}^i \frac{da^j}{ds_a} \frac{da^k}{ds_a} \right) = e_a F^i{}_j \frac{da^j}{ds_a} \,, \tag{4.35}$$

we find the inertial mass m_a^{inertial} multiplying the acceleration term in the equation to give a force on the right that is determined by an external field F_{ij} and a coupling constant e_a characterising the particle. This is a typical equation of motion when there is some non-gravitational force (in this case electromagnetic) acting on the particle. Now in Newtonian gravitational theory, if the external field happens to be a gravitational potential Φ, one gets an equation of motion like this:

$$m_a^{\text{inertial}} \left(\frac{d^2 a^i}{ds_a^2} + \overline{\Gamma}^i_{jk} \frac{da^j}{ds_a} \frac{da^k}{ds_a} \right) = m_a^{\text{pg}} h^{ij} \Phi_{,j} , \qquad (4.36)$$

where m_a^{pg} is the passive gravitational mass of particle a, $(h^{ij}) = \text{diag}\,(0,1,1,1)$, and $\overline{\Gamma}^i_{jk}$ is the connection appropriate to Newtonian spacetime and relative to whatever coordinates we have chosen to describe it. But due to the observed equality of inertial mass and passive gravitational mass, viz., $m_a^{\text{pg}} = m_a^{\text{inertial}}$, the coupling factor on the right-hand side is just the same factor as we have on the left-hand side. The only relevant characteristic of our point particle thus cancels out.

This explains how the inertial mass disappears from the equation, precisely because of the observed equality of inertial mass and passive gravitational mass, but how do we get rid of the gravitational potential we have just introduced? Of course, we can absorb it into the connection, following the much more detailed account of all this in [18]. We now have a new connection

$$\Gamma^i_{jk} := \overline{\Gamma}^i_{jk} + h^{il} \Phi_{,l} t_j t_k , \qquad (4.37)$$

where $(t_i) := (1,0,0,0)$, so that $t_i = \partial t / \partial x^i$. Equation (4.36) becomes

$$\frac{d^2 a^i}{ds_a^2} + \Gamma^i_{jk} \frac{da^j}{ds_a} \frac{da^k}{ds_a} = 0 , \qquad (4.38)$$

still in this Newtonian context. So the equality of inertial and passive gravitational mass allows us to treat the trajectories of particles subjected only to gravitational effects as geodesics of a non-flat connection, because we do expect this new connection in (4.37) to be non-flat in general.

As Friedman says in [18], the equality of inertial and passive gravitational mass implies the existence of a connection Γ such that freely falling objects follow geodesics of Γ. This does not work for other types of interaction, where the ratio of $m_a := m_a^{\text{inertial}}$ to the coupling factor, e.g., e_a for a charged particle, is not the same for all bodies. The worldlines of charged particles in an EM field cannot be construed as the geodesics of any single connection, because m_a/e_a in (4.33) varies from one particle to another.

Put another way, the equality of inertial and passive gravitational mass must be true if any theory of gravitation like general relativity, in which gravitational interaction is explained by the dependence of a nonflat connection on the distribution of matter, is to be possible. Note in passing that general relativity is not the only theory of this type. Classical gravitational theory can also be formulated in this way by taking advantage of the very same equivalence of inertial and passive gravitational mass. Friedman's book [18] is recommended for anyone who thinks that Newtonian gravitational theory cannot be given a fully covariant and totally geometric treatment. The essential difference with general relativity is that, in this treatment of Newtonian gravity, there is a flat connection $\overline{\Gamma}$ living alongside the non-flat connection Γ of (4.37). The deep fact here is that, in general relativity, the non-flat connection is the *only* connection of the spacetime.

So what of the first point mentioned at the beginning of this section, namely that the very possibility of a theory like GR implies the equality of inertial and passive gravitational mass? Might this not undermine the result of Sect. 4.2 which used GR and SEP to show that at least the self-force contributions to inertial mass are likely to equal the self-force contributions to passive gravitational mass? Put like this, we appear to be assuming the result in order to demonstrate it.

Looking back at Sect. 4.2, what we proposed was a new law (4.25), viz.,

$$\boxed{\sum_{\text{fields}} F_{\text{self}} + F_{\text{supp}} = 0} \tag{4.39}$$

which would replace Newton's second law $\mathbf{F} = m\mathbf{a}$ and its direct extensions to GR with the help of SEP. Newton's second law in its usual form follows from (4.39) by analysing the self-forces into some multiple of the four-acceleration, and as mentioned earlier, the whole problem of the research program suggested in this book is to show that this is always possible, not just for EM forces, but for the other forces too, and then to show that there is no other mechanical mass. So a dynamical law, viz., (4.39), is still necessary here, but from it we can deduce results that were merely imposed previously, at least in the case where the inertia is entirely due to self-force effects.

For one thing, we understand physically why a supporting force is needed, namely to balance self-forces. In GR as it is usually presented, the supporting force is needed because the particle has non-zero four-acceleration, but we do not know why a non-zero four-acceleration should require a (supporting) force any more than we know why an acceleration should require a force in Newtonian physics.

Another point is that self-forces make a distinction between uniform velocities and changing velocities. The self-force is zero when the particle has a uniform velocity, and only becomes non-zero when the particle velocity is changing. So we understand from (4.39) why no force F_{supp} is required on the particle to keep it in free fall. (The terminology 'supporting force' is something of a misnomer here!) And we understand the contrast between Newton's first and second laws, in the same way as the self-force idea explains this contrast in Newtonian physics.

And finally, although the equality of inertial and passive gravitational mass was crucial to the very existence of a theory like GR, we do have a mechanism here to explain why this should be the case. It is not easy to formulate this mechanism in the context of GR, because there is no such thing as passive gravitational mass once one has adopted general relativity as the fundamental theory. The best we can do is an argument of the form leading up to (4.24) on p. 52, which was based on the claim that the passive gravitational mass m_{PG} would be defined by the proportionality between the four-acceleration and F_{supp}. This argument, like all others presented, should be regarded critically by the reader.

4.3.3 Do Einstein's Equations Explain Inertia?

To come now to the second point listed at the beginning of this section, it turns out that the geodesic principle is not a principle at all, and neither is it likely to be any better than an approximation for a real particle that cannot be treated as a mathematical point. For the fact is that the geodesic principle follows from Einstein's equations in general relativity, provided that we also have SEP and *provided that we can make suitable assumptions about the particle*. Here follows a proof of sorts.

Recall first that Einstein's equations can be written

$$G_{ij} = -\kappa T_{ij} , \qquad (4.40)$$

where

$$G_{ij} := R_{ij} - \frac{1}{2} g_{ij} R \qquad (4.41)$$

is the Einstein tensor expressed in terms of the Ricci tensor R_{ij} and curvature scalar R, κ is a constant that turns out to be expressible as

$$\kappa = \frac{8\pi G}{c^4} , \qquad (4.42)$$

and T_{ij} is the energy–momentum tensor expressing the distribution of mass and energy in the spacetime.

Now the covariant divergence of the Einstein tensor is zero in many circumstances, in particular when something called the torsion is zero. But the torsion is indeed often zero. In fact, it is sourced by the spin currents of matter in such a way that, in contrast to curvature, it does not propagate in spacetime, so it could only be nonzero in regions where there is matter or energy with some rotational property. A very clear, though somewhat sophisticated account of all this can be found in [13, Chap. 5]. Anyway, in a region where there is no spinning matter, Einstein's equation (4.40) implies that the covariant divergence of the energy–momentum tensor is zero. This is what we shall use to derive the geodesic 'principle'.

To do this, we shall consider an almost-pointlike particle. So when almost-point particles are not acted upon by forces (apart from gravitational effects), we would like to show that their trajectories take the form

$$\frac{d^2 x^m}{ds^2} + \Gamma^m_{kj} \frac{dx^k}{ds} \frac{dx^j}{ds} = 0 . \qquad (4.43)$$

We consider a small blob of dustlike (i.e., zero pressure) matter with density ρ and velocity field

$$v^i = \frac{dx^i}{ds} . \qquad (4.44)$$

This equation expresses the fact that we view each component dust particle as having its own worldline $x^i(s)$. The energy–momentum tensor for this matter is then

$$T^{ik} = \rho \frac{dx^i}{ds}\frac{dx^k}{ds} \tag{4.45}$$

and we are saying that Einstein's field equation (4.40) implies that

$$T^{ik}_{\;\;;k} = 0 . \tag{4.46}$$

We analyse (4.46) by inserting (4.45) and the result is the geodesic equation (4.26). For completeness, here is the argument. We have

$$\rho_{,k}\frac{dx^i}{ds}\frac{dx^k}{ds} + \rho\left(\frac{\partial}{\partial x^k}\frac{dx^i}{ds} + \Gamma^i_{km}\frac{dx^m}{ds}\right)\frac{dx^k}{ds} + \rho\frac{dx^i}{ds}\left(\frac{\partial}{\partial x^k}\frac{dx^k}{ds} + \Gamma^k_{km}\frac{dx^m}{ds}\right) = 0 . \tag{4.47}$$

If we did not have the idea of a velocity field v^i, it would be difficult to interpret partial derivatives of dx^i/ds with respect to the coordinates. But as things are, we can say

$$\frac{dx^k}{ds}\frac{\partial}{\partial x^k}\frac{dx^i}{ds} = \frac{dx^k}{ds}\frac{\partial v^i}{\partial x^k} = \frac{dv^i}{ds} = \frac{d^2x^i}{ds^2} . \tag{4.48}$$

The terms in the second bracket of (4.47) are

$$\frac{\partial v^k}{\partial x^k} + \Gamma^k_{km}v^m = \mathrm{div}\,v , \tag{4.49}$$

and the whole thing can now be expressed by

$$\mathrm{div}(\rho v)\frac{dx^i}{ds} + \rho\left(\frac{d^2x^i}{ds^2} + \Gamma^i_{km}\frac{dx^k}{ds}\frac{dx^m}{ds}\right) = 0 . \tag{4.50}$$

By mass conservation,

$$\mathrm{div}(\rho v) = 0 , \tag{4.51}$$

and the result follows.

This proof purports to show that each constitutive particle of the blob follows a geodesic. But then we did not allow these particles to jostle one another. For example, we have zero pressure, as attested by the form of the energy–momentum tensor in (4.45). And we did not allow the particles to generate any torsion by revolving about the center of energy of the blob. And neither did we endow them with electric charge. It is in this sense that the geodesic 'principle' is in fact just an approximation, unless the test particle is not a blob, but a mathematical point.

Some argue that inertia is explained in general relativity, precisely because of the above proof (or better variants of it). This point of view is expressed in the philoso-

phical study by Brown [14, p. 141]. He claims that GR is the first in the long line of
dynamical theories, based on the Aristotelian distinction between natural and forced
motions of bodies, that *explains* inertial motion. This is not the view taken here, for
reasons to be explained shortly. However, other issues discussed in Brown's book, in
particular what he refers to as the dynamical approach to spacetime structure, should
mark a turning point in our understanding of relativity theories that is exactly in line
with the approach advocated in the present book, and in particular with the issues
discussed in Bell's paper [5] and extensions of those points to GR [12].

Concerning the putative explanation of inertial motion according to Brown [14,
Sect. 9.3], inertia in GR is just as much a consequence of the field equations as
gravitational waves, i.e., inertial motion of test particles is just part of the dynamics.
Here is an argument against that view.

Recall the discussion just after (4.33) on p. 55. It was pointed out that actions
like (4.30) and (4.32) are designed to give appropriate field equations, and that the
appropriate field equation for the particle labelled by a is a geodesic equation, or an
equation like (4.33). Now in GR, one adds a gravitational part to the action, viz.,

$$\mathscr{A}_{\mathrm{grav}} := \frac{c^3}{16\pi G} \int R(-g)^{1/2} \mathrm{d}^4 x . \tag{4.52}$$

Some textbooks motivate this as follows. When the metric is varied in $\mathscr{A}_{\mathrm{grav}}$, a
constant multiple of the Einstein tensor pops out. The point about this is the observation that, when the metric is varied in an action like (4.32), the energy–momentum
tensor T_{ij} pops out. One gets a sum of contributions to this tensor from the matter
as encapsulated in the action term

$$-\sum_a c m_a \int \mathrm{d}s_a , \tag{4.53}$$

and from the EM fields as encapsulated in the action term

$$-\frac{1}{16\pi c} \int F_{ik} F^{ik} (-g)^{1/2} \mathrm{d}^4 x .$$

Setting the variation of the full action with respect to the metric equal to zero, one
then obtains the Einstein equations, with the Einstein tensor on one side and the
total energy–momentum on the other side.

Now the covariant divergence of the Einstein tensor is zero (assuming zero torsion) and this could in principle be worrying, because the Einstein equation then
implies that the covariant divergence of the total energy–momentum is zero. However, there is a general result that the energy–momentum tensor derived from an
action of the form

$$\int L(-g)^{1/2} \mathrm{d}^4 x \tag{4.54}$$

by varying the metric always has zero covariant divergence when L is a scalar, and
more sophisticated versions, e.g., invariance of the matter action under the group of

diffeomorphisms is sufficient to guarantee zero covariant divergence of the corresponding energy–momentum tensor on shell if the torsion is zero [13, Sect. 6.5]. (As mentioned above, if the torsion is not zero, the covariant divergence of the Einstein tensor is not zero either. This case is not considered here.) Of course, the action $\mathscr{A}$ in (4.53) does not have the form (4.54), but one expects some general theorem to ensure that the resulting energy–momentum tensor will have zero covariant divergence on shell, i.e., when the field equations, that is, the geodesic equations, are satisfied.

So it looks as though the geodesic equations, and their variants with a force on one side, are built in by construction of the action. It is no surprise therefore that they should pop out again when we set the covariant divergence of the energy–momentum tensor equal to zero. Perhaps one should be more suspicious of arguments from actions. They are neat, and bring a level of unity in the sense that one can derive several dynamical equations from the same action by varying different items. On the other hand, we are only getting out what we put in somewhere else.

A recent commentator [16] asserts that the motion of massive test particles is independent of SEP. This refers to the above idea that geodesic motion follows by conservation of energy–momentum, which in turn follows from Einstein's equations, whence the inertia of massive objects is supposed to be explained by the theory. Here we argue against both conclusions:

(i)   independence from SEP, and
(ii)   insofar as geodesic motion is a consequence of Einstein's equations, this explains inertia.

Note, however, that the paper [16] is highly recommended for a clear account of the idea advocated by Brown, and also in this book, that the metric tensor in relativity theories gets its geometric significance through detailed physical arguments.

Regarding (ii), we have just seen a counter-argument. When we wanted to deduce geodesic motion from Einstein's equations, we decreed that the energy–momentum tensor of the test particle was that of a very small cloud of dust, then reasoned heuristically and took a point-particle limit at the end of a short calculation. Further, we assumed that the component matter was not spinning in any way, and that the component particles carried no electric charge. Put like that, one sees how limited the proof is. No real particle with spatial extent could be like this, and indeed, no real particle would actually free fall along a geodesic, and nor even would its center of energy.

In any case, there is another aspect of the model that one ought to justify – one ought really to explain the choice of energy–momentum tensor. Another proof of the geodesic principle from Einstein's equation due to DeWitt is more sophisticated, using distributions [15]. There one derives the energy–momentum tensor (distribution) for a point particle by varying the metric in the usual action $S = -m \int d\tau$ for a point particle in relativity theory. One then shows that conservation implies the geodesic equation. But it would be a circular argument to say that this proves that the geodesic equation follows from Einstein's equation, because the action $S = -m \int d\tau$

is designed to deliver the geodesic equation when one varies the particle worldline of which it is a functional.

In fact, the variational method itself shows what is happening here. The point is that the action S is invariant under coordinate changes. This alone implies that the covariant divergence of the energy–momentum tensor derived from it by varying the metric will be zero if and only if the particle follows a geodesic. It is difficult to believe that one would really explain inertial behaviour solely by the coordinate invariance of the action. But of course, as mentioned above, the action is designed to yield geodesic motion.

Interestingly, and probably significantly, all this assumes there is no torsion, because it uses the Levi-Civita connection. But torsion is generated by spinning matter as mentioned above [13, Chap. 5]. This brings us to the claim (i) above that SEP would not be needed to show that test particles follow geodesics.

In the above demonstration of the geodesic principle, we require the particle to be moving in a region of spacetime where the torsion is zero, because this is a sufficient condition for the covariant divergence of the Einstein tensor to be zero. One often forgets torsion outright and just decrees the connection coefficients to be symmetric in their two lower indices. However, it is interesting to draw attention to torsion here because it is precisely spinning matter that generates torsion. Since torsion does not propagate beyond its sources, one only needs to assume that the test particle (blob) moves in empty spacetime. But what if the blob is itself spinning?

Now it is known that a spinning blob of matter will not free fall along a geodesic. The spin angular momentum of the blob couples with the curvature and tweaks it off the geodesic. How does one show this? One begins with a Lagrangian which treats the blob as an ensemble of particles, then expands everything about the center of energy [15]. The best thing would be to include all the electromagnetic forces holding the particles together, but fortunately one can just make a quasi-rigidity assumption and go from there. The latter assumption avoids talking about, but nevertheless embodies, non-gravitational forces. It thus assumes, in a very hidden way admittedly, the strong principle of equivalence.

It is the center of energy of the blob that approximately (but not quite) follows a geodesic. It seems a remarkable achievement just to get this. But it is not so remarkable, because that Lagrangian mentioned in the last paragraph is precisely the one that is designed to deliver geodesic motion for the constituent particles of the blob, were they not constrained by quasi-rigidity.

In fact, Butterfield does specify that he is talking about non-rotating test particles in [16]. But the point remains that one really must ask what is meant by a test particle. It is supposed to be a mathematical point, but that is an approximation. And the fact is that all test particles are going to involve non-gravitational forces. Of course, it is precisely for the blob of dust that one gets one of the derivations, taking a limit in the end as the size of the blob goes to zero. On the other hand, any realistic test particle (even with a limit at the end) is going to involve non-gravitational forces, and will in general be spinning (not spinning would be a very special and improbable case). Even if not spinning, the general Lagrangian approach including EM forces is going to predict deviations from the geodesic.

Note in this context that photons have to follow null curves because of Maxwell's equations [17], and this brings in a need for SEP. Now it would be a strange thing in a way if SEP were required to show that photons have to follow null curves, but massive particles could get away without having to obey any vestige of the laws of any of the forces governing their make-up, and hence avoid any need for SEP. Put another way, if we say that there are no non-gravitational effects to be taken into account when considering our test particle, then since SEP deals only in non-gravitational effects, it cannot be needed to say *anything* about the test particle. This is pure logic. There is no physics at all in it.

4.4 Active Gravitational Mass

Before leaving the subject of general relativity, it is important to mention the notion of active gravitational mass. In Newtonian theory, the gravitational force of the Sun, with mass $M_\odot$, on the Earth, with mass M_{Earth}, is of course

$$F = \frac{GM_\odot M_{\text{Earth}}}{R^2} , \tag{4.55}$$

where R is the distance separating the two bodies and G is the gravitational constant. If one is considering the path of the Earth through space as a result of the gravitational force on it due to the Sun, then $M_\odot$ is referred to as the active gravitational mass of the Sun in Newtonian theory, because this is what is supposed to have an attractive effect on the Earth as a test particle.

Of course, with the advent of relativity theory, we know that even the radiation in the Sun adds to the total value of $M_\odot$, as does anything in the star that increases its energy. And the view of what happens here is very different from the view in Newtonian physics as described by (4.55). In GR, the active gravitational mass is anything that curves spacetime, i.e., anything that contributes to the energy–momentum tensor of matter, radiation, or anything else there might be out there.

The point about this is that, in Newtonian physics, one has another great, unexplained equality, namely the equality of passive and active gravitational mass, both of these being equal also to the inertial mass. In this chapter we have indicated a way one might understand why the passive gravitational mass might be equal to the inertial mass, but we have not mentioned the other equality. In the view described here, the active gravitational masses have to do their work before we apply our derivation of the passive gravitational mass. Indeed they have to curve the spacetime, or at least alter the metric from the Minkowski form. In this sense, active and passive gravitational mass are not on the same footing in the theory outlined in this chapter. That aspect of things therefore remains a mystery.

4.5 The Machian Program and the Brans–Dicke Theory

An interesting discussion of Mach's principle can be found in Narlikar's action-packed book on cosmology [19]. This is a potted version of that account. One can measure the spin of the Earth about its polar axis either by observing the rising and setting of the stars, or by observing the gradual rotation of a Foucault pendulum. It is not obvious that the two methods should give the same result, but they do. The first method defines a frame relative to the background of distant stars, while the second is based on the existence of absolute space according to Newton, or the existence of inertial frames, so that when one is using a rotating frame like the one fixed on the surface of the Earth, Newton's laws of motion require us to take the rotation of the frame into account.

One is thus assuming that the background of distant stars coincides with Newton's absolute space. Realising this, Mach reasoned that Newton's postulate of absolute space, relative to which Newton's laws take their simplest form, might be related in some way to the large scale distribution of matter in the Universe. Actually, there has been much debate about what Mach actually meant, and what various authors, such as Einstein himself, took it to mean [18,20], but the basic idea is that the inertia of a massive particle must somehow originate in an interaction between that particle and all the other matter in the Universe.

Of course, the absolute space is a problem in itself. How does this frame of reference obtain its special status in which no inertial forces (as Newton called them) are required? And how could this frame be identified without recourse to Newton's second law, which is based on it? There is the possibility that it is the Universe itself that provides a background reference frame that may be identified with Newton's special frame. It remains of course to formulate this relationship.

Narlikar provides the following idea [19]. He imagines a single body in an otherwise empty universe. When there are no forces on it, Newton's second law gives

$$m\mathbf{a} = 0,$$

where m is the inertial mass of the body and $\mathbf{a}$ its acceleration. Newton would deduce that $\mathbf{a} = 0$, i.e., that the body was moving with uniform velocity relative to absolute space. But in this empty universe, there is no background relative to which the velocity could be measured. Saying that $\mathbf{a} = 0$ has no operational meaning here. In fact, it would be better if $\mathbf{a}$ were indeterminate, and the perfect way to obtain that would be to deduce rather that

$$m = 0.$$

The idea is that our measure of inertia might depend on the existence of the background in such a way that, in the absence of the background, the measure is zero.

The Newtonian view that inertia is a property of the matter alone is replaced by the idea that it is a property also of the background provided by the distribution of matter and energy throughout the whole Universe. On the face of things, one might wonder whether general relativity achieves this. After all, the curvature at any space-time event depends on the energy–momentum tensor in the neighbourhood of that event, and this tensor is generated by the distribution of matter and energy throughout the Universe. But the fact is that GR does not achieve the Machian program. There seem to be three points:

- We do not see how to deduce the inertial mass of a particle from the local values of the energy–momentum tensor.
- We do not explain why the law of motion of a particle should be basically just an extension of $\mathbf{F} = m\mathbf{a}$.
- Absolute rotation is independent of external masses because, if two spheres S_1 and S_2 were alone in the Universe, in relative rotation about the axis joining their centers, it would still be possible for one and only one of them to experience distorting differential effects.

A deeper discussion of the third point can be found in [18, Sect. V.5].

The Brans–Dicke theory tries to implement the idea of an inertial mass that can vary as a test particle moves through spacetime, not just as a result of its changing velocity, but due to the presence of a scalar field ϕ that permeates spacetime and results somehow from the overall distribution of matter. It will be interesting to review some simple features of this theory, while a more detailed account can be obtained from [19].

In fact, Brans and Dicke wanted a framework in which the gravitational constant G would arise from the structure of the Universe, whence a changing G would be viewed as a Machian consequence of a changing Universe. For various reasons, they postulated that G should behave as the reciprocal of a scalar field ϕ, viz., $G \sim \phi^{-1}$. They then wrote down an adapted version of the Einstein–Hilbert action (4.52) and derived a variant of Einstein's equation for the curvature (now depending on ϕ as well as the energy–momentum tensor), and a field equation for ϕ in which it is sourced by the trace of the energy–momentum tensor.

Solar System observations require the proposed variation of G to be very small, and Narlikar considers cosmological implications which can nevertheless be significant. We cannot go into the details here, but will concentrate on certain qualitative features of the theory. It turns out that the Brans–Dicke theory can be expressed as a theory in which G is constant but a particle's mass varies depending on the spacetime event at which it is located. This is what we would like to sketch here.

If g is the metric relative to which the Brans–Dicke theory has been formulated, we consider a conformally equivalent metric $\bar{g}$ defined by

$$\bar{g} := \Omega^2 g, \tag{4.56}$$

for some non-zero scalar field Ω, and consider the possibility that a particle might move through the $\bar{g}$ spacetime with a variable mass that is some scalar function of its location in spacetime. Let us find the equation for its worldline by the usual variation of the action

$$\int_{\bar{s}_0}^{\bar{s}_1} mc\,d\bar{s}\,,$$

where $\bar{s}$ is proper time along the worldline, as defined by the barred metric.

As usual, we vary the worldline $x(\bar{s})$, keeping the endpoints $x_0 = x(\bar{s}_0)$ and $x_1 = x(\bar{s}_1)$ fixed. We must remember to keep m inside the integral for this exercise, since it is a function of space and time coordinates. We shall carry out the derivation without mentioning the explicit form of this function. One set of terms comes from

$$c\int_{\bar{s}_0}^{\bar{s}_1} m\delta\left(\bar{g}_{ij}\frac{dx^i}{d\bar{s}}\frac{dx^j}{d\bar{s}}\right)^{1/2} d\bar{s}\,,$$

which is just

$$c\int_{\bar{s}_0}^{\bar{s}_1} \frac{m}{2}\left(\bar{g}_{ij}\frac{dx^i}{d\bar{s}}\frac{dx^j}{d\bar{s}}\right)^{-1/2}\left[\delta\bar{g}_{ij}\frac{dx^i}{d\bar{s}}\frac{dx^j}{d\bar{s}}+2\bar{g}_{ij}\delta\left(\frac{dx^i}{d\bar{s}}\right)\frac{dx^j}{d\bar{s}}\right] d\bar{s}\,.$$

This in turn can be written

$$\frac{c}{2}\int_{\bar{s}_0}^{\bar{s}_1} m\left[\bar{g}_{ij,k}\delta x^k(\bar{s})\frac{dx^i}{d\bar{s}}\frac{dx^j}{d\bar{s}}+2\bar{g}_{ij}\left(\frac{d}{d\bar{s}}\delta x^i\right)\frac{dx^j}{d\bar{s}}\right] d\bar{s}\,. \tag{4.57}$$

The second term gets integrated by parts in quite the usual way, so that the whole thing becomes

$$c\int_{\bar{s}_0}^{\bar{s}_1} m\left(\frac{1}{2}\bar{g}_{kj,i}\frac{dx^k}{d\bar{s}}\frac{dx^j}{d\bar{s}}-\bar{g}_{ij,k}\frac{dx^k}{d\bar{s}}\frac{dx^j}{d\bar{s}}-\bar{g}_{ij}\frac{d^2x^j}{d\bar{s}^2}\right)\delta x^i(\bar{s})\,d\bar{s}\,,$$

terms arising in the standard constant mass situation, plus another term arising because the mass gets caught up in the integration by parts, viz.,

$$-c\int_{\bar{s}_0}^{\bar{s}_1} \frac{dm}{d\bar{s}}\bar{g}_{ij}\delta x^i\frac{dx^j}{d\bar{s}}\,d\bar{s}\,. \tag{4.58}$$

This new term can be reexpressed as

$$-c\int_{\bar{s}_0}^{\bar{s}_1} \frac{\partial m}{\partial x^k}\bar{g}_{ij}\frac{dx^k}{d\bar{s}}\frac{dx^j}{d\bar{s}}\delta x^i\,d\bar{s}\,. \tag{4.59}$$

But there is a further new term arising from

$$c\int_{\bar{s}_0}^{\bar{s}_1} d\bar{s}\,\delta m = c\int_{\bar{s}_0}^{\bar{s}_1} \frac{\partial m}{\partial x^i}\delta x^i\,d\bar{s}\,.$$

This can also be written

$$c\int_{\bar{s}_0}^{\bar{s}_1} d\bar{s}\,\delta m = c\int_{\bar{s}_0}^{\bar{s}_1} m(\ln m)_{,i}\,\delta x^i\,d\bar{s}\,.$$

We now have all the terms from the variation in the right form to deduce the equation of motion of the variable mass particle, viz.,

$$\frac{1}{2}\bar{g}_{kj,i}\frac{\mathrm{d}x^k}{\mathrm{d}\bar{s}}\frac{\mathrm{d}x^j}{\mathrm{d}\bar{s}} - \bar{g}_{ij,k}\frac{\mathrm{d}x^k}{\mathrm{d}\bar{s}}\frac{\mathrm{d}x^j}{\mathrm{d}\bar{s}} - \bar{g}_{ij}\frac{\mathrm{d}^2x^j}{\mathrm{d}\bar{s}^2} = \bar{g}_{ij}\frac{\mathrm{d}x^k}{\mathrm{d}\bar{s}}\frac{\mathrm{d}x^j}{\mathrm{d}\bar{s}}(\ln m)_{,k} - (\ln m)_{,i}\;.$$

After a slight reorganisation,

$$\boxed{\frac{\mathrm{d}^2x^m}{\mathrm{d}\bar{s}^2} + \bar{\Gamma}^m_{kj}\frac{\mathrm{d}x^k}{\mathrm{d}\bar{s}}\frac{\mathrm{d}x^j}{\mathrm{d}\bar{s}} = \left(\bar{g}^{mk} - \frac{\mathrm{d}x^m}{\mathrm{d}\bar{s}}\frac{\mathrm{d}x^k}{\mathrm{d}\bar{s}}\right)(\ln m)_{,k}}\tag{4.60}$$

where we have used the usual definition of the Levi-Civita connection $\bar{\Gamma}$ for the barred metric, viz.,

$$\bar{\Gamma}^m_{kj} := \frac{1}{2}\bar{g}^{mi}\left(\frac{\partial\bar{g}_{ik}}{\partial x^j} + \frac{\partial\bar{g}_{ij}}{\partial x^k} - \frac{\partial\bar{g}_{kj}}{\partial x^i}\right)\;.\tag{4.61}$$

This is still completely general, in the sense that we have not yet specified the space-time dependence of m. The varying mass acts rather like a perturbing force through the terms on the right-hand side of (4.60).

Let us now compare (4.60) with the geodesic equation relative to the unbarred metric, viz.,

$$\frac{\mathrm{d}^2x^m}{\mathrm{d}s^2} + \Gamma^m_{kj}\frac{\mathrm{d}x^k}{\mathrm{d}s}\frac{\mathrm{d}x^j}{\mathrm{d}s} = 0\;.\tag{4.62}$$

The idea is, of course, to rewrite this in terms of barred quantities, using the fact that $\mathrm{d}s = \Omega^{-1}\mathrm{d}\bar{s}$. To begin with,

$$\frac{\mathrm{d}^2x^m}{\mathrm{d}s^2} = \Omega\frac{\mathrm{d}}{\mathrm{d}s}\left(\Omega\frac{\mathrm{d}}{\mathrm{d}s}x^m\right) = \Omega^2\frac{\mathrm{d}^2x^m}{\mathrm{d}\bar{s}^2} + \Omega\Omega_{,k}\frac{\mathrm{d}x^k}{\mathrm{d}\bar{s}}\frac{\mathrm{d}x^m}{\mathrm{d}\bar{s}}\;.$$

Now the barred and unbarred Levi-Civita connections, as calculated from (4.61) and a similar relation for the unbarred metric, are related by

$$\bar{\Gamma}^i_{jk} = \frac{\Omega_{,k}}{\Omega}\delta^i_j + \frac{\Omega_{,j}}{\Omega}\delta^i_k - \frac{\Omega_{,m}}{\Omega}g_{jk}g^{im} + \Gamma^i_{jk}\;,$$

so

$$\Gamma^m_{kj}\frac{\mathrm{d}x^k}{\mathrm{d}s}\frac{\mathrm{d}x^j}{\mathrm{d}s} = \Omega^2\bar{\Gamma}^m_{kj}\frac{\mathrm{d}x^k}{\mathrm{d}\bar{s}}\frac{\mathrm{d}x^j}{\mathrm{d}\bar{s}} - 2\Omega^2\frac{\mathrm{d}x^m}{\mathrm{d}\bar{s}}\frac{\mathrm{d}x^j}{\mathrm{d}\bar{s}}(\ln\Omega)_{,j} + g^{mn}(\ln\Omega)_{,n}\;.$$

Putting this together,

$$\frac{\mathrm{d}^2x^m}{\mathrm{d}s^2} + \Gamma^m_{kj}\frac{\mathrm{d}x^k}{\mathrm{d}s}\frac{\mathrm{d}x^j}{\mathrm{d}s} = \Omega^2\left(\frac{\mathrm{d}^2x^m}{\mathrm{d}\bar{s}^2} + \bar{\Gamma}^m_{kj}\frac{\mathrm{d}x^k}{\mathrm{d}\bar{s}}\frac{\mathrm{d}x^j}{\mathrm{d}\bar{s}}\right) - \Omega\Omega_{,k}\frac{\mathrm{d}x^k}{\mathrm{d}\bar{s}}\frac{\mathrm{d}x^m}{\mathrm{d}\bar{s}} + g^{mn}(\ln\Omega)_{,n}\;.$$

In other words, the worldline with equation (4.62) relative to metric g has equation

$$\frac{d^2 x^m}{d\bar{s}^2} + \bar{\Gamma}^m_{kj} \frac{dx^k}{d\bar{s}} \frac{dx^j}{d\bar{s}} = \frac{\Omega_{,k}}{\Omega} \frac{dx^k}{d\bar{s}} \frac{dx^m}{d\bar{s}} - \bar{g}^{mn} (\ln \Omega)_{,n} ,$$

relative to the metric $\bar{g}$. Alternatively, this may be written

$$\frac{d^2 x^m}{d\bar{s}^2} + \bar{\Gamma}^m_{kj} \frac{dx^k}{d\bar{s}} \frac{dx^j}{d\bar{s}} = \left(\frac{dx^m}{d\bar{s}} \frac{dx^k}{d\bar{s}} - \bar{g}^{mk} \right) (\ln \Omega)_{,k} \qquad (4.63)$$

This is actually quite general, for any conformal transformation

$$\bar{g}_{mn} = \Omega^2 g_{mn} .$$

Suppose now that we want this to look (relative to $\bar{g}$) like the trajectory of a variable mass particle. Comparing (4.60) and (4.63), we must have

$$\left(\bar{g}^{mk} - \frac{dx^m}{d\bar{s}} \frac{dx^k}{d\bar{s}} \right) (\ln m)_{,k} = \left(\frac{dx^m}{d\bar{s}} \frac{dx^k}{d\bar{s}} - \bar{g}^{mk} \right) (\ln \Omega)_{,k} .$$

We conclude that this is the case if and only if

$$(\ln m)_{,k} = -(\ln \Omega)_{,k} \qquad \forall k .$$

This implies that $m\Omega$ is constant.

We thus take the following view. Relative to the metric $\bar{g}$, a particle has mass $m = m_0/\Omega$, for some constant m_0, and follows the strange worldline (4.60), which becomes the worldline

$$\frac{d^2 x^m}{ds^2} + \Gamma^m_{kj} \frac{dx^k}{ds} \frac{dx^j}{ds} = 0 , \qquad (4.64)$$

when expressed relative to g. It remains only to choose a scalar function Ω with some relevance to the Brans–Dicke formulation, but it is worth stressing that this choice is not essential to the arguments in this section. Indeed, the above arguments are completely general for all conformal transformations. Relative to one metric g, massive particles and light follow certain paths. Relative to a conformally equivalent metric $\bar{g}$, the light paths are the same, and the paths of massive particles look as though their masses change in such a way that $m\Omega$ is constant.

Now it turns out [19] that, if we consider a conformally equivalent metric $\bar{g}$ defined by

$$\bar{g} := \Omega^2 g , \qquad \Omega^2 := \frac{\phi}{\bar{\phi}} , \qquad (4.65)$$

where $\bar{\phi}$ is a constant and ϕ is the special scalar field introduced originally by Brans and Dicke to obtain a varying gravitational 'constant', then the Brans–Dicke universe would appear to be general relativistic in the barred view, in the sense that we retrieve something resembling Einstein's equation for the barred curvature. But what we have shown above is that, in the $\bar{g}$ view, particles have to be attributed masses varying in the way suggested, viz.,

$$m = m_0 \left(\frac{\bar{\phi}}{\phi} \right)^{1/2} = \frac{m_0}{\Omega} , \tag{4.66}$$

if their worldlines are to remain geodesic for g. The logic is therefore this: we observe that particles follow geodesic worldlines in the Brans–Dicke universe with metric g, but we prefer to use a conformally equivalent metric $\bar{g}$ in which the gravitational coupling between massive objects is constant, and relative to this metric, the particles would appear to move as if they had a varying mass.

Is there any way of understanding the $\bar{g}$ model physically? Here is a rather heuristic argument. In the Newtonian approximation for gravitational effects, the force between two active gravitational masses m_1 and m_2 is proportional to Gm_1m_2. Now we have made the hypothesis that G is actually a function of spacetime, going as ϕ^{-1}. We might imagine taking a new view of this situation in which

$$G = G_0 \frac{\bar{\phi}}{\phi} ,$$

for some spatiotemporal constants G_0 and $\bar{\phi}$. We then absorb ϕ equitably into the two masses, so that we now have spatiotemporally varying active gravitational masses. Functionally, they go as $m_1(\bar{\phi}/\phi)^{1/2}$ and $m_2(\bar{\phi}/\phi)^{1/2}$. At this point we would make precisely the choice in (4.66). Of course, there is a more sophisticated GR version of this involving the Brans–Dicke extension of Einstein's equation [19].

The reason for discussing all this was just to see whether any progress could be made toward understanding inertia by this kind of theory. A very interesting point about the above discussion is that we have a clear link between two quantities which are only coincidentally related in general relativity, viz., the inertial and active gravitational masses. However, we may make a serious criticism of the idea that this method somehow explains inertial mass as being caused by the large scale distribution of matter and energy in the universe, in other words, of the idea that it implements a Machian program. We have seen that there is a view of things, the $\bar{g}$ view, in which particles move around as though their inertial masses were a function of their space and time coordinates. But this does not change the fact that their trajectories are given by an equation of the type

$$m \left(\frac{\mathrm{d}^2 x^m}{\mathrm{d}\bar{s}^2} + \bar{\Gamma}^m_{kj} \frac{\mathrm{d}x^k}{\mathrm{d}\bar{s}} \frac{\mathrm{d}x^j}{\mathrm{d}\bar{s}} \right) = \left(\bar{g}^{mk} - \frac{\mathrm{d}x^m}{\mathrm{d}\bar{s}} \frac{\mathrm{d}x^k}{\mathrm{d}\bar{s}} \right) m_{,k} .$$

All we have changed from the standard situation, where the equation would be

$$m \left(\frac{\mathrm{d}^2 x^m}{\mathrm{d}\bar{s}^2} + \bar{\Gamma}^m_{kj} \frac{\mathrm{d}x^k}{\mathrm{d}\bar{s}} \frac{\mathrm{d}x^j}{\mathrm{d}\bar{s}} \right) = 0 , \tag{4.67}$$

is that the otherwise unexplained inertial mass is now allowed to vary! For example, we have not explained the origin of m_0 in (4.66), and the equation (4.67) is still basically the same, with standard second order terms. It is the fact that it is second order with this particular structure which means that accelerations cannot just be viewed as a relative thing, and which compels us to introduce inertial forces when we do not take normal coordinates at the point of observation. When there are real physical forces (apart from gravity, which has been geometrised away and is no longer considered to be a force), we still basically have Newton's equation $\mathbf{F} = m\mathbf{a}$. Nothing really seems to have changed.

In fact, a Machian program needs to explain why we have this type of equation at all, a much bigger project than has been undertaken here! But what hypothesis led us to introduce this equation for particle paths? We originally said that particles had to follow geodesics in the g view, so that when they were not acted upon by forces (apart from gravitational effects), their trajectories would take the form

$$\frac{d^2x^m}{ds^2} + \Gamma^m_{kj}\frac{dx^k}{ds}\frac{dx^j}{ds} = 0 . \tag{4.68}$$

Why did we make this our hypothesis? Certainly, it arises from a variation of the action, but that action was specifically designed to produce this result. Is there some better explanation?

What we must remember here is that the theory is not an application of general relativity, although it could be said to be a variant of general relativity. We started from an action, but we did not say what the point of contact would be with physical reality. Let us suppose we impose a strong principle of equivalence, that is, we say that any physical interaction other than gravitation behaves in a locally inertial frame as though gravity were absent. We impose this principle relative to the g metric. Relative to such a frame, any particle will then move in a straight line with uniform velocity, i.e., it will follow the trajectory described by

$$\frac{dv^i}{ds} = 0 , \tag{4.69}$$

where v^i is its 4-velocity. This is expressed covariantly through (4.68). If there are non-gravitational forces, we start with $\mathbf{F} = m\mathbf{a}$ in the locally inertial frame and we find (4.68) with a force term on the right-hand side. It is quite clear that we still have $\mathbf{F} = m\mathbf{a}$. In other words, we may have found a way for the inertial mass to vary, but we have not explained inertia and inertial effects. The basic equation is still Newton's equation.

4.6 Conclusion

Of course the view advocated here is that one might make better progress in explaining inertia by paying more attention to the fact that test particles are not likely to be well modelled by mathematical points. We have the concrete example of the spinning particle and the effect of curvature on its motion. A similar effect occurs when the particle is a source of some classical force field, typically electromagnetic. The spatially extended particle then exerts forces on itself and in simple cases it can be shown that these forces oppose acceleration in flat spacetime and explain why a force is needed to keep the particle off a geodesic in curved spacetime. If all inertia were due to these self-forces, the geodesic equation, or relevant extension of $\mathbf{F} = m\mathbf{a}$, would then be replaced by an equation of the form $\sum F = 0$, where the F summed over include self-forces.

Treating elementary particles like electrons as spatially extended will not make it easier to model them physically, and the point particle approximation has proven its worth in many ways. The idea in this book is not to reject all the successes of point particle models. And furthermore, we shall only be considering classical theory here, so the wonderfully successful world of quantum theory will barely get a mention. But the origin of inertia is nevertheless worth the detour, and once a classical explanation is found, there is no obvious reason why a quantum version of it should not be constructed.

Chapter 5
Momentum and Energy in the EM Fields of a Charge Dumbbell

In this chapter we are going to carry out similar calculations to those in Sects. 3.1 and 3.2, but this time we treat the electron as consisting of two spherical shells of charge, each of radius a, whose centers are separated by a distance $d \gg a$ when the system is at rest in an inertial frame. Each charge shell contains a total charge $q_e/2$, uniformly distributed over the shell.

This is not quite the toy electron of Sect. 3.7, which consisted of two point charges $q_e/2$, separated by distance d when the system is at rest in an inertial frame. The idea in this book is that one should do away with point particles. The point particle idea is just a useful approximation. In fact, we apply it in Sect. 5.1 to obtain the energy in the fields of the charge dumbbell due to bringing together the two component charges. Of course, there is more energy in the fields due to assembling the 'point' charge components, and we obtained that in Sect. 3.1 under the assumption that they were actually spherical charge shells.

In Sect. 5.2, the charge dumbbell is set in motion along its axis, with a constant velocity, and we obtain the momentum in the EM fields in order to deduce a momentum-derived EM mass as in Sect. 3.2. In Sect. 5.3, the charge dumbbell is set in uniform velocity motion perpendicularly to its axis for the same purpose, and we find a *different* momentum-derived EM mass! This is one of the interesting results with the toy electron: it does not have spherical symmetry like the spherical charge shell, and we find that the momentum-derived EM mass depends on the direction in which the system is moving in relation to its own structure.

5.1 Energy Considerations

When we calculated the energy in the electromagnetic field surrounding a stationary electron, treated as a spherical shell of charge of radius a, we found (see Sect. 3.1)

$$U_{\text{electron}} = \frac{1}{2}\frac{e^2}{a} \, , \tag{5.1}$$

Lyle, S.N.: *Momentum and Energy in the EM Fields of a Charge Dumbbell*. Lect. Notes Phys. **796**, 73–91 (2010)
DOI 10.1007/978-3-642-04785-5_5

having made the definition $e^2 := q_e^2/4\pi\varepsilon_0$. If we divide this by c^2 we get the mass equivalent [the energy-derived EM mass of (3.26) on p. 40]

$$m_{\text{EM}}^{\text{EDM}} = \frac{1}{2}\frac{e^2}{ac^2} .$$ (5.2)

This is not the same as the momentum-derived electromagnetic mass for the same system, which is [see (3.10) on p. 35]

$$m_{\text{EM}}^{\text{MDM}} = \frac{2}{3}\frac{e^2}{ac^2} .$$ (5.3)

The momentum-derived electromagnetic mass is bigger than the energy-derived electromagnetic mass by a factor of 4/3.

A similar thing happens with the two-point model. The energy required to bring the second charge concentration from infinity to a distance d from the first is

$$U(d) = -\int_\infty^d \frac{q_e^2}{16\pi\varepsilon_0}\frac{1}{x^2}\,\mathrm{d}x = \frac{e^2}{4d} .$$ (5.4)

The mass equivalent is $e^2/4dc^2$. Of course, a lot more energy would be required to assemble the much smaller charge shells, as can be seen from (5.2). What we find here is a hierarchy of contributions to the energy-derived EM mass, inversely proportional to the spatial dimensions of the construction level. We shall express this by writing

$$\boxed{m_{\text{EM}}^{\text{EDM}}(\text{level } d) = \frac{e^2}{4dc^2}}$$ (5.5)

but bearing in mind that there are other contributions to $m_{\text{EM}}^{\text{EDM}}$ from any smaller charge assemblies making up the system.

Now we shall find that, when the system moves along its axis, the contribution to the momentum-derived electromagnetic mass from the construction level with spatial dimension d is twice the value in (5.5) (see Sect. 5.2), but when the system moves perpendicularly to its axis, we do get the same value for the two ways of deriving this contribution to the EM mass (see Sect. 5.3).

According to Feynman, the Poincaré stresses have to be involved in the model to get agreement between energy-derived and momentum-derived contributions to the EM mass [2, Sect. 28.4]. Naturally, the Poincaré stresses, which we shall usually call binding forces, will themselves be related to fields, and these fields will also contribute to the system's inertia. Feynman says that the complexity of the approach rules it out as a viable alternative. However, this amounts to keeping the point particle model of the electron, which is what we hope to contest in this book. We shall come back to the notorious discrepancy between the energy- and momentum-derived EM masses in Chap. 11.

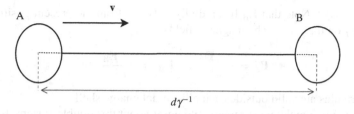

Fig. 5.1 Charge dumbbell in motion along its axis with uniform velocity **v**, hence contracted by a factor γ^{-1}. The charge shells are ellipsoidal

5.2 Longitudinal Motion of the Dumbbell

We assume that the electron is made up of two shells of charge, each being a uniform distribution of charge $q_e/2$. One is taken to be centered on $x(t) = vt$ and the other on $x(t) = vt + d/\gamma$ (see Fig. 5.1). Note that the rest length of the system is d, so we are assuming a FitzGerald contraction.

Let us first set up a suitable notation. The charge concentrations are labelled A and B, and the line joining them, which is also the axis of motion, will be the x axis. Each charge concentration produces electric and magnetic fields, denoted $\mathbf{E}_A$, $\mathbf{B}_A$, $\mathbf{E}_B$, and $\mathbf{B}_B$. The total fields are

$$\mathbf{E} = \mathbf{E}_A + \mathbf{E}_B , \quad \mathbf{B} = \mathbf{B}_A + \mathbf{B}_B . \tag{5.6}$$

We know what all these fields are [see (2.73) and (2.74) on p. 28]. For instance, we have

$$E_{Ax} = \frac{q\gamma(v)}{4\pi\varepsilon_0} \frac{x - vt}{[\gamma^2(x - vt)^2 + y^2 + z^2]^{3/2}} , \tag{5.7}$$

$$E_{Ay} = \frac{q\gamma(v)}{4\pi\varepsilon_0} \frac{y}{[\gamma^2(x - vt)^2 + y^2 + z^2]^{3/2}} , \tag{5.8}$$

$$E_{Az} = \frac{q\gamma(v)}{4\pi\varepsilon_0} \frac{z}{[\gamma^2(x - vt)^2 + y^2 + z^2]^{3/2}} , \tag{5.9}$$

and likewise

$$E_{Bx} = \frac{q\gamma(v)}{4\pi\varepsilon_0} \frac{x - vt - d/\gamma}{[\gamma^2(x - vt - d/\gamma)^2 + y^2 + z^2]^{3/2}} , \tag{5.10}$$

$$E_{By} = \frac{q\gamma(v)}{4\pi\varepsilon_0} \frac{y}{[\gamma^2(x - vt - d/\gamma)^2 + y^2 + z^2]^{3/2}} , \tag{5.11}$$

$$E_{Bz} = \frac{q\gamma(v)}{4\pi\varepsilon_0} \frac{z}{[\gamma^2(x - vt - d/\gamma)^2 + y^2 + z^2]^{3/2}} , \tag{5.12}$$

where $q = q_e/2$. Note that $\mathbf{E}_B$ lies radially outward from the present position of B at $vt + d/\gamma$ on the x axis. The magnetic fields are

$$\mathbf{B}_A = \frac{\mathbf{v} \times \mathbf{E}_A}{c^2} , \quad \mathbf{B}_B = \frac{\mathbf{v} \times \mathbf{E}_B}{c^2} . \tag{5.13}$$

These formulas are valid outside each ellipsoidal charge shell.

We now come to the most remarkable point about this double system. Each pair $\mathbf{E}_A$, $\mathbf{B}_A$, and $\mathbf{E}_B$, $\mathbf{B}_B$ has an associated momentum density $\mathbf{g}_A$, $\mathbf{g}_B$ given by [see (2.39) on p. 12]

$$\mathbf{g}_A = \varepsilon_0 \mathbf{E}_A \times \mathbf{B}_A , \quad \mathbf{g}_B = \varepsilon_0 \mathbf{E}_B \times \mathbf{B}_B . \tag{5.14}$$

But the momentum density of the combined system is not at all the mere sum of these. There is interference, just as in quantum theory, and this is going to alter the electromagnetic contribution to the inertial mass when it is derived from momentum considerations! So it is interference between the fields that leads to a different electromagnetic contribution to inertial mass. Let us investigate this quantitatively.

For the combined system,

$$\mathbf{g} = \varepsilon_0 \mathbf{E} \times \mathbf{B}$$
$$= \mathbf{g}_A + \mathbf{g}_B + \varepsilon_0 \mathbf{E}_A \times \mathbf{B}_B + \varepsilon_0 \mathbf{E}_B \times \mathbf{B}_A . \tag{5.15}$$

We are thus going to get the following momentum in the electromagnetic fields:

$$\mathbf{p} = \mathbf{p}_A + \mathbf{p}_B + \varepsilon_0 \int_{\text{all space}} [\mathbf{E}_A \times \mathbf{B}_B + \mathbf{E}_B \times \mathbf{B}_A] \, dV , \tag{5.16}$$

where, in the non-relativistic approximation, we have already found [see (3.9) on p. 35]

$$\mathbf{p}_A = \mathbf{p}_B = \frac{2}{3} \frac{(e/2)^2}{ac^2} \mathbf{v} . \tag{5.17}$$

Note that the sum of $\mathbf{p}_A$ and $\mathbf{p}_B$ comes to only half what we had when all the charge was evenly distributed around a single spherical shell. This has no particular interpretation. For example, it does not mean that we have gained or lost momentum. Each spatial distribution taken to represent the electron gives a different model. For example, we could halve the value of a in the present case to get back the total spherical shell contribution to the momentum-derived electromagnetic mass.

Our task is thus to evaluate the integral in (5.16). We will then find out how the momentum-derived electromagnetic mass contribution from a spatial structure of dimension d compares with that from spatial structures of dimension a within the same system.

It turns out that the geometry of the situation is particularly simple. Note that, as far as the fields outside the charge shells are concerned, we may treat the latter as point charges at vt and $vt + d/\gamma$. Then it is easy to see that the two magnetic field

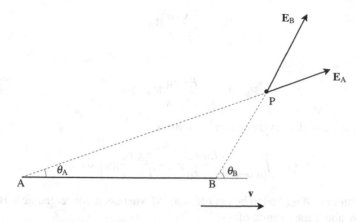

Fig. 5.2 Charge dumbbell in motion along its axis with uniform speed v. The fields outside the charge shells are the same as if the latter were replaced by point charges $q_e/2$ (A and B). The diagram shows why the magnetic field vectors at the given field point must be parallel. In fact, both $\mathbf{B}_A = \mathbf{v} \times \mathbf{E}_A / c^2$ and $\mathbf{B}_B = \mathbf{v} \times \mathbf{E}_B / c^2$ point up out of the page. This is because, for any choice of field point P, the three vectors $\mathbf{E}_A$, $\mathbf{E}_B$, and $\mathbf{v}$ are coplanar. Note that the electric fields $\mathbf{E}_A$ and $\mathbf{E}_B$ at P lie along AP and BP, respectively, i.e., they point from the current positions of the charges, because the latter have uniform velocities

vectors at a given field point outside the charge shells are parallel (see Fig. 5.2). Here is a mathematical proof. Using the fact that $\mathbf{E}_B$ is radial from the current position of charge B,

$$\mathbf{E}_B = E_B \frac{\mathbf{r}_B}{r_B} = \frac{E_B}{r_B} [\mathbf{r}_A - d\mathbf{v}/\gamma v] \ ,$$

where $\mathbf{r}_B$ is the vector from B to the field point, and $\mathbf{r}_A$ the vector from A to the field point (see Fig. 5.2). Now we have

$$\mathbf{B}_B = \frac{\mathbf{v} \times \mathbf{E}_B}{c^2} = \frac{E_B}{r_B c^2} \mathbf{v} \times \mathbf{r}_A$$

$$= \frac{E_B}{r_B c^2} \mathbf{v} \times \mathbf{E}_A \frac{r_A}{E_A}$$

$$= \frac{E_B}{E_A} \frac{r_A}{r_B} \mathbf{B}_A \ .$$

An interesting consequence of this is that

$$\mathbf{E}_A \times \mathbf{B}_B = \frac{E_B}{E_A} \frac{r_A}{r_B} \mathbf{E}_A \times \mathbf{B}_A \ . \tag{5.18}$$

Likewise,

$$\mathbf{B}_A = \frac{E_A}{E_B}\frac{r_B}{r_A}\mathbf{B}_B \, , \tag{5.19}$$

and hence,

$$\mathbf{E}_B \times \mathbf{B}_A = \frac{E_A}{E_B}\frac{r_B}{r_A}\mathbf{E}_B \times \mathbf{B}_B \, . \tag{5.20}$$

The integral we seek to evaluate can be written

$$I := \int_{\text{all space}} \left[\frac{E_B}{E_A}\frac{r_A}{r_B}\mathbf{g}_A + \frac{E_A}{E_B}\frac{r_B}{r_A}\mathbf{g}_B \right] dV \, . \tag{5.21}$$

Now the integral of $\mathbf{g}_A$ had to be cut off near A, whereas it converged near B. What can we say about the integral of

$$\frac{E_B}{E_A}\frac{r_A}{r_B}\mathbf{g}_A$$

at these points? It will turn out that there is no singularity at either A or B and we shall be able to simplify the integration by integrating over all space, neglecting the contributions from the excluded spheres (ellipsoids) at A and B.

The time has come to put in explicit formulas. Consider the plane containing A, B, and the field point. Taking the axis of motion as the x axis,

$$\mathbf{v} = \begin{pmatrix} v \\ 0 \end{pmatrix}, \quad \mathbf{r}_A = \begin{pmatrix} r_A \cos\theta_A \\ r_A \sin\theta_A \end{pmatrix}, \quad \mathbf{r}_B = \begin{pmatrix} r_A \cos\theta_A - d/\gamma \\ r_A \sin\theta_A \end{pmatrix}, \tag{5.22}$$

where θ_A is the angle shown in Fig. 5.2. In this notation, A is at $(vt,0,0)$ and B is at $(vt+d/\gamma,0,0)$. We can also write

$$\mathbf{r}_A = \begin{pmatrix} x - vt \\ y \\ z \end{pmatrix}, \quad \mathbf{r}_B = \begin{pmatrix} x - vt - d/\gamma \\ y \\ z \end{pmatrix} = \mathbf{r}_A - \begin{pmatrix} d/\gamma \\ 0 \\ 0 \end{pmatrix} \, . \tag{5.23}$$

The magnetic field $\mathbf{B}_B$ sticks vertically out of the plane of Fig. 5.2, and has magnitude

$$B_B = \frac{vE_B}{c^2} \sin\theta_B \, , \tag{5.24}$$

where θ_B is the angle between the x direction (along $\mathbf{v}$) and the vector $\mathbf{r}_B$ from B to the field point, as shown in Fig. 5.2. The magnitude of $\varepsilon_0 \mathbf{E}_A \times \mathbf{B}_B$ is then

$$|\varepsilon_0 \mathbf{E}_A \times \mathbf{B}_B| = \frac{\varepsilon_0 v}{c^2} E_A E_B \sin\theta_B \, . \tag{5.25}$$

This vector points down from the field point obliquely towards the axis of motion, in fact, making an angle of θ_A to the vertical, as can be seen from Fig. 5.2. When we integrate around the axis of motion, only the components along that axis actually contribute, the others cancelling by symmetry. We therefore multiply the above magnitude by a factor of $\sin \theta_A$ to pick up this component, and just integrate that. This is a key point again. The resulting contribution to the field momentum is along the direction of motion.

This brings us to the following contribution to the integrand:

$$\frac{\varepsilon_0 v}{c^2} E_A E_B \sin \theta_A \sin \theta_B \ . \tag{5.26}$$

We could have deduced this from (5.18), viz.,

$$\mathbf{E}_A \times \mathbf{B}_B = \frac{E_B}{E_A} \frac{r_A}{r_B} \mathbf{E}_A \times \mathbf{B}_A \ . \tag{5.27}$$

We have already seen in Sect. 3.2 that $\varepsilon_0 \mathbf{E}_A \times \mathbf{B}_A$ leads to an integrand

$$\frac{\varepsilon_0 v E_A^2}{c^2} \sin^2 \theta_A \ .$$

Hence, $\varepsilon_0 \mathbf{E}_A \times \mathbf{B}_B$ is going to give an integrand

$$\frac{\varepsilon_0 v E_A E_B}{c^2} \sin^2 \theta_A \frac{r_A}{r_B} \ .$$

However, it is obvious from Fig. 5.2 that

$$r_A \sin \theta_A = r_B \sin \theta_B \ .$$

We arrive at the same answer.

Note that (5.26) is perfectly symmetrical in A and B. Checking carefully, we find that $\varepsilon_0 \mathbf{E}_B \times \mathbf{B}_A$ contributes exactly the same quantity to the integrand. We now have

$$I = \frac{2\varepsilon_0 \mathbf{v}}{c^2} \int_{\text{all space}} E_A E_B \sin \theta_A \sin \theta_B \, dV \ . \tag{5.28}$$

Note that this result holds just as well in both the relativistic case and the non-relativistic case.

5.2.1 Non-Relativistic Calculation

We shall now assume that $v \ll c$, whence $\gamma \approx 1$, and take

$$E_A \approx \frac{q_e/2}{4\pi\varepsilon_0 r_A^2} \ , \quad E_B \approx \frac{q_e/2}{4\pi\varepsilon_0 r_B^2} \ . \tag{5.29}$$

We shall use integration variables $r := r_A$ and $\theta := \theta_A$. The volume element is

$$dV = 2\pi r^2 \sin\theta \, d\theta \, dr \,.$$

We can write

$$E_B = \frac{q_e/2}{4\pi\varepsilon_0 \left[(r\cos\theta - d)^2 + r^2 \sin^2\theta \right]} \,, \tag{5.30}$$

and since $r_B \sin\theta_B = r\sin\theta$,

$$\sin\theta_B = \frac{r\sin\theta}{\left[(r\cos\theta - d)^2 + r^2 \sin^2\theta \right]^{1/2}} \,. \tag{5.31}$$

Putting all these ingredients together, we find

$$I = \frac{q_e^2 \mathbf{v}}{16\pi\varepsilon_0 c^2} \int_{\text{all space}} \frac{r\sin^3\theta}{\left[r^2 - 2dr\cos\theta + d^2 \right]^{3/2}} \, d\theta \, dr \,. \tag{5.32}$$

We can now justify a claim made earlier, which allows us to integrate over all space, even the interiors of the two small spheres A and B. Note that the integrand has no singularities. When $r = 0$ this is clear. When $r = d$ and $\theta = 0$, the denominator looks problematic, but we have a term going as θ^3 in the numerator.

We treat the integral over r as having the form

$$I_1 = \int_0^\infty \frac{Ar}{\left[r^2 - 2Br + C \right]^{3/2}} \, dr \,, \tag{5.33}$$

where $A := \sin^3\theta$, $B := d\cos\theta$ and $C := d^2$. The denominator is a power of

$$(r - B)^2 + D^2 \,, \quad \text{where} \quad D := d\sin\theta \,, \quad D^2 = C - B^2 \,.$$

Put $X := r - B$ so that $dr = dX$ and

$$I_1 = \int_{-B}^\infty \frac{A(X + B)}{\left[X^2 + D^2 \right]^{3/2}} \, dX$$

$$= A\left[-(X^2 + D^2)^{-1/2} \right]_{-B}^\infty + AB \int_{-B}^\infty \frac{1}{\left[X^2 + D^2 \right]^{3/2}} \, dX \,.$$

We now substitute $X = D\tan\alpha$ so that

$$dX = D(1 + \tan^2\alpha) \, d\alpha \,.$$

What looks as though it will become rather complicated soon simplifies to give

$$I_1 = \frac{A}{\left[B^2 + D^2 \right]^{1/2}} + \frac{AB}{D^2} \left[1 + \frac{B/D}{\sqrt{1 + B^2/D^2}} \right] \,. \tag{5.34}$$

Decoding this back to the original variables, we have

$$I = \frac{q_e^2 \mathbf{v}}{16\pi\varepsilon_0 c^2 d} \int_0^\pi \left[\sin\theta + \frac{1}{2}\sin 2\theta \right] d\theta . \qquad (5.35)$$

Note that we now have the $1/d$ dependence. Finally,

$$I = \frac{q_e^2 \mathbf{v}}{8\pi\varepsilon_0 c^2 d} . \qquad (5.36)$$

Replacing $q_e^2/4\pi\varepsilon_0$ by e^2, we have the momentum in the electromagnetic fields according to (5.16), viz.,

$$\mathbf{p} = \mathbf{p}_A + \mathbf{p}_B + \frac{e^2 \mathbf{v}}{2c^2 d} , \qquad (5.37)$$

where

$$\mathbf{p}_A = \mathbf{p}_B = \frac{2}{3}\frac{(e/2)^2}{ac^2}\mathbf{v} . \qquad (5.38)$$

Hence,

$$\mathbf{p} = \frac{4}{3}\frac{(e/2)^2}{ac^2}\mathbf{v} + \frac{e^2 \mathbf{v}}{2c^2 d} . \qquad (5.39)$$

The momentum-derived electromagnetic mass has a component from each level of structure, increasing as the dimension of the structure decreases. Here we note that

$$\boxed{m_{\text{EM}}^{\text{MDM}}(\text{level } d) = \frac{e^2}{2dc^2} \quad \text{(longitudinal motion)}} \qquad (5.40)$$

Comparing with (5.5) on p. 74, we find once again that we have a discrepancy between the momentum- and energy-derived EM masses. This will be explained in Chap. 11.

5.2.2 Relativistic Calculation

We now return to (5.28) and attempt to integrate in the general case, where v may be similar to c. Given the simplicity of the last result, we expect to obtain the electromagnetic mass contribution from the d structure with a factor of $\gamma(v)$, rather as happened in (3.24) on p. 37.

In the relativistic case we still have (5.16), viz.,

$$\mathbf{p} = \mathbf{p}_A + \mathbf{p}_B + \varepsilon_0 \int_{\text{all space}} [\mathbf{E}_A \times \mathbf{B}_B + \mathbf{E}_B \times \mathbf{B}_A]\, dV , \qquad (5.41)$$

but now

$$\mathbf{p}_A = \mathbf{p}_B = \frac{2}{3} \frac{(e/2)^2}{ac^2} \gamma(v)\mathbf{v} . \tag{5.42}$$

Likewise, we still have (5.28), viz.,

$$I = \frac{2\varepsilon_0 v}{c^2} \int_{\text{all space}} E_A E_B \sin\theta_A \sin\theta_B \, dV , \tag{5.43}$$

where I is the integral in (5.41).

At the beginning of this section, we already mentioned the problem of the system length. Just as we allowed the charge spheres to FitzGerald contract, we must expect the present system to contract in motion. This is an assumption, but a very natural one. It amounts to saying that the binding forces (or Poincaré stresses) holding the system together despite the electromagnetic repulsion are Lorentz covariant. In other words, we make the physical hypothesis that, if d is the rest length of the system, then its length in motion will be d/γ. Of course, for the non-relativistic calculation, this had no effect on the final answer, because we set $\gamma = 1$ in all formulas.

By (5.7–5.12), the magnitudes of the electric fields are

$$E_A = \frac{\gamma q_e/2}{4\pi\varepsilon_0} \frac{\left[(x-vt)^2 + y^2 + z^2\right]^{1/2}}{\left[\gamma^2(x-vt)^2 + y^2 + z^2\right]^{3/2}} , \tag{5.44}$$

$$E_B = \frac{\gamma q_e/2}{4\pi\varepsilon_0} \frac{\left[(x-vt-d/\gamma)^2 + y^2 + z^2\right]^{1/2}}{\left[\gamma^2(x-vt-d/\gamma)^2 + y^2 + z^2\right]^{3/2}} . \tag{5.45}$$

We are going to need suitable integration variables, and we also need to express $\sin\theta_A$ and $\sin\theta_B$. One piece of good fortune that carries over from the non-relativistic case is that the integrand has no singularities and we do not need the cutoffs around A and B. If we assume $a \ll d$, we can just integrate over all space, including the points $x = vt$ and $x = vt + d/\gamma$ where A and B happen to be situated at time t.

Let us introduce a new variable

$$X := \gamma(x-vt) , \quad dX = \gamma dx . \tag{5.46}$$

Then

$$dV = \frac{dX \, dy \, dz}{\gamma} . \tag{5.47}$$

We also introduce

$$R := (X^2 + y^2 + z^2)^{1/2} , \quad X = R\cos\theta' , \quad (y^2 + z^2)^{1/2} = R\sin\theta' . \tag{5.48}$$

As before, R and θ' are length and angle in the (X, y, z) space. The volume element in this space is

$$dX dy dz = 2\pi R^2 \sin\theta' d\theta' dR . \tag{5.49}$$

Now back in the (x, y, z) space,

$$\sin\theta_A = \frac{(y^2 + z^2)^{1/2}}{[(x - vt)^2 + y^2 + z^2]^{1/2}} . \tag{5.50}$$

Likewise,

$$\sin\theta_B = \frac{(y^2 + z^2)^{1/2}}{[(x - vt - d/\gamma)^2 + y^2 + z^2]^{1/2}} . \tag{5.51}$$

When we work out $E_A E_B \sin\theta_A \sin\theta_B$, some of the awkward factors cancel, and we find

$$E_A E_B \sin\theta_A \sin\theta_B =$$
$$\frac{\gamma^2 q_e^2/4}{16\pi^2 \varepsilon_0^2} \frac{y^2 + z^2}{[\gamma^2(x - vt)^2 + y^2 + z^2]^{3/2} [\gamma^2(x - vt - d/\gamma)^2 + y^2 + z^2]^{3/2}} . \tag{5.52}$$

Note that

$$\gamma^2(x - vt - d/\gamma)^2 = (X - d)^2 .$$

Hence,

$$E_A E_B \sin\theta_A \sin\theta_B = \frac{\gamma^2 q_e^2/4}{16\pi^2 \varepsilon_0^2} \frac{\sin^2\theta'}{R[R^2 - 2dX + d^2]^{3/2}} , \tag{5.53}$$

and we soon deduce that

$$I = \frac{q_e^2 \gamma(v) \mathbf{v}}{16\pi\varepsilon_0 c^2} \int_{\text{all space}} \frac{R \sin^3\theta'}{[R^2 - 2dR\cos\theta + d^2]^{3/2}} d\theta' dR . \tag{5.54}$$

This should be compared with (5.32). We have exactly the same integral, even though the meaning of the dummy variables has changed, with the introduction of a factor of $\gamma(v)$.

We conclude that in the general case where v may attain relativistic values,

$$\mathbf{p} = \frac{4}{3} \frac{(e/2)^2}{ac^2} \gamma(v)\mathbf{v} + \frac{e^2 \gamma(v)\mathbf{v}}{2c^2 d} . \tag{5.55}$$

As before, the electromagnetic mass has a component from each level of structure, increasing as the dimension of the structure decreases. Note also that we have found the same contribution $e^2/2c^2 d$ as in (5.40), but now including the usual Lorentz factor:

$$\boxed{m_{\text{EM}}^{\text{MDM}}(\text{level } d) = \frac{e^2}{2dc^2} \gamma(v) \qquad \text{(longitudinal motion)}} \tag{5.56}$$

5.3 Transverse Motion of the Dumbbell

There is an obvious question concerning our two-point system, since it picks out a
direction in space, unlike the small spherical shells. Up to now we have considered motion along the line joining the 'point' charges. We might wonder, however,
whether we would get the same momentum in the fields if the system were to move
in a direction perpendicular to the system axis. If not, it would seem that electromagnetic mass could vary with direction for non-spherically symmetrical charge
distributions. Let us therefore carry out such a calculation. We will treat only the
relativistic case.

We place A at $(vt, 0, 0)$ and B at $(vt, 0, d)$ (see Fig. 5.3). The vectors from A and
B to the field point are

$$\mathbf{r}_A = \begin{pmatrix} x - vt \\ y \\ z \end{pmatrix}, \quad \mathbf{r}_B = \begin{pmatrix} x - vt \\ y \\ z - d \end{pmatrix}, \quad \text{so that} \quad \mathbf{r}_A - \mathbf{r}_B = \begin{pmatrix} 0 \\ 0 \\ d \end{pmatrix}. \quad (5.57)$$

Note that the electric fields are still radial from the point where the charge is located
now, i.e.,

$$\mathbf{E}_A = \frac{E_A}{r_A} \mathbf{r}_A, \quad \mathbf{E}_B = \frac{E_B}{r_B} \mathbf{r}_B. \quad (5.58)$$

As before, the magnetic fields are given by

$$\mathbf{B}_A = \frac{\mathbf{v} \times \mathbf{E}_A}{c^2}, \quad \mathbf{B}_B = \frac{\mathbf{v} \times \mathbf{E}_B}{c^2}, \quad (5.59)$$

where $\mathbf{v}$ is still taken to be in the x direction. This time, however, the two magnetic
fields are no longer parallel and it is preferable to adopt a purely vectorial analysis.
Indeed, it is this fact that compels us to carry out the ϕ integration around the axis
of motion, rather than just appeal to some axial symmetry.

We have

$$\mathbf{B}_B = \frac{E_B}{r_B c^2} \mathbf{v} \times \mathbf{r}_B$$

$$= \frac{E_B}{r_B c^2} \mathbf{v} \times \left[\mathbf{r}_A - \begin{pmatrix} 0 \\ 0 \\ d \end{pmatrix} \right]$$

$$= \frac{E_B}{E_A} \frac{r_A}{r_B} \mathbf{B}_A + \frac{E_B v d}{r_B c^2} \begin{pmatrix} 0 \\ 1 \\ 0 \end{pmatrix}. \quad (5.60)$$

This also implies the symmetrical result

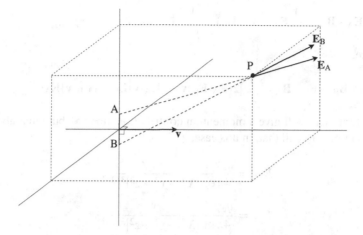

Fig. 5.3 Charge dumbbell in motion perpendicular to its axis with uniform speed v. The fields outside the charge shells are the same as if the latter were replaced by point charges $q_e/2$ (A and B). The diagram shows why the magnetic field vectors $\mathbf{B}_A = \mathbf{v} \times \mathbf{E}_A/c^2$ and $\mathbf{B}_B = \mathbf{v} \times \mathbf{E}_B/c^2$ at the given field point are not generally parallel. Note that the electric field vectors $\mathbf{E}_A$ and $\mathbf{E}_B$ lie along AP and BP, respectively, i.e., they point from the current positions of the charges, because the latter have uniform velocities. They are clearly not generally coplanar with $\mathbf{v}$

$$\mathbf{B}_A = \frac{E_A}{E_B} \frac{r_B}{r_A} \mathbf{B}_B - \frac{E_A v d}{r_A c^2} \begin{pmatrix} 0 \\ 1 \\ 0 \end{pmatrix} . \tag{5.61}$$

The momentum in the fields is calculated as usual from

$$\mathbf{p} = \varepsilon_0 \int_{\text{all space}} (\mathbf{E}_A + \mathbf{E}_B) \times (\mathbf{B}_A + \mathbf{B}_B) \, dV \tag{5.62}$$

$$= \mathbf{p}_A + \mathbf{p}_B + \varepsilon_0 \int_{\text{all space}} [\mathbf{E}_A \times \mathbf{B}_B + \mathbf{E}_B \times \mathbf{B}_A] \, dV ,$$

where

$$\mathbf{p}_A = \varepsilon_0 \int_{\text{all space}} \mathbf{E}_A \times \mathbf{B}_A \, dV , \quad \mathbf{p}_B = \varepsilon_0 \int_{\text{all space}} \mathbf{E}_B \times \mathbf{B}_B \, dV , \tag{5.63}$$

and we find as before that

$$\mathbf{p}_A = \mathbf{p}_B = \frac{2}{3} \frac{(e/2)^2}{ac^2} \gamma(v) \mathbf{v} . \tag{5.64}$$

Note that the integration is not quite over all of space, since we omit the interiors of the small spherical (ellipsoidal) shells of charge.

Let us now calculate one of the terms in the remaining integral. We have

$$\mathbf{E_A}\times\mathbf{B_B} = \frac{1}{c^2}\mathbf{E_A}\times(\mathbf{v}\times\mathbf{E_B}) = \frac{1}{c^2}\left[(\mathbf{E_A}\cdot\mathbf{E_B})\mathbf{v} - (\mathbf{E_A}\cdot\mathbf{v})\mathbf{E_B}\right] , \tag{5.65}$$

and hence,

$$\mathbf{E_A}\times\mathbf{B_B} + \mathbf{E_B}\times\mathbf{B_A} = \frac{1}{c^2}\left[2(\mathbf{E_A}\cdot\mathbf{E_B})\mathbf{v} - (\mathbf{E_A}\cdot\mathbf{v})\mathbf{E_B} - (\mathbf{E_B}\cdot\mathbf{v})\mathbf{E_A}\right] . \tag{5.66}$$

The first term here will give a momentum contribution along $\mathbf{v}$, but what about the other two terms? Recall that, in this case,

$$E_{Ax} = \frac{q_e\gamma/2}{4\pi\varepsilon_0}\frac{X/\gamma}{[X^2+y^2+z^2]^{3/2}} , \tag{5.67}$$

$$E_{Ay} = \frac{q_e\gamma/2}{4\pi\varepsilon_0}\frac{y}{[X^2+y^2+z^2]^{3/2}} , \tag{5.68}$$

$$E_{Az} = \frac{q_e\gamma/2}{4\pi\varepsilon_0}\frac{z}{[X^2+y^2+z^2]^{3/2}} , \tag{5.69}$$

and likewise

$$E_{Bx} = \frac{q_e\gamma/2}{4\pi\varepsilon_0}\frac{X/\gamma}{[X^2+y^2+(z-d)^2]^{3/2}} , \tag{5.70}$$

$$E_{By} = \frac{q_e\gamma/2}{4\pi\varepsilon_0}\frac{y}{[X^2+y^2+(z-d)^2]^{3/2}} , \tag{5.71}$$

$$E_{Bz} = \frac{q_e\gamma/2}{4\pi\varepsilon_0}\frac{z-d}{[X^2+y^2+(z-d)^2]^{3/2}} , \tag{5.72}$$

where we define $X := \gamma(x-vt)$.

The x component of the integral of $-(\mathbf{E_A}\cdot\mathbf{v})\mathbf{E_B} - (\mathbf{E_B}\cdot\mathbf{v})\mathbf{E_A}$ can now be calculated. We find

$$-2v\left(\frac{q_e\gamma/2}{4\pi\varepsilon_0}\right)^2\int\frac{\mathrm{d}X\mathrm{d}y\mathrm{d}z}{\gamma}\frac{X^2/\gamma^2}{[X^2+y^2+z^2]^{3/2}[X^2+y^2+(z-d)^2]^{3/2}} .$$

The y component is

$$-2v\left(\frac{q_e\gamma/2}{4\pi\varepsilon_0}\right)^2\int\frac{\mathrm{d}X\mathrm{d}y\mathrm{d}z}{\gamma}\frac{Xy/\gamma}{[X^2+y^2+z^2]^{3/2}[X^2+y^2+(z-d)^2]^{3/2}} .$$

Now when this is integrated over either X or y, we get zero because it constitutes an odd function in both variables. Likewise for the z component,

$$-2v\left(\frac{q_e\gamma/2}{4\pi\varepsilon_0}\right)^2\int\frac{\mathrm{d}X\mathrm{d}y\mathrm{d}z}{\gamma}\frac{X(z-d)/\gamma+Xz/\gamma}{[X^2+y^2+z^2]^{3/2}[X^2+y^2+(z-d)^2]^{3/2}} ,$$

the integral over X will be zero.

We have reached a very important conclusion here. The field momentum is still along the axis of motion of the charge system. The last term we require is the integral of $2(\mathbf{E_A \cdot E_B})\mathbf{v}$, given by

$$2\mathbf{v}\left(\frac{q_e\gamma/2}{4\pi\varepsilon_0}\right)^2 \int \frac{dXdydz}{\gamma} \frac{X^2/\gamma^2+y^2+z(z-d)}{[X^2+y^2+z^2]^{3/2}[X^2+y^2+(z-d)^2]^{3/2}} \,.$$

Finally the interference contribution to the field momentum is

$$2\mathbf{v}\frac{\varepsilon_0}{c^2}\left(\frac{q_e\gamma/2}{4\pi\varepsilon_0}\right)^2 \int \frac{dXdydz}{\gamma} \frac{y^2+z(z-d)}{[X^2+y^2+z^2]^{3/2}[X^2+y^2+(z-d)^2]^{3/2}} \,. \qquad (5.73)$$

Grouping the constants, we have

$$\mathbf{P}_{\text{int}} = \mathbf{v}\gamma\frac{q_e^2}{32\pi^2\varepsilon_0 c^2} \int dXdydz\frac{y^2+z(z-d)}{[X^2+y^2+z^2]^{3/2}[X^2+y^2+(z-d)^2]^{3/2}} \,. \qquad (5.74)$$

We even have the Lorentz factor.

Our immediate concern is whether or not the integral converges without the cutoff at A and B. If so, we can integrate over the otherwise excluded spherical (ellipsoidal) regions at these points, neglecting the small contribution they make to the integral. The best approach is still to move to spherical coordinates

$$X = R\sin\theta\cos\phi\,, \quad y = R\sin\theta\sin\phi\,, \quad z = R\cos\theta\,.$$

Then

$$X^2+y^2+z^2 = R^2\,, \quad X^2+y^2+(z-d)^2 = R^2 - 2dR\cos\theta + d^2\,,$$

and

$$y^2+z(z-d) = R^2\sin^2\theta\sin^2\phi + R\cos\theta[R\cos\theta - d]\,.$$

Using the standard result

$$dX\,dy\,dz = R^2\sin\theta dR d\theta\,d\phi\,,$$

we now have

$$\mathbf{P}_{\text{int}} = \mathbf{v}\gamma\frac{q_e^2}{32\pi^2\varepsilon_0 c^2} \int dRd\theta d\phi\frac{R\sin^3\theta\sin^2\phi + \sin\theta\cos\theta[R\cos\theta - d]}{[R^2 - 2dR\cos\theta + d^2]^{3/2}} \,. \qquad (5.75)$$

We are now in a position to assess the convergence of the integral at the two sensitive points where the integrand is potentially singular (i.e., at A and B).

From the symmetry between the two potential singularities, we need only consider one of them, viz., A, since it is mathematically the simplest in these coordinates. Now what happens when R goes to zero? We find that there is no longer any singularity in these coordinates, because the volume measure has absorbed it. Needless to say, however, it is best to carry out the R integration before the θ integration. Before either, we shall carry out the ϕ integration. It is obvious since $\sin^2 \phi + \cos^2 \phi = 1$ and cosine is just sine translated by $\pi/2$ that the integral of $\sin^2 \phi$ over a whole cycle must be half of 2π. We now have

$$\mathbf{p}_{\text{int}} = \mathbf{v}\gamma \frac{q_e^2}{32\pi\varepsilon_0 c^2} \int dRd\theta \frac{R\sin^3\theta + 2\sin\theta\cos\theta[R\cos\theta - d]}{[R^2 - 2dR\cos\theta + d^2]^{3/2}} . \qquad (5.76)$$

Replacing $\cos^2\theta$ by $1 - \sin^2\theta$ in the numerator of the integrand, we can write the integral I as a sum $I = I_1 + I_2$, where

$$I_1 = -\int dRd\theta \frac{R\sin^3\theta}{[R^2 - 2dR\cos\theta + d^2]^{3/2}} \qquad (5.77)$$

and

$$I_2 = \int dRd\theta \frac{2\sin\theta[R - d\cos\theta]}{[R^2 - 2dR\cos\theta + d^2]^{3/2}} . \qquad (5.78)$$

We have already met I_1 in (5.32), where we found

$$I_1 = -\frac{2}{d} . \qquad (5.79)$$

The second integral can be immediately integrated with respect to R to give

$$I_2 = \int_{\theta=0}^{\pi} d\theta \sin\theta \left[\frac{-2}{[R^2 - 2dR\cos\theta + d^2]^{1/2}} \right]_0^{\infty} = \frac{2}{d}\int_0^{\pi}\sin\theta\, d\theta = \frac{4}{d} . \qquad (5.80)$$

The final result is

$$\mathbf{p}_{\text{int}} = \frac{q_e^2}{16\pi\varepsilon_0 c^2 d}\gamma(v)\mathbf{v} . \qquad (5.81)$$

Replacing $q_e^2/4\pi\varepsilon_0$ by e^2, this gives

$$\mathbf{p}_{\text{int}} = \frac{e^2}{4dc^2}\gamma(v)\mathbf{v} . \qquad (5.82)$$

We can conclude that the momentum-derived electromagnetic mass does indeed depend on the orientation of the spatial charge structure with respect to the axis of motion:

$$\boxed{m_{\text{EM}}^{\text{MDM}}(\text{level } d) = \frac{e^2}{4dc^2}\gamma(v) \qquad \text{(transverse motion)}} \qquad (5.83)$$

The value obtained for a transverse motion is only half the value (5.56) for the longitudinal motion studied previously.

5.4 Neutral Particle

We can consider a radically different situation with the same structure. Let us suppose the charge concentrations at A and B have opposite signs so that the whole system is neutral. This is thus an electric dipole. We can calculate the momentum in the fields when the system is moving along at constant velocity in our frame. It is very easy to adapt the above arguments and reach a conclusion. It should be noted that this kind of study was not possible with an extended charge distribution in the form of a spherical shell, where all the charge had to be like charge.

To begin with, if B now has positive charge, its electric field is reversed compared with previous calculations, i.e., $\mathbf{E_B} \rightarrow -\mathbf{E_B}$. But then its magnetic field is also reversed, being calculated from

$$\mathbf{B_B} = \frac{1}{c^2}\mathbf{v} \times \mathbf{E_B} \ .$$

This means that $\mathbf{g_B} = \varepsilon_0 \mathbf{E_B} \times \mathbf{B_B}$ is unchanged. Hence the spatial structure of dimension a is going to contribute precisely as before, regardless of sign. But look at the cross terms $\varepsilon_0 \mathbf{E_B} \times \mathbf{B_A}$ and $\varepsilon_0 \mathbf{E_A} \times \mathbf{B_B}$. These both merely change sign. This time the d-level structure is going to reduce electromagnetic mass by exactly the amounts we had to add previously, in our various calculations. In other words, the interference contribution to electromagnetic mass now becomes a reduction of order $1/d$. Note that the structure on the smaller scale introduces terms of order $1/a$ and these clearly dominate.

What is the significance for a quark model of mesons, for example? The pion π^+ is supposed to be $u\bar{d}$, where u has charge $+2/3$ and $\bar{d}$ has charge $+1/3$. Note that the binding forces (Poincaré stresses) are now provided by the strong force between the quarks. In our model, we would expect a certain electromagnetic mass from whatever tiny spatial structure the individual quarks might have, and a much smaller addition to this value due to the presence of two charge centers with like charge values. Contrast with the neutral pion π^0 which is supposed to be a superposition of states like $u\bar{u}$, $d\bar{d}$ and so on. In this case, the contribution from the larger spatial structure is a reduction. The neutral π meson has mass 135.0 MeV, whilst the charged mesons have the same, higher mass of 139.6 MeV, so this does seem to confirm the idea that the mass difference at least arises from the presence of electromagnetic effects (see also the discussion in Sect. 13.3.3). Note also that one would expect the strong force to contribute to bootstrap effects. Has this ever been seriously investigated? More about that in Chap. 13.

It is interesting to note a slight difference between the toy model presented here and the discussion by Feynman in [2, Chap. 28]. In the latter, it is merely noted that the charged mesons are charged whilst the neutral meson is not. (What else could one do with a spherical charge shell?) Then the former can be attributed to an electromagnetic contribution to their mass, whilst the latter cannot. In the toy model, although inexcusably simplistic, all the particles are considered to be made up of charged subparticles.

We should be able to estimate the size d of the quark–antiquark bound systems from the toy model. It is easy to adjust our model to the case where the charge concentrations have different magnitudes, in this case 1/3 and 2/3. In the formulas (5.44) and (5.45) for the magnitudes of E_A and E_B to be fed into (5.43), we simply replace $q/2$ by $q_1 = 1/3$ in one and $q_2 = 2/3$ in the other. The result is that we have to replace e^2 in (5.55) by $8e^2/9$. These differences are actually rather cosmetic given the crude spatial structure we have attributed to the quark–antiquark relationship! The neutral π meson is a superposition of $d\bar{d}$ and $u\bar{u}$ so we presumably need to take some kind of statistical average of the corresponding charge factors. Once again, this is probably not justified in the circumstances of such a crude model.

If we nevertheless include all the factors and merely average over the $u\bar{u}$ and $d\bar{d}$ combinations, we predict a mass difference of e^2/d. This also assumes that u and d have the same mass, which is not realistic. If we equate e^2/d in suitable units with 4.6 MeV, the mass (times c^2) difference between the neutral and charged π mesons, we find that $d \sim 10^{-14}$ cm. This is apparently what would be expected from experimental cross-section determinations.

Note that Feynman's less elaborate considerations lead to the same estimate of pion dimensions. He merely argues that the charged pions are more massive than the neutral one by 4.6 MeV, so that according to his spherical shell model, they consist of uniform distributions of charge over a spherical shell of radius 10^{-14} cm. This says nothing about the size of the neutral pion! In the present version, the pions are made of quark–antiquark pairs and it is assumed that all have the same 'length' d. It is then predicted that d must be 10^{-14} cm in order to explain the mass difference as having electromagnetic origin.

Another neutral particle with charged counterpart is the neutron, now considered to be udd (see also the discussion in Sect. 13.3.3). The charged counterpart is uud, the proton. The neutron can be thought to look something like a proton with a negative pion cloud around it, a kind of sum of uud and $d\bar{u}$. Neutron decay occurs when the pion turns into an electron and electron antineutrino via W^- production. It is not entirely surprising that the neutron is heavier, since any spatial structure here is bound to be more complicated than the one we have assumed. We may wonder how far these spatial structure models might be taken. Methods would have to be developed to handle the integrals that arise in calculating the electromagnetic field momentum.

5.5 Spread of Field Momentum after an Acceleration

One might make the criticism that we consider an ideal case when we treat an electron that has always had some constant velocity, wherein we may integrate a rather simple field quantity over the whole of space. But the idea that the electron must in reality have had some other velocity at some finite time in the past does not indicate that the inertial mass must somehow build up during or after a velocity change. In actual fact, any build-up is only due to the fact that the self-force within the electron has to come into effect, which could only take something like a time d_e/c, where d_e is the electron diameter.

It can be shown explicitly that field momentum gradually fills space after a short and sharp acceleration [8, Chap. 4]. Much later, i.e., after a long coordinate time, the field momentum will have spread to fill a very large sphere. The charge itself will then have been moving at its new velocity for that same coordinate time. If we now calculate the total field momentum in the relevant sphere at any given coordinate time, we will always obtain the same answer, and it will be the result that we have found for an electron that has always had this velocity when we integrate over the whole of space.

The idea is just that during the acceleration, we must impart some momentum to the system just to ensure that the fields can obtain their rightful dose! If there are other contributions to the electron inertial mass, i.e., other reasons for it to oppose acceleration, then more work will be required from the force imparting the acceleration. But what we can understand from this is that a force is required right from the start of the acceleration, to overcome the self-force that the electron imposes on itself to oppose the change of motion. There may be a slight delay of the order of the time it takes a light signal to cross the electron, but there is no delay of any kind due to the fact that the field momentum has to spread out across the whole of space, which will of course take an infinite time.

Chapter 6
Self-Force for Transverse Linear Acceleration

6.1 Setting the Scene

We consider the dumbbell charge system under a linear acceleration perpendicular to its axis as shown in Fig. 6.1. So we have two like charges $q_e/2$, treated as points and separated by a constant distance d. Let us formulate this. We shall take the dumbbell axis to lie along the z direction, and the motion of each charge to be solely in the x direction. Then the position vectors of A and B will have the form

$$\mathbf{r}^A(t) = \begin{pmatrix} x(t) \\ 0 \\ 0 \end{pmatrix}, \qquad \mathbf{r}^B(t) = \begin{pmatrix} x(t) \\ 0 \\ d \end{pmatrix}, \tag{6.1}$$

where $x(t)$ describes the motion of each of A and B through its time derivatives

$$v(t) := \dot{x}(t), \qquad a(t) := \ddot{x}(t), \tag{6.2}$$

and so on.

Now the idea is to use the formulas for the electromagnetic fields due to A to calculate the electric and magnetic forces of A on B, then to use the formulas for the electromagnetic fields due to B to calculate the electric and magnetic forces of B on A. We then simply add the two forces to see if there is a net electromagnetic force of the system on itself as it were. Naturally, A and B repel one another electrically. Indeed, we shall find something of the electromagnetic forces they exert on one another, and it is clear that we are considering a very ideal situation when we assume that they are able to remain at a constant separation d in the z direction. We are not going to worry about the binding force required to maintain this constraint. Ultimately, it too may be considered to come from some fundamental force of which A and B are the sources, and the claim made implicitly throughout this book is that this will lead to other self-forces that must ultimately be taken into consideration.

But for the moment, we are concerned only with the EM self-force. Now the formula (2.61) given on p. 21 in Chap. 2 for the electric field due to a point charge

Lyle, S.N.: *Self-Force for Transverse Linear Acceleration.* Lect. Notes Phys. **796**, 93–113 (2010)
DOI 10.1007/978-3-642-04785-5_6   © Springer-Verlag Berlin Heidelberg 2010

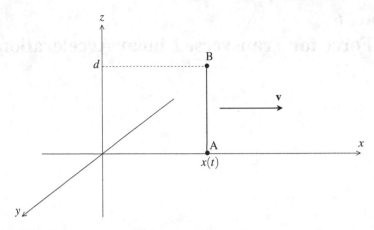

Fig. 6.1 Dumbbell charge system under linear acceleration perpendicular to its axis

like A or B, with charge $q_e/2$ and arbitrary worldline given by functions $x^\mu(\tau)$ of the proper time τ in Minkowski spacetime, is

$$
\mathbf{E} = \frac{q_e}{8\pi\varepsilon_0} \frac{\left(\mathbf{r}_{01} - \frac{r_{01}\mathbf{v}}{c}\right)\left(1 - \frac{v^2}{c^2}\right) + \frac{\mathbf{r}_{01}}{c^2} \times \left[\left(\mathbf{r}_{01} - \frac{r_{01}\mathbf{v}}{c}\right) \times \frac{d\mathbf{v}}{dt}\right]}{(r_{01} - \mathbf{r}_{01}\cdot\mathbf{v}/c)^3} , \qquad (6.3)
$$

where (t_0, x_0, y_0, z_0) is the field point. Then

$$
\mathbf{r}_{01} = \mathbf{x}_0 - \mathbf{x}(\tau_+) = \begin{pmatrix} x_0 - x^1(\tau_+) \\ y_0 - x^2(\tau_+) \\ z_0 - x^3(\tau_+) \end{pmatrix} , \qquad (6.4)
$$

where τ_+ is the retarded time. In words, $\mathbf{r}_{01}$ is the vector from the relevant retarded point to the field point. $\mathbf{v}$ is the coordinate velocity of the source at the retarded time, and $d\mathbf{v}/dt$ is the coordinate acceleration at the retarded time.

Consider first the fields due to A. In the present situation, we are only concerned with the fields at B, so it is worth introducing a special notation. We replace $\mathbf{r}_{01}$ by

$$
\mathbf{r}_+^{AB} := \mathbf{r}^B(t) - \mathbf{r}^A(t_+) = \begin{pmatrix} x(t) - x(t_+^A) \\ 0 \\ d \end{pmatrix} , \qquad (6.5)
$$

where t_+^A is the coordinate retarded time, rather than the proper retarded time, in whatever inertial frame we have selected to view things from. So this is the vector from A at the appropriate retarded time t_+^A to B at the time t we have chosen to consider. Let r_+^{AB} be the length of this vector, viz.,

$$r_+^{AB} = \left\{ \left[x(t) - x(t_+^A) \right]^2 + d^2 \right\}^{1/2} . \tag{6.6}$$

Then the condition determining t_+^A is

$$c(t - t_+^A) = r_+^{AB} , \tag{6.7}$$

which just says that the light travel time from A at the retarded time t_+^A to B at the time t is just right for a signal from A at time t_+^A to arrive at B at time t. Explicitly then, the condition defining t_+^A is

$$c(t - t_+^A) = \left\{ \left[x(t) - x(t_+^A) \right]^2 + d^2 \right\}^{1/2} . \tag{6.8}$$

The reader should be warned that the innocuous looking relation (6.7) lies at the heart of all self-force calculations, and is the source of all the difficulties. It is easy enough to write down, but it is not at all designed to facilitate the task of expanding in powers of d. But nevertheless, what we shall do is consider $t_+^A = t_+^A(t,d)$ as a function of t and d and expand it as a power series in d, to whatever order is required.

However, before doing that, it is worth noting the following, completely erroneous intuitive analysis of how a self-force might come about. In the figure, the system is shown moving in the positive x direction. Now, the electric field due to B affecting A at a certain time originates from a point slightly to the left of the current position of B, in the sense that B was slightly to the left of its current position at the relevant retarded time. Likewise for the electric field due to A affecting B. One might think that, when the charges have the same sign, the mutual repulsion of A and B would have a rightward component that would thus tend to assist rather than hinder the acceleration.

This argument is totally invalid. The situation is much more delicate than this. Note, for example, that when the system is moving inertially, with constant velocity, the electric field due to B is radial from its *current* position! This is something discussed in detail in Chap. 2. We have to be talking about an effect due to the acceleration, not just the velocity, when we produce any explanation. Let us therefore carry out a more accurate calculation, just to check that the self-force is to the left when a system of like charges accelerates to the right.

6.2 Electric Self-Force

Before making expansions in powers of d, we shall obtain the exact expression for the electrical self-force, according to the formula (6.3). First, however, we can make a notational simplification due to the symmetry of the situation. We note that the condition for the retarded time t_+^B when we consider the fields at A due to B is

$$c(t - t_+^B) = \left\{ [x(t) - x(t_+^B)]^2 + d^2 \right\}^{1/2} , \tag{6.9}$$

precisely the same as the condition (6.8) for the retarded time t_+^A. Of course, by symmetry, we have

$$t_+^B(t,d) = t_+^A(t,d) =: t_+ , \tag{6.10}$$

dropping reference to A or B in the new symbol.

To express the electric field at B due to A as given by (6.3), one of the terms we require is

$$\mathbf{r}_+^{AB} - r_+^{AB} \mathbf{v}_+^A / c = \begin{pmatrix} x(t) - x(t_+) - (t - t_+) v(t_+) \\ 0 \\ d \end{pmatrix} , \tag{6.11}$$

where $\mathbf{v}_+^A$ is the velocity of A at the retarded time t_+, viz.,

$$\mathbf{v}_+^A = \begin{pmatrix} v(t_+) \\ 0 \\ 0 \end{pmatrix} . \tag{6.12}$$

Given the type of notation arising here, it will be useful to shorten things still further by simply writing $x_+ := x(t_+)$ and $x := x(t)$ (an abuse of notation), together with $v_+ := v(t_+)$ and $a_+ := a(t_+)$. Then the last two relations become

$$\mathbf{r}_+^{AB} - r_+^{AB} \mathbf{v}_+^A / c = \begin{pmatrix} x - x_+ - (t - t_+) v_+ \\ 0 \\ d \end{pmatrix} , \qquad \mathbf{v}_+^A = \begin{pmatrix} v_+ \\ 0 \\ 0 \end{pmatrix} . \tag{6.13}$$

We also have

$$\left. \frac{d\mathbf{v}^A}{dt} \right|_{t=t_+} = \begin{pmatrix} a_+ \\ 0 \\ 0 \end{pmatrix} . \tag{6.14}$$

Hence,

$$\left(\mathbf{r}_+^{AB} - r_+^{AB} \mathbf{v}_+^A / c \right) \times \left. \frac{d\mathbf{v}^A}{dt} \right|_{t=t_+} = \mathbf{r}_+^{AB} \times \left. \frac{d\mathbf{v}^A}{dt} \right|_{t=t_+}$$

$$= \begin{pmatrix} 0 \\ 0 \\ d \end{pmatrix} \times \begin{pmatrix} a_+ \\ 0 \\ 0 \end{pmatrix} = \begin{pmatrix} 0 \\ da_+ \\ 0 \end{pmatrix} ,$$

using the fact that $\mathbf{v}^A$ and $\mathbf{a}^A$ are always parallel. Finally, the double vector product term in (6.3) becomes

$$
\frac{\mathbf{r}_+^{AB}}{c^2} \times \left[\left(\mathbf{r}_+^{AB} - \frac{r_+^{AB} \mathbf{v}_+^A}{c} \right) \times \frac{d\mathbf{v}^A}{dt} \Big|_{t=t_+} \right] = \frac{da_+}{c^2} \begin{pmatrix} x - x_+ \\ 0 \\ d \end{pmatrix} \times \begin{pmatrix} 0 \\ 1 \\ 0 \end{pmatrix}
$$

$$
= \frac{da_+}{c^2} \begin{pmatrix} -d \\ 0 \\ x - x_+ \end{pmatrix}.
$$

For the denominator of (6.3), we require

$$
\mathbf{r}_+^{AB} \cdot \mathbf{v}_+^A = \begin{pmatrix} x - x_+ \\ 0 \\ d \end{pmatrix} \cdot \begin{pmatrix} v_+ \\ 0 \\ 0 \end{pmatrix} = v_+(x - x_+),
$$

whence

$$
r_+^{AB} - \mathbf{r}_+^{AB} \cdot \mathbf{v}_+^A/c = c(t - t_+) - v_+(x - x_+)/c.
$$

Putting the pieces together, the electric field at B due to A is

$$
\mathbf{E}^A(B) = \frac{q_e}{8\pi\varepsilon_0} \frac{\dfrac{1}{\gamma_+^2} \begin{pmatrix} x - x_+ - (t - t_+)v_+ \\ 0 \\ d \end{pmatrix} + \dfrac{da_+}{c^2} \begin{pmatrix} -d \\ 0 \\ x - x_+ \end{pmatrix}}{\left[c(t - t_+) - v_+(x - x_+)/c \right]^3}, \tag{6.15}
$$

with the shorthand

$$
\gamma := \gamma(v(t)), \qquad \gamma_+ := \gamma(v(t_+)),
$$

whence

$$
\frac{1}{\gamma_+^2} = 1 - \frac{v_+^2}{c^2}.
$$

Note that we have been using $c(t - t_+)$ as a placeholder for r_+^A, given the relation (6.7). We now have the electric force on B due to A from

$$
\mathbf{F}^{\text{elec}}(\text{on B due to A}) = \frac{q_e}{2} \mathbf{E}^A(B),
$$

whereupon

$$\mathbf{F}^{\text{elec}}(\text{on B due to A}) = \frac{e^2}{4} \frac{\dfrac{1}{\gamma_+^2}\begin{pmatrix} x-x_+ - (t-t_+)v_+ \\ 0 \\ d \end{pmatrix} + \dfrac{da_+}{c^2}\begin{pmatrix} -d \\ 0 \\ x-x_+ \end{pmatrix}}{\left[c(t-t_+) - v_+(x-x_+)/c \right]^3},$$

(6.16)

having made the replacement $e^2 := q_e^2/4\pi\varepsilon_0$. Note that there is a component in the z direction. In a moment we shall see that it perfectly cancels the z component of the electrical force of B on A.

At any time t, knowing the position $\mathbf{r}^A(t)$ of A, the vector from B at the appropriate retarded time to A at time t is

$$\mathbf{r}_+^{BA} = \begin{pmatrix} x - x_+ \\ 0 \\ -d \end{pmatrix},$$

which is the same as $\mathbf{r}_+^{AB}$ but with d replaced by $-d$. Of course, by symmetry,

$$r_+^{BA} = \left[(x-x_+)^2 + d^2 \right]^{1/2} = r_+^{AB}.$$

Furthermore,

$$\mathbf{v}^B = \begin{pmatrix} v_+ \\ 0 \\ 0 \end{pmatrix} = \mathbf{v}^A,$$

whence

$$\mathbf{r}_+^{BA} - r_+^{BA}\mathbf{v}_+^B/c = \begin{pmatrix} x - x_+ - (t - t_+)v_+ \\ 0 \\ -d \end{pmatrix},$$

the same as $\mathbf{r}_+^{AB} - r_+^{AB}\mathbf{v}_+^A/c$ but with d replaced by $-d$. Since

$$\left.\frac{d\mathbf{v}^B}{dt}\right|_{t=t_+} = \begin{pmatrix} a_+ \\ 0 \\ 0 \end{pmatrix} = \left.\frac{d\mathbf{v}^A}{dt}\right|_{t=t_+},$$

we then have

$$\frac{\mathbf{r}_+^{BA}}{c^2} \times \left[\left(\mathbf{r}_+^{BA} - \frac{r_+^{BA}\mathbf{v}_+^B}{c} \right) \times \left.\frac{d\mathbf{v}^B}{dt}\right|_{t=t_+} \right] = -\frac{da_+}{c^2}\begin{pmatrix} d \\ 0 \\ x - x_+ \end{pmatrix},$$

as before, but with d replaced by $-d$. The denominator of (6.3) is unchanged, i.e.,

$$r_+^{BA} - \mathbf{r}_+^{BA} \cdot \mathbf{v}_+^B/c = c(t-t_+) - v_+(x-x_+)/c = r_+^{AB} - \mathbf{r}_+^{AB} \cdot \mathbf{v}_+^A/c \,.$$

The electric field at A due to B is thus

$$\mathbf{E}^B(A) = \frac{q_e}{8\pi\varepsilon_0} \frac{\frac{1}{\gamma_+^2}\begin{pmatrix} x-x_+ - (t-t_+)v_+ \\ 0 \\ -d \end{pmatrix} + \frac{da_+}{c^2}\begin{pmatrix} -d \\ 0 \\ -(x-x_+) \end{pmatrix}}{\left[c(t-t_+) - v_+(x-x_+)/c\right]^3}, \qquad (6.17)$$

and the electric force of B on A is

$$\mathbf{F}^{elec}(\text{on A due to B}) = \frac{e^2}{4} \frac{\frac{1}{\gamma_+^2}\begin{pmatrix} x-x_+ - (t-t_+)v_+ \\ 0 \\ -d \end{pmatrix} + \frac{da_+}{c^2}\begin{pmatrix} -d \\ 0 \\ -(x-x_+) \end{pmatrix}}{\left[c(t-t_+) - v_+(x-x_+)/c\right]^3}.$$

$$(6.18)$$

We observe that the x components of $\mathbf{F}^{elec}$(on A due to B) and $\mathbf{F}^{elec}$(on B due to A) are the same, while their z components have opposite sign and simply cancel if we make a sum of the two forces. Note that $x-x_+$ and a_+ may or may not be positive, so their product may have either sign. These forces along the axis of the system (in the z direction) can thus tend to stretch or to compress the dumbbell. However, we are assuming a (rather astonishing) binding force that just manages to keep the two charges A and B at exactly the same distance apart whatever happens!

We are concerned about the electric self-force in the x direction, the axis of acceleration, and simply take the sum of the x components to get

$$\boxed{(F_x)_{\text{self}}^{\text{elec}} = \frac{e^2}{2} \frac{\left[x-x_+ - (t-t_+)v_+\right]\left(1-v_+^2/c^2\right) - d^2 a_+/c^2}{\left[c(t-t_+) - v_+(x-x_+)/c\right]^3}} \qquad (6.19)$$

Note that this result is still exact. An immediate question is this: does this self-force act to the left or to the right? We shall find that, to highest order in d, it acts to the left if and only if a is positive. But before carrying out the power series expansion (in powers of d), let us consider the magnetic self-force.

6.3 Magnetic Self-Force

According to the formula (2.62) given on p. 21 of Chap. 2, the magnetic field at B
due to A is given by

$$\mathbf{B}^A(B) = \frac{\mathbf{r}_+^{AB} \times \mathbf{E}^A(B)}{cr_+^{AB}} , \tag{6.20}$$

where $\mathbf{E}^A(B)$ is given by (6.15), and the magnetic force of A on B is then

$$\mathbf{F}^{mag}(\text{on B due to A}) = \frac{q_e}{2} \mathbf{v}^B(t) \times \mathbf{B}^A(B) . \tag{6.21}$$

This leads to

$$\mathbf{B}^A(B) = \frac{1}{cr_+^{AB}} \begin{pmatrix} x - x_+ \\ 0 \\ d \end{pmatrix} \times \begin{pmatrix} E_1^A(B) \\ 0 \\ E_3^A(B) \end{pmatrix} = \frac{1}{cr_+^{AB}} \begin{pmatrix} 0 \\ dE_1^A(B) - (x - x_+)E_3^A(B) \\ 0 \end{pmatrix} ,$$

which is in the y direction, perpendicular to the plane of the page in Fig. 6.1. When
we take the vector product with $\mathbf{v}^B(t)$, which is solely in the x direction, we get a
force in the z direction:

$$\mathbf{F}^{mag}(\text{on B due to A}) = \frac{q_e}{2} \frac{v}{cr_+^{AB}} \left[dE_1^A(B) - (x - x_+)E_3^A(B) \right] \begin{pmatrix} 0 \\ 0 \\ 1 \end{pmatrix} . \tag{6.22}$$

To find $\mathbf{F}^{mag}(\text{on A due to B})$, we note that $E_1^B(A) = E_1^A(B)$ is unchanged, while
$E_3^B(A) = -E_3^A(B)$ has changed sign. When we put $d \rightarrow -d$ to obtain the magne-
tic self-force $\mathbf{F}^{mag}(\text{on A due to B})$ from $\mathbf{F}^{mag}(\text{on B due to A})$, we thus get the same
quantity with the opposite sign. This had to happen by symmetry, because the ma-
gnetic force does not lie along the axis of symmetry for this scenario (the x axis).
The result is therefore

$$\boxed{\mathbf{F}_{self}^{mag} = 0} \tag{6.23}$$

Note that the magnetic forces tend to either stretch or compress our dumbbell along
its axis. However, once again, the miraculous binding force sees to it that the dumbbell remains unchanged.

6.4 Constant Velocity Case

So we have shown that the net electromagnetic self-force of the system on itself is
purely electric, the magnetic effects summing to zero, and that it lies along the line
of acceleration (the x axis). Let us now take the total self-force as given by (6.19),
viz.,

$$F_x^{\text{self}} = \frac{e^2}{2} \frac{\left[x - x_+ - (t - t_+)v_+\right]\left(1 - v_+^2/c^2\right) - d^2 a_+/c^2}{\left[c(t - t_+) - v_+(x - x_+)/c\right]^3}$$

(6.24)

and expand as a power series in d to $O(d^0)$. Before beginning though, this is a good point to note what happens when the system is moving along at constant velocity. In this case, $v_+ = v$, $a_+ = 0 = a$, and $v_+(t - t_+) = x - x_+$, so

$$\mathbf{F}_{\text{self}} = 0 \quad \text{(constant velocity case)}$$

(6.25)

Note that this is an exact result, not depending on any power series expansion.

It would have been interesting if the self-force had been non-zero for a constant velocity of the system. In such a world, Newton's first law would not hold, because one would have to deal with this kind of effect, in bound particle systems at least, if not also in all elementary particles, in order to get the relevant object just to move along at constant velocity. One would have to oppose this reluctance to move, which we are used to finding only when we wish something to move with an acceleration, not when the motion is merely a constant velocity.

This particular self-force thus makes a clear distinction between constant velocity and acceleration. In a way, it is a step toward understanding Newton's second law. Since self-forces do indeed contribute to the inertia of bound particle systems, here we have a (pre-quantum) explanation for why accelerations should be different from constant velocities in this respect.

6.5 Power Series Expansion of Self-Force

The time has come to examine the self-force in more detail. The first task is to analyse the retarded time $t_+ = t_+(t,d)$, defined by (6.8) on p. 95, viz.,

$$c(t - t_+) = \left\{ \left[x(t) - x(t_+)\right]^2 + d^2 \right\}^{1/2}.$$

(6.26)

As mentioned there, this simple relation lies at the heart of the self-force effect, and in the more complex scenarios considered later, it is the source of all the difficulty in the calculation.

However, in the present scenario, things are simple enough. It will be worth taking the expansion of t_+ to third order in d so that we may expand $\mathbf{F}_{\text{self}}$ to $O(d^0)$. We shall then find that there is an $O(d^{-1})$ term in $\mathbf{F}_{\text{self}}$, which obviously diverges if we let the system dimension tend to zero, and can be considered to contribute to the inertial mass of the system, and an $O(d^0)$ term, i.e., independent of the system dimension, which in fact explains why our system is able to radiate electromagnetic energy when accelerated.

6.5.1 Expansions for Retarded Time and Retarded Point

We seek an expansion of the form

$$
t_+(t,d) = t_+(t,0) + \left.\frac{\partial t_+}{\partial d}\right|_{d=0} d + \frac{1}{2}\left.\frac{\partial^2 t_+}{\partial d^2}\right|_{d=0} d^2 + \frac{1}{6}\left.\frac{\partial^3 t_+}{\partial d^3}\right|_{d=0} d^3 + O(d^4) .
$$

(6.27)

We obviously have $t_+(t,0) = t$, i.e., the retarded time is just t itself when $d = 0$, because A and B lie at the same point in space. For notational purposes in what follows, we shall define

$$
U := \left.\frac{\partial t_+}{\partial d}\right|_{d=0} , \quad V := \left.\frac{\partial^2 t_+}{\partial d^2}\right|_{d=0} , \quad W := \left.\frac{\partial^3 t_+}{\partial d^3}\right|_{d=0} ,
$$

so that the above expansion becomes

$$
t_+(t,d) = t + Ud + \frac{1}{2}Vd^2 + \frac{1}{6}Wd^3 + O(d^4) .
$$

(6.28)

Taking the partial derivative of (6.26) with respect to d, we obtain

$$
-c\frac{\partial t_+}{\partial d} = \frac{-(x-x_+)v_+ \dfrac{\partial t_+}{\partial d} + d}{\left[(x-x_+)^2 + d^2\right]^{1/2}} ,
$$

whence

$$
\frac{\partial t_+}{\partial d} = \frac{d}{(x-x_+)v_+ - c^2(t-t_+)} .
$$

(6.29)

Since

$$
x = x_+ + \left.\frac{dx}{dt}\right|_{t=t_+} (t-t_+) + O(t-t_+)^2 ,
$$

we have

$$
x - x_+ = v_+(t-t_+) + O(d^2) = -v_+ Ud + O(d^2) .
$$

Hence,

$$
U = \lim_{d\to 0} \frac{d}{-v_+ Udv_+ + c^2 Ud + O(d^2)} = \frac{1}{(c^2 - v^2)U} = \frac{\gamma^2}{c^2 U} ,
$$

and finally,

$$\boxed{U = \frac{\partial t_+}{\partial d}\bigg|_{d=0} = -\frac{\gamma}{c}} \tag{6.30}$$

Note that U is expected to be negative, because t_+ should get further behind t as d increases.

Differentiating (6.29) with respect to d, we now obtain

$$\frac{\partial^2 t_+}{\partial d^2} = \frac{(x-x_+)v_+ - c^2(t-t_+) + \left[v_+^2 - (x-x_+)a_+ - c^2\right]d\dfrac{\partial t_+}{\partial d}}{\left[(x-x_+)v_+ - c^2(t-t_+)\right]^2}. \tag{6.31}$$

We have to take the limit of both sides as $d \to 0$ in order to find V. On the right-hand side, both numerator and denominator go as d^2, so we need to expand each to that order. (The method is beginning to creak, and we shall use a more efficient one to obtain W!)

The denominator is easy, since we have

$$x - x_+ = \frac{v\gamma}{c}d + O(d^2), \quad v_+ = v + O(d), \quad t - t_+ = \frac{\gamma}{c}d + O(d^2),$$

whence

$$\begin{aligned}
\left[(x-x_+)v_+ - c^2(t-t_+)\right]^2 &= \left(\frac{v^2\gamma}{c} - c^2\frac{\gamma}{c}\right)^2 d^2 + O(d^3) \\
&= c^2 d^2 \gamma^2 \left(1 - \frac{v^2}{c^2}\right)^2 + O(d^3) \\
&= \frac{c^2}{\gamma^2}d^2 + O(d^3).
\end{aligned}$$

The numerator is more delicate. We require $x - x_+$ to $O(d^2)$, v_+ to $O(d)$, $t - t_+$ to $O(d^2)$, v_+^2 to $O(d)$, and $\partial t_+/\partial d$ to $O(d)$. Now

$$t_+ = t - \frac{\gamma}{c}d + \frac{1}{2}Vd^2 + O(d^3),$$

so

$$t - t_+ = \frac{\gamma}{c}d - \frac{1}{2}Vd^2 + O(d^3), \qquad \frac{\partial t_+}{\partial d} = -\frac{\gamma}{c} + Vd + O(d^2).$$

Further

$$x(t_+) = x(t + t_+ - t)$$

$$= x(t) + v(t)(t_+ - t) + \frac{1}{2}a(t)(t_+ - t)^2 + O(t_+ - t)^3$$

$$= x + v\left(-\frac{\gamma}{c}d + \frac{1}{2}Vd^2\right) + \frac{1}{2}a\frac{\gamma^2}{c^2}d^2 + O(d^3).$$

Hence,

$$x - x_+ = \frac{\gamma v}{c}d - \left(\frac{1}{2}vV + \frac{a\gamma^2}{2c^2}\right)d^2 + O(d^2).$$

Finally,

$$v(t_+) = v(t + t_+ - t)$$

$$= v(t) + a(t)(t_+ - t) + O(t_+ - t)^2$$

$$= v + a\left(-\frac{\gamma}{c}d\right) + O(d^2),$$

so that

$$v_+ = v - \frac{a\gamma}{c}d + O(d^2), \qquad v_+^2 = v^2 - \frac{2av\gamma}{c}d + O(d^2).$$

The numerator in (6.31) is then given by

$$(x - x_+)v_+ - c^2(t - t_+) + \left[v_+^2 - (x - x_+)a_+ - c^2\right]d\frac{\partial t_+}{\partial d}$$

$$= \left[\frac{3a\gamma^2 v}{2c^2} + \frac{1}{2}(v^2 - c^2)V\right]d^2 + O(d^3).$$

Returning to (6.31) and taking the limit of both sides as $d \to 0$, we obtain

$$V = \left.\frac{\partial^2 t_+}{\partial d^2}\right|_{d=0} = \lim_{d \to 0} \frac{(x - x_+)v_+ - c^2(t - t_+) + \left[v_+^2 - (x - x_+)a_+ - c^2\right]d\frac{\partial t_+}{\partial d}}{\left[(x - x_+)v_+ - c^2(t - t_+)\right]^2}$$

$$= \lim_{d \to 0} \frac{\left[\frac{3a\gamma^2 v}{2c^2} + \frac{1}{2}(v^2 - c^2)V\right]d^2 + O(d^3)}{\frac{c^2}{\gamma^2}d^2 + O(d^3)}$$

$$= \frac{\gamma^2}{c^2}\left[\frac{3a\gamma^2 v}{2c^2} + \frac{1}{2}(v^2 - c^2)V\right]$$

$$= \frac{3a\gamma^4 v}{2c^4} - \frac{1}{2}V.$$

At last we have the result

$$V = \frac{\partial^2 t_+}{\partial d^2}\bigg|_{d=0} = \frac{av\gamma^4}{c^4} \tag{6.32}$$

So far we have shown that

$$t_+ = t - \frac{\gamma}{c}d + \frac{av\gamma^4}{2c^4}d^2 + \frac{1}{6}Wd^3 + O(d^4), \tag{6.33}$$

or

$$t - t_+ = \frac{\gamma}{c}d - \frac{av\gamma^4}{2c^4}d^2 - \frac{1}{6}Wd^3 + O(d^4) \tag{6.34}$$

This is enough to obtain $\mathbf{F}_{\text{self}}$ to $O(d^{-1})$, the lowest order term in the self-force, contributing to the inertial mass of the system. In the three other scenarios, we shall only go this far, but in the present case, things are simple enough to justify going for the radiation reaction term at $O(d^0)$. To do so, we must find W.

Taking another derivative of (6.26) with respect to d and using the above method is much too long-winded, hence open to error, so we use another approach, viz., expand both $(x - x_+)^2$ and $(t - t_+)^2$ to suitable orders, substitute the expansions into

$$c^2(t - t_+)^2 = (x - x_+)^2 + d^2, \tag{6.35}$$

then equate coefficients of powers of d on either side. We could have used this simple idea to obtain U and V above, and will opt for this method in the other three scenarios.

Squaring (6.34) and multiplying by c^2, we obtain

$$c^2(t - t_+)^2 = \gamma^2 d^2 - \frac{av\gamma^5}{c^3}d^3 + \left(\frac{a^2v^2\gamma^8}{4c^6} - \frac{W\gamma c}{3}\right)d^4 + O(d^5), \tag{6.36}$$

which is indeed accurate up to $O(d^4)$. Next we need x_+ to $O(d^3)$, using

$$x_+ = x + \frac{\partial x_+}{\partial d}\bigg|_{d=0} d + \frac{1}{2}\frac{\partial^2 x_+}{\partial d^2}\bigg|_{d=0} d^2 + \frac{1}{6}\frac{\partial^3 x_+}{\partial d^3}\bigg|_{d=0} d^3 + O(d^4),$$

where

$$\frac{\partial x_+}{\partial d} = v_+\frac{\partial t_+}{\partial d}, \qquad \frac{\partial^2 x_+}{\partial d^2} = a_+\left(\frac{\partial t_+}{\partial d}\right)^2 + v_+\frac{\partial^2 t_+}{\partial d^2},$$

and

$$\frac{\partial^3 x_+}{\partial d^3} = \dot{a}_+ \left(\frac{\partial t_+}{\partial d} \right)^3 + 3a_+ \frac{\partial t_+}{\partial d} \frac{\partial^2 t_+}{\partial d^2} + v_+ \frac{\partial^3 t_+}{\partial d^3} .$$

It follows that

$$\left. \frac{\partial x_+}{\partial d} \right|_{d=0} = v \left. \frac{\partial t_+}{\partial d} \right|_{d=0} = -\frac{\gamma v}{c} ,$$

using (6.30) for $\partial t_+ / \partial d |_{d=0}$, and

$$\left. \frac{\partial^2 x_+}{\partial d^2} \right|_{d=0} = \frac{a\gamma^2}{c^2} + \frac{av^2\gamma^4}{c^4} = \frac{a\gamma^4}{c^2} ,$$

using (6.32) for $\partial^2 t_+ / \partial d^2 |_{d=0}$. Finally,

$$\left. \frac{\partial^3 x_+}{\partial d^3} \right|_{d=0} = \dot{a} \left(-\frac{\gamma}{c} \right)^3 + 3a \left(-\frac{\gamma}{c} \right) \frac{av\gamma^4}{c^4} + vW$$

$$= vW - \frac{\dot{a}\gamma^3}{c^3} - \frac{3a^2\gamma^5 v}{c^5} .$$

We thus obtain

$$\boxed{ x - x_+ = \frac{\gamma v}{c} d - \frac{a\gamma^4}{2c^2} d^2 + \frac{1}{6} \left(\frac{\dot{a}\gamma^3}{c^3} + \frac{3a^2\gamma^5 v}{c^5} - vW \right) d^3 + O(d^4) } \qquad (6.37)$$

Squaring this, we now have

$$(x - x_+)^2 = \frac{\gamma^2 v^2}{c^2} d^2 - \frac{av\gamma^5}{c^3} d^3 + \left(\frac{\dot{a}v\gamma^4}{3c^4} + \frac{a^2 v^2\gamma^6}{c^6} - \frac{\gamma v^2 W}{3c} + \frac{a^2\gamma^8}{4c^4} \right) d^4 + O(d^5) ,$$

which is indeed correct to $O(d^4)$, and hence

$$(x - x_+)^2 + d^2 = \gamma^2 d^2 - \frac{av\gamma^5}{c^3} d^3 + \left(\frac{\dot{a}v\gamma^4}{3c^4} + \frac{a^2 v^2\gamma^6}{c^6} - \frac{\gamma v^2 W}{3c} + \frac{a^2\gamma^8}{4c^4} \right) d^4 + O(d^5) .$$

$$(6.38)$$

Comparing (6.36) term by term with (6.38), as dictated by the defining condition (6.35) for t_+, we find that the coefficients of d^2 and d^3 do agree, something we carefully arranged when we calculated U in (6.30) and V in (6.32). Requiring the coefficients of the $O(d^4)$ terms to agree implies the relation

$$\frac{a^2 v^2\gamma^8}{4c^6} - \frac{W\gamma c}{3} = \frac{\dot{a}v\gamma^4}{3c^4} + \frac{a^2 v^2\gamma^6}{c^6} - \frac{\gamma v^2 W}{3c} + \frac{a^2\gamma^8}{4c^4} .$$

Solving for W, we conclude this analysis with the result

$$W = \frac{\partial^3 t_+}{\partial d^3}\bigg|_{d=0} = -\frac{\gamma^5}{4c^5}\left(3a^2\gamma^2 + \frac{12a^2v^2\gamma^2}{c^2} + 4\dot{a}v\right) \tag{6.39}$$

We substitute this into (6.34) to obtain

$$t - t_+ = \frac{\gamma}{c}d - \frac{av\gamma^4}{2c^4}d^2 + \frac{\gamma^5}{24c^5}\left(3a^2\gamma^2 + \frac{12a^2v^2\gamma^2}{c^2} + 4\dot{a}v\right)d^3 + O(d^4) \tag{6.40}$$

and into (6.37) to obtain, after a short calculation,

$$x - x_+ = \frac{\gamma v}{c}d - \frac{a\gamma^4}{2c^2}d^2 + \frac{\gamma^5}{6c^3}\left(\dot{a} + \frac{15a^2v\gamma^2}{4c^2}\right)d^3 + O(d^4) \tag{6.41}$$

6.5.2 Expansion for Total Self-Force

We return now to the exact formula (6.19) for the total self-force on p. 99, viz.,

$$F_x^{\text{self}} = \frac{e^2}{2}\frac{\left[x - x_+ - (t - t_+)v_+\right](1 - v_+^2/c^2) - d^2 a_+/c^2}{\left[c(t - t_+) - v_+(x - x_+)/c\right]^3} \tag{6.42}$$

Since the denominator goes as d^3, we must expand the numerator to $O(d^3)$ in order to get F_x^{self} to $O(d^0)$. It turns out that the $O(d)$ terms one might expect in the numerator actually cancel, so that we shall obtain terms in d^2 and d^3. When we then come to divide by the denominator, we shall require terms in d^3 and d^4.

Starting with the denominator, we note that we require $c(t - t_+) - v_+(x - x_+)/c$ to $O(d^2)$ to fulfill the above program. This means that we require v_+ to $O(d)$. However, later we shall require v_+ to $O(d^2)$. Now

$$v_+ = v + \frac{\partial v_+}{\partial d}\bigg|_{d=0}d + +\frac{1}{2}\frac{\partial^2 v_+}{\partial d^2}\bigg|_{d=0}d^2 + O(d^3),$$

and

$$\frac{\partial v_+}{\partial d} = a_+\frac{\partial t_+}{\partial d} \implies \frac{\partial v_+}{\partial d}\bigg|_{d=0} = -\frac{a\gamma}{c},$$

by (6.30) on p. 103, while

$$\frac{\partial^2 v_+}{\partial d^2} = \dot{a}_+\left(\frac{\partial t_+}{\partial d}\right)^2 + a_+\frac{\partial^2 t_+}{\partial d^2} \implies \frac{\partial^2 v_+}{\partial d^2}\bigg|_{d=0} = \frac{\dot{a}\gamma^2}{c^2} + \frac{a^2v\gamma^4}{c^4},$$

by (6.32) on p. 105. Hence,

$$v_+ = v - \frac{\gamma a}{c}d + \frac{\gamma^2}{2c^2}\left(\dot{a} + \frac{a^2 v \gamma^2}{c^2}\right)d^2 + O(d^3) \qquad (6.43)$$

Now (6.40) and (6.41) imply that

$$c(t - t_+) - v_+(x - x_+)/c = \frac{d}{\gamma} + \frac{av\gamma^2}{c^3}d^2 + O(d^3),$$

whence

$$\left[c(t - t_+) - v_+(x - x_+)/c\right]^{-3} = \frac{\gamma^3}{d^3\left[1 + \frac{3av\gamma^3}{c^3}d + O(d^2)\right]}. \qquad (6.44)$$

Now consider the numerator of (6.42). To begin with,

$$a_+ = a + \dot{a}\left.\frac{\partial t_+}{\partial d}\right|_{d=0}d + O(d^2) = a - \frac{\gamma \dot{a}}{c}d + O(d^2),$$

whence

$$\frac{d^2 a_+}{c^2} = \frac{a}{c^2}d^2 - \frac{\gamma \dot{a}}{c^3}d^3 + O(d^4). \qquad (6.45)$$

Putting together (6.40) for $t - t_+$, (6.41) for $x - x_+$, and (6.43) for v_+, keeping the terms to $O(d^2)$ in the latter for this calculation, we now have

$$x - x_+ - v_+(t - t_+) = \frac{\gamma^2 a}{2c^2}d^2 - \left(\frac{\dot{a}\gamma^3}{3c^3} + \frac{a^2 v \gamma^5}{2c^5}\right)d^3 + O(d^4). \qquad (6.46)$$

This has to be multiplied by $1 - v_+^2/c^2$, keeping terms to $O(d)$. Equation (6.43) implies

$$1 - \frac{v_+^2}{c^2} = \gamma^{-2} + \frac{2a\gamma v}{c^3}d + O(d^2). \qquad (6.47)$$

This implies that

$$\left[x - x_+ - v_+(t - t_+)\right]\left(1 - \frac{v_+^2}{c^2}\right) = \frac{a}{2c^2}d^2 + \left(\frac{a^2 \gamma^3 v}{2c^5} - \frac{\gamma \dot{a}}{3c^3}\right)d^3 + O(d^4).$$

The numerator of (6.42) is thus

$$\left[x - x_+ - v_+(t - t_+)\right]\left(1 - \frac{v_+^2}{c^2}\right) - \frac{d^2 a_+}{c^2} = -\frac{a}{2c^2}d^2 + \left(\frac{a^2 \gamma^3 v}{2c^5} + \frac{2\gamma \dot{a}}{3c^3}\right)d^3 + O(d^4).$$

We now have the expansion of the total electromagnetic self-force in the x direction, using the last relation and (6.44) to calculate

$$F_x^{\text{self}} = \frac{e^2}{2} \frac{\gamma^3}{d^3} \left[1 - \frac{3av\gamma^3}{c^3}d + O(d^2) \right] \left[-\frac{a}{2c^2}d^2 + \left(\frac{a^2\gamma^3 v}{2c^5} + \frac{2\gamma\dot{a}}{3c^3} \right) d^3 + O(d^4) \right],$$

whence

$$F_x^{\text{self}} = -\frac{e^2}{4c^2 d}a\gamma^3 + \frac{e^2\gamma^3\dot{a}}{3c^3} + \frac{e^2 va^2\gamma^6}{c^5} + O(d) \tag{6.48}$$

6.6 Interpreting the Expansion of the Self-Force

The first observation is that there is no term going as d^{-2}. However, there is a term going as d^{-1}. If we let the system size go to zero, the self-force diverges. This is a classical version of the problem that requires mass renormalisation in quantum field theories of elementary particles. But there is also a term that is independent of the system size and remains unchanged when we let the system size go to zero. This is the radiation reaction term. We shall consider each of these terms separately in the next two sections.

But first the reader should also note what would happen if the two charges in the dumbbell system were opposite in sign. It is very easy to check that every term in F_x^{self} simply changes sign (replacing e^2 by $-e^2$). In such a system, the EM self-force will actually help the acceleration along, i.e., the EM contribution to the inertial mass will be negative. This fits perfectly with the idea that a negative binding energy in a bound state particle should decrease its inertial mass.

6.6.1 Divergent Term in the Self-Force and Mass Renormalisation

The divergent term in (6.48) (as $d \to 0$) is negative when a is positive, and positive when a is negative. This is a very significant fact. It means that, when the acceleration is instantaneously to the right (positive x axis), the self-force acts instantaneously to the left (negative x axis), and vice versa. It is in this sense that the self-force contributes to the inertia of the system. Whatever inertial masses one may attribute to the two point charges separately, with total m_0 say, there will be another contribution from this self-force effect.

Let us see how one might renormalise the inertial mass of the system, i.e., absorb the self-force contribution into a total inertial mass. Recall the Lorentz force law, or at least, the three spatial components of it:

$$\frac{\mathrm{d}}{\mathrm{d}t}(m_0\gamma\mathbf{v}) = e(\mathbf{E} + \mathbf{v} \times \mathbf{B}) + \text{EM self-force}, \tag{6.49}$$

where $\mathbf{v}$ is the coordinate 3-velocity of the system and $\mathbf{E}$ and $\mathbf{B}$ are some external EM fields. The self-force

$$\text{EM self-force} = -\frac{e^2}{4c^2 d}\gamma(v)^3 \ddot{x} \tag{6.50}$$

in the x direction, where $\ddot{x} = a$, has been added to the other terms on the right-hand side of (6.49), which are due to the external electromagnetic fields. But note that

$$\frac{\mathrm{d}}{\mathrm{d}t}\left[\gamma(v)v\right] = \gamma(v)^3 \ddot{x} .$$

This means that the divergent $O(d^{-1})$ contribution to the self-force has exactly the form required for it to contribute to inertial mass, even under relativistic conditions, where that mass varies with speed.

This is quite a remarkable situation. To begin with it allows us to replace the bare mass m_0 by the renormalised mass

$$m_{\text{ren}} := m_0 + \frac{e^2}{4c^2 d} , \tag{6.51}$$

which would be infinite if the system had no spatial extension, and then rewrite the Lorentz force law in the form

$$\frac{\mathrm{d}}{\mathrm{d}t}(m_{\text{ren}}\gamma\mathbf{v}) = e(\mathbf{E} + \mathbf{v} \times \mathbf{B}) , \tag{6.52}$$

ignoring the possible presence of any self-forces, as one usually does for the electron, for example.

But better than that, it shows that the self-force contribution to the inertial mass expressed in the renormalisation equation (6.51) will actually vary the way inertial mass is supposed to vary with speed in the special theory of relativity. This was discovered before the advent of the fully fledged relativity theory. With hindsight one might say that this had to happen because Maxwell's theory, from which this contribution was derived, is Lorentz symmetric (or relativistically invariant, or however one would like to put that), and of course it does not matter whether one is aware of that when deriving this result. But the reader is advised to look at the complexity of the calculation, and indeed the arbitrariness of the structure of our toy electron, before taking this situation as obvious.

To put it another way, one might say that we have here an explanation as to why inertial mass should increase as predicted by relativity theory when a system moves faster: it is because the self-forces within the system increase the way they do, at least as far as the self-force contributions to the inertial mass are concerned.

6.6.2 Constant Term in the Self-Force and Radiation Reaction

We said after (6.52), viz.,

$$\frac{\mathrm{d}}{\mathrm{d}t}(m_{\text{ren}}\gamma\mathbf{v}) = e(\mathbf{E} + \mathbf{v} \times \mathbf{B}) , \tag{6.53}$$

that one could ignore the possible presence of any self-force terms, but this is not quite true. For we have another term in the self-force which cannot be absorbed into the inertial mass of the system, and this term remains the same when we allow $d \to 0$. If we add in this term on the right-hand side of (6.53), we obtain something very close to the Lorentz–Dirac equation, which is discussed in great detail in [4]. The point is that this term is a radiation reaction, i.e., a force back on the system due to the fact that it radiates EM energy.

Now a charged particle like an electron will radiate EM energy whenever it is accelerated, so in any situation where it is accelerated, there will be a radiation reaction force on it. This implies that its equation of motion will never be just the Lorentz force law (6.53). Dirac was the first to obtain an adjusted equation [3], which is thus known as the Lorentz–Dirac equation. His and subsequent derivations treat the electron as a point particle and derive this equation from conservation of energy and momentum in a delicate limiting process. (The process also throws up the mass renormalisation term mentioned in the last section, this being infinite in that case, since the electron has no spatial extent.)

The fact that accelerating charges radiate is merely a consequence of Maxwell's equations. These imply the famous Larmor formula (2.63) on p. 22 for the rate P at which energy is radiated away by a charge q with acceleration a (further discussion can be found in [1, 4]):

$$P = \frac{2}{3} \frac{q^2}{c^3} a^2 . \tag{6.54}$$

Let us apply this to the present system. It will be an opportunity to discover something rather remarkable.

Our system comprises two point charges $e/2$, each with acceleration a. Now when we wrote down (6.48), we only included the interaction self-forces between the two charges. But there will also be a self-force of each component charge on itself. The latter will contribute infinitely to mass renormalisation since the component charges are mathematical points, but we shall suppose that renormalisation has been carried out in the usual way. Concerning the next term, independent of the particle dimensions, each point component will exert a self-force

$$\frac{2}{3} \frac{(e/2)^2}{c^3} \dddot{x}$$

on itself, according to the general result (3.28) on p. 41 [see also Sect. 3.5 and (3.30) on p. 43]. Adding twice this (once for each point component) to the term in the interparticle self-force (6.48) that is independent of the interparticle distance, we have a total for this particular term equal to

$$\frac{2}{3} \frac{(e/2)^2}{c^3} \dddot{x} + \frac{2}{3} \frac{(e/2)^2}{c^3} \dddot{x} + \frac{1}{3} \frac{e^2}{c^3} \dddot{x} = \frac{2}{3} \frac{e^2}{c^3} \dddot{x} ,$$

which is exactly equal to the corresponding term in the general result (3.28) for a particle with total charge e [see also (3.30) on p. 43].

To see that this reasoning is correct, consider how all that works out when the two component charges have opposite signs. We still get

$$\frac{2}{3}\frac{(e/2)^2}{c^3}\dddot{x}$$

for the term in the self-force of each component on itself that is independent of its dimensions (since each component is viewed as being made up of like charges, even though it is taken to be a point particle). But as mentioned at the beginning of this section, all the terms in the interparticle self-force (6.48) are reversed in sign, so the total for the self-force term that is independent of the spatial dimensions of the system is now

$$\frac{2}{3}\frac{(e/2)^2}{c^3}\dddot{x} + \frac{2}{3}\frac{(e/2)^2}{c^3}\dddot{x} - \frac{1}{3}\frac{e^2}{c^3}\dddot{x} = 0\,.$$

We are saying therefore that the radiation reaction is precisely zero in this case. That agrees perfectly with the corresponding term in the general result (3.28) in the case where the total charge of the system is zero.

Consider how this works out in terms of the Larmor radiation formula, first for the case where the component point particles in our system have like charges. According to (6.54), the power radiated away by each considered as a point charge is

$$\frac{2}{3}\left(\frac{e}{2}\right)^2\frac{1}{c^3}\ddot{x}^2\,,$$

giving a total

$$P(\text{components}) = \frac{2}{3}\left(\frac{e}{2}\right)^2\frac{1}{c^3}\ddot{x}^2 + \frac{2}{3}\left(\frac{e}{2}\right)^2\frac{1}{c^3}\ddot{x}^2 = \frac{1}{3}\frac{e^2}{c^3}\ddot{x}^2\,. \qquad (6.55)$$

But there is also the interaction force between the two components, i.e., the $O(d^0)$ self-force term

$$F_{\text{rad}}^{\text{self}} = \frac{e^2\gamma^3\dot{a}}{3c^3} + \frac{e^2va^2\gamma^6}{c^5}\,,$$

in (6.48). First note that the curious second term will be negligible in the non-relativistic case, so let us consider that case, setting also $\gamma \approx 1$. Then the rate of doing work dW/dt on our system against the main radiation term in the interaction self-force would be

$$\frac{dW}{dt} = -\frac{e^2\dddot{x}}{3c^3}\dot{x} = -\frac{1}{3}\frac{e^2}{c^3}\frac{d}{dt}(\dot{x}\ddot{x}) + \frac{1}{3}\frac{e^2}{c^3}\ddot{x}^2\,.$$

If the motion of the system is periodic, with $x(t) \propto \cos\omega t$, then

$$\dot{x}\ddot{x} \propto \sin 2\omega t\,,$$

so the first term in dW/dt will average to zero. But the second term is always positive, and must be added to the total (6.55) from the components. What we get then is

$$\frac{1}{3}\frac{e^2}{c^3}\ddot{x}^2 + \frac{1}{3}\frac{e^2}{c^3}\ddot{x}^2 = \frac{2}{3}\frac{e^2}{c^3}\ddot{x}^2 \,,$$

which is precisely the Larmor radiation term predicted in (6.54) for the dumbbell system when it has a total charge e.

When the component point charges in our dumbbell have opposite signs, we still have (6.55), but now the sign changes in $F_{\text{rad}}^{\text{self}}$, so the sign changes in the relevant term of the rate of doing work dW/dt on our system against the main radiation term in the interaction self-force. What we get now for the total radiated power is

$$\frac{1}{3}\frac{e^2}{c^3}\ddot{x}^2 - \frac{1}{3}\frac{e^2}{c^3}\ddot{x}^2 = 0 \,,$$

which is precisely the Larmor radiation term predicted in (6.54) for the dumbbell system when it has a total charge 0.

This is another point in favour of the idea that one should consider all charged particles to have spatial extent, and to constitute some kind of spatial distribution of charge (although obviously not the very artificial one chosen here, which in a sense exacerbates the problem, by comprising two point charges instead of just one). The point here is that otherwise one has no explanation as to why the electron should radiate energy, or what force one is working against in order to drive energy off into the surrounding space in the form of EM radiation. As pointed out by Feynman [2], when a radio antenna is radiating, the forces come from the influence of one part of the antenna current on another. In the case of a single accelerating electron radiating into otherwise empty space, there is only one place the force could come from, namely, the action of one part of the electron on another.

Chapter 7
Self-Force for Axial Linear Acceleration

7.1 Setting the Scene

We now consider the dumbbell charge system under a linear acceleration along its axis as shown in Fig. 7.1. So we have two like charges $q_e/2$, labelled A and B, and each is moving along the x axis. We shall say that A has trajectory given by $x_A(t) =: x(t)$ in whatever inertial frame we have selected, with speed $v_A(t) =: v(t) = \dot{x}(t)$ and acceleration $a_A(t) =: a(t) = \ddot{x}(t)$, these being in the x direction. But this time we have a problem, because we know that objects contract along the direction of motion in the special theory of relativity (FitzGerald contraction). And if our system is contracting as it moves along, this means that its length will not generally be d, and the speed $v_B(t)$ of B will not just be $v(t)$.

This is clearly a physical problem. Mathematically speaking, we could insist that the left-hand charge A have motion $x_A(t)$ and the right-hand charge have motion $x_B(t) = d + x_A(t)$. The binding forces would have to do some clever work to make this happen, because the electromagnetic forces go into strange contortions, as we shall soon see. It is much more likely that our electron will FitzGerald contract, just as the proverbial rigid rod in relativity theory will do when moved. The rigid rod is basically the way it is because of the electromagnetic forces set up between its constituent atoms [5]. It is more satisfying to think that the binding forces in our electron will not cleverly try to counteract this natural kind of contraction. We should expect them to obey some relativistic theory, that is, a theory which is Lorentz symmetric. Then our electron will contract. This is therefore a hypothesis that we shall make concerning the binding forces (Poincaré stresses).

For the moment, let us not go into too much detail about this, relegating the discussion of rigidity to its own chapter (see Chap. 12). It is an interesting subject in itself. Let us just note that, when Lorentz himself calculated the self-force on a spherical charged sphere under acceleration along a straight line, his sphere was assumed to contract in the direction of motion, becoming an ellipsoid. It did this in a rather special way, which in fact constitutes the very definition of rigidity as it is usually given [6]: the sphere always looked exactly spherical in its instantaneous

Lyle, S.N.: *Self-Force for Axial Linear Acceleration*. Lect. Notes Phys. **796**, 115–133 (2010)
DOI 10.1007/978-3-642-04785-5_7   © Springer-Verlag Berlin Heidelberg 2010

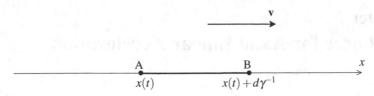

Fig. 7.1 Dumbbell charge system under linear acceleration along its axis. d is fixed, being the rest frame length of the system, but γ is a function of t via the speed $v(t)$ of A

rest frame [7]. This assumption actually makes the calculation tractable for the case of the sphere. Suffice it to say for the moment that we do not find much about the length of accelerating rigid rods in books about relativity theory (but see [5]).

There is an obvious physical point here concerning the time required for the separation between A and B to adjust to its FitzGerald-contracted value. Physically there is sure to be a small time lag before the separation between A and B assumes the correct FitzGerald-contracted length. If their speeds change too quickly, the approximations we make here are likely to be inaccurate. The time lag should be something like d/c, the time required for the electromagnetic effects, and presumably also the binding force effects, to travel across the electron.

Furthermore, charges A and B are moving at different speeds at any given time. This means that there is some question about where we should say the electron is located, and what speed it is going at. Rather than taking averages over the two component charges, let us just say that the electron is at

$$x(t) := x^A(t) , \quad \text{so that} \quad x^B(t) = x(t) + \sqrt{1 - \frac{v(t)^2}{c^2}}\, d , \tag{7.1}$$

where

$$v(t) := \dot{x}(t) , \quad \text{and hence} \quad v^B(t) = v(t) - \frac{\gamma(v)v(t)\ddot{x}(t)d}{c^2} , \tag{7.2}$$

where

$$\gamma(v) = \left(1 - \frac{v^2}{c^2}\right)^{-1/2} .$$

This means that we are assuming the instantaneous FitzGerald contraction of the separation by feeding in the speed of the left-hand charge. The position vectors of A and B will thus have the form

$$\mathbf{r}^A(t) = \begin{pmatrix} x^A(t) \\ 0 \\ 0 \end{pmatrix} , \quad \mathbf{r}^B(t) = \begin{pmatrix} x^B(t) \\ 0 \\ 0 \end{pmatrix} , \tag{7.3}$$

where $x(t) = x^A(t)$ effectively determines the motion of each of A and B through its time derivatives

$$v(t) := \dot{x}(t) , \qquad a(t) := \ddot{x}(t) , \qquad (7.4)$$

and the relations (7.1) and (7.2).

As in Chap. 6, we use the formulas for the electromagnetic fields due to A to calculate the electric and magnetic forces of A on B, then use the formulas for the electromagnetic fields due to B to calculate the electric and magnetic forces of B on A. We then simply add the two forces to see if there is a net electromagnetic force of the system on itself as it were. Naturally, A and B repel one another electrically. As in the scenario of Chap. 6, there is a potentially misleading argument as to why there should be a self-force on this system when it is moving. It goes as follows. Suppose the system is moving to the right. Charge A repels charge B with a force inversely proportional to the square of the separation. However, the force felt by B at some given time t in our inertial frame was the one generated by A at a retarded time $t_+^A < t$. Now A was slightly further to the left then than it is now. This tends to reduce the repulsion of B. But charge B repels charge A likewise. It too was slightly further to the left at the relevant retarded time t_+^B. This tends to increase its repulsive force on A. One would like to conclude that there will be a net force to the left.

It turns out that there is, but only if the instantaneous acceleration of the system is to the right. The situation is much more delicate than the above argument would suggest. When the system is moving inertially, with constant velocity, the electric field due to B is radial from its *current* position! This is something discussed in detail in Chap. 2. We have to be talking about an effect due to the acceleration, not just the velocity, when we produce any explanation. We shall thus carry out a more accurate calculation, to check that the self-force really is to the left when a system of like charges accelerates to the right.

Now the formula (2.61) given on p. 21 of Chap. 2 for the electric field due to a point charge like A or B, with charge $q_e/2$ and arbitrary worldline given by functions $x^\mu(\tau)$ of the proper time τ in Minkowski spacetime, is

$$\mathbf{E} = \frac{q_e}{8\pi\varepsilon_0} \frac{\left(\mathbf{r}_{01} - \dfrac{r_{01}\mathbf{v}}{c}\right)\left(1 - \dfrac{v^2}{c^2}\right) + \dfrac{r_{01}}{c^2} \times \left[\left(\mathbf{r}_{01} - \dfrac{r_{01}\mathbf{v}}{c}\right) \times \dfrac{d\mathbf{v}}{dt}\right]}{\left(r_{01} - \mathbf{r}_{01}\cdot\mathbf{v}/c\right)^3} , \qquad (7.5)$$

where (t_0, x_0, y_0, z_0) is the field point. Then

$$\mathbf{r}_{01} = \mathbf{x}_0 - \mathbf{x}(\tau_+) = \begin{pmatrix} x_0 - x^1(\tau_+) \\ y_0 - x^2(\tau_+) \\ z_0 - x^3(\tau_+) \end{pmatrix} , \qquad (7.6)$$

where τ_+ is the retarded time. In words, $\mathbf{r}_{01}$ is the vector from the relevant retarded point to the field point. $\mathbf{v}$ is the coordinate velocity of the source at the retarded time, and $d\mathbf{v}/dt$ is the coordinate acceleration at the retarded time.

Consider first the fields due to A. We are only concerned with the fields it produces at B, so once again we introduce a special notation. We replace $\mathbf{r}_{01}$ by the vector from A at the retarded time t_+^A to B at the time t we have chosen to consider, viz.,

$$\mathbf{r}_+^{AB} := \mathbf{r}^B(t) - \mathbf{r}^A(t_+^A) = \begin{pmatrix} x^B(t) - x^A(t_+^A) \\ 0 \\ 0 \end{pmatrix} . \qquad (7.7)$$

Note that t_+^A is the coordinate retarded time, rather than the proper retarded time, in whatever inertial frame we have selected to view things from. Let r_+^{AB} be the length of this vector, viz.,

$$r_+^{AB} = x^B(t) - x^A(t_+^A) . \qquad (7.8)$$

Then the condition determining t_+^A is

$$c(t - t_+^A) = r_+^{AB} , \qquad (7.9)$$

which just says that the light travel time from A at the retarded time t_+^A to B at the time t is just right for a signal from A at time t_+^A to arrive at B at time t. Explicitly then, the condition defining t_+^A is

$$\boxed{c(t - t_+^A) = x^B(t) - x^A(t_+^A)} \qquad (7.10)$$

When we consider the fields due to B, we are only concerned with the fields it produces at A, and we thus replace $\mathbf{r}_{01}$ in (7.5) by the vector from B at the retarded time t_+^B to A at the time t we have chosen to consider, viz.,

$$\mathbf{r}_+^{BA} := \mathbf{r}^A(t) - \mathbf{r}^B(t_+^B) = \begin{pmatrix} x^A(t) - x^B(t_+^B) \\ 0 \\ 0 \end{pmatrix} . \qquad (7.11)$$

Let r_+^{BA} be the length of this vector, viz.,

$$r_+^{BA} = \left| x^A(t) - x^B(t_+^B) \right| = x^B(t_+^B) - x^A(t) . \qquad (7.12)$$

Then the condition determining t_+^B is

$$c(t - t_+^B) = r_+^{BA} , \qquad (7.13)$$

which just says that the light travel time from B at the retarded time t_+^B to A at the time t is just right for a signal from B at time t_+^B to arrive at A at time t. Explicitly then, the condition defining t_+^B is

$$c(t - t_+^B) = x^B(t_+^B) - x^A(t) \qquad (7.14)$$

Once again, the innocuous looking relations (7.10) and (7.14) lie at the heart of the self-force calculation, and are the source of all the difficulties. Although they are simple conditions, they were not designed to facilitate the task of expanding in powers of d. However, this is precisely what we shall do, treating $t_+^A = t_+^A(t, d)$ and $t_+^B = t_+^B(t, d)$ as functions of t and d, and expanding them as power series in d, to whatever order is required.

Many things about the present scenario are more complicated than in the one discussed in Chap. 6. In particular, there are now two different retarded time functions t_+^A and t_+^B to be expanded. Some things are simpler, however. In the present case, all vectors arising in the required applications of (7.5) have only one nonzero component, in the x direction. This means that the vector product term in the numerator of (7.5) will be zero for both the electric field $\mathbf{E}^B(A)$ at A due to B and the electric field $\mathbf{E}^A(B)$ at B due to A.

Furthermore, the magnetic field $\mathbf{B}^B(A)$ at A due to B as given by (2.62) on p. 21 is

$$\mathbf{B}^B(A) = \frac{\mathbf{r}_+^{BA} \times \mathbf{E}^B(A)}{c r_+^{BA}} , \qquad (7.15)$$

and the two vectors in the numerator are parallel (in the x direction) so their vector product is zero. Likewise, the magnetic field $\mathbf{B}^A(B)$ at B due to A, viz.,

$$\mathbf{B}^A(B) = \frac{\mathbf{r}_+^{AB} \times \mathbf{E}^A(B)}{c r_+^{AB}} , \qquad (7.16)$$

will be zero. (We are not saying that A and B do not generate magnetic fields. They do. It is just that neither can ever generate a magnetic field where the other happens to be.) So in this scenario, there will be no magnetic self-force at all, not even one that might try to stretch or compress the system, as we found in Chap. 6.

7.2 Self-Force

Before making expansions in powers of d, let us write down an exact expression for the self-force on the system. The first thing is to express the electric field at B due to A, viz.,

$$\mathbf{E}^A(B) = \frac{q_e}{8\pi\varepsilon_0} \frac{\left(\mathbf{r}_+^{AB} - \dfrac{r_+^{AB} \mathbf{v}_+^A}{c} \right) \left[1 - \dfrac{(\mathbf{v}_+^A)^2}{c^2} \right]}{\left(r_+^{AB} - \mathbf{r}_+^{AB} \cdot \mathbf{v}_+^A/c \right)^3} , \qquad (7.17)$$

and the electric field at A due to B, viz.,

$$\mathbf{E}^B(A) = \frac{q_e}{8\pi\varepsilon_0} \frac{\left(\mathbf{r}_+^{BA} - \frac{r_+^{BA}\mathbf{v}_+^B}{c}\right)\left[1 - \frac{(\mathbf{v}_+^B)^2}{c^2}\right]}{\left(r_+^{BA} - \mathbf{r}_+^{BA}\cdot\mathbf{v}_+^B/c\right)^3}, \tag{7.18}$$

in which $\mathbf{v}_+^A$ and $\mathbf{v}_+^B$ are the velocities of A at the retarded time t_+^A and B at the retarded time t_+^B, respectively.

Now by (7.7), we have

$$r_+^{AB} - \frac{\mathbf{r}_+^{AB}\cdot\mathbf{v}_+^A}{c} = r_+^{AB}\left(1 - \frac{v_+^A}{c}\right)$$

and

$$\mathbf{r}_+^{AB} - \frac{r_+^{AB}\mathbf{v}_+^A}{c} = r_+^{AB}\left(1 - \frac{v_+^A}{c}\right)\begin{pmatrix} 1 \\ 0 \\ 0 \end{pmatrix}.$$

Hence,

$$\mathbf{E}^A(B) = \frac{q_e}{8\pi\varepsilon_0} \frac{1 + \frac{v_+^A}{c}}{(r_+^{AB})^2\left(1 - \frac{v_+^A}{c}\right)}\begin{pmatrix} 1 \\ 0 \\ 0 \end{pmatrix}. \tag{7.19}$$

Now the electric force on B due to A is

$$\mathbf{F}^{\text{elec}}(\text{on B due to A}) = \frac{q_e}{2}\mathbf{E}^A(B) = \frac{q_e^2}{16\pi\varepsilon_0} \frac{1 + \frac{v_+^A}{c}}{(r_+^{AB})^2\left(1 - \frac{v_+^A}{c}\right)}\begin{pmatrix} 1 \\ 0 \\ 0 \end{pmatrix}. \tag{7.20}$$

Likewise, by (7.11) and (7.12),

$$r_+^{BA} - \frac{\mathbf{r}_+^{BA}\cdot\mathbf{v}_+^B}{c} = r_+^{BA}\left(1 + \frac{v_+^B}{c}\right)$$

and

$$\mathbf{r}_+^{BA} - \frac{r_+^{BA}\mathbf{v}_+^B}{c} = -r_+^{BA}\left(1 + \frac{v_+^B}{c}\right)\begin{pmatrix} 1 \\ 0 \\ 0 \end{pmatrix}.$$

Hence,

$$\mathbf{E}^{B}(A) = -\frac{q_e}{8\pi\varepsilon_0} \frac{1 - \frac{v_+^B}{c}}{(r_+^{BA})^2 \left(1 + \frac{v_+^B}{c}\right)} \begin{pmatrix} 1 \\ 0 \\ 0 \end{pmatrix}. \tag{7.21}$$

Now the electric force on A due to B is

$$\mathbf{F}^{\mathrm{elec}}(\text{on A due to B}) = \frac{q_e}{2}\mathbf{E}^{B}(A) = -\frac{q_e^2}{16\pi\varepsilon_0} \frac{1 - \frac{v_+^B}{c}}{(r_+^{BA})^2 \left(1 + \frac{v_+^B}{c}\right)} \begin{pmatrix} 1 \\ 0 \\ 0 \end{pmatrix}. \tag{7.22}$$

Finally, making the replacement $e^2 := q_e^2/4\pi\varepsilon_0$, the total electric, and hence also the total electromagnetic force of the system on itself is

$$\mathbf{F}^{\mathrm{self}} := \mathbf{F}^{\mathrm{elec}}(\text{on B due to A}) + \mathbf{F}^{\mathrm{elec}}(\text{on A due to B}),$$

which gives the result

$$\boxed{\mathbf{F}^{\mathrm{self}} = \frac{e^2}{4}\left[\frac{1 + \frac{v_+^A}{c}}{(r_+^{AB})^2 \left(1 - \frac{v_+^A}{c}\right)} - \frac{1 - \frac{v_+^B}{c}}{(r_+^{BA})^2 \left(1 + \frac{v_+^B}{c}\right)}\right] \begin{pmatrix} 1 \\ 0 \\ 0 \end{pmatrix}} \tag{7.23}$$

By virtue of the relations (7.9) and (7.13) defining the retarded times t_+^A and t_+^B, respectively, this can also be written in the form

$$\boxed{\mathbf{F}^{\mathrm{self}} = \frac{e^2}{4c^2}\left[\frac{1 + \frac{v_+^A}{c}}{(t - t_+^A)^2 \left(1 - \frac{v_+^A}{c}\right)} - \frac{1 - \frac{v_+^B}{c}}{(t - t_+^B)^2 \left(1 + \frac{v_+^B}{c}\right)}\right] \begin{pmatrix} 1 \\ 0 \\ 0 \end{pmatrix}} \tag{7.24}$$

7.3 Constant Velocity Case

It was mentioned above that there would be no EM self-force in the case where the system has uniform velocity $\mathbf{v}$. Note that, when A has uniform velocity, one expects B also to have uniform velocity, and one expects the separation of the two charges to be the FitzGerald-contracted distance $d\sqrt{1 - v^2/c^2}$, as given in this case by (7.1) on p. 116. Note that this separation is constant when v is constant.

It is not immediately obvious that $\mathbf{F}^{\mathrm{self}}$ in (7.24) will be zero here. Let us see why it is in fact zero. In the constant velocity case, $v_+^A = v = v_+^B$, so

$$\mathbf{F}^{\text{self}}(\text{constant velocity}) = \frac{e^2}{4c^2}\left[\frac{1+\dfrac{v}{c}}{(t-t_+^A)^2\left(1-\dfrac{v}{c}\right)} - \frac{1-\dfrac{v}{c}}{(t-t_+^B)^2\left(1+\dfrac{v}{c}\right)}\right]\begin{pmatrix}1\\0\\0\end{pmatrix},$$

and it only remains to find $t - t_+^A$ and $t - t_+^B$ from the defining conditions

$$c(t-t_+^A) = x^B(t) - x^A(t_+^A)\,, \qquad c(t-t_+^B) = x^B(t_+^B) - x^A(t)\,.$$

Now

$$x^B(t) = x(t) + d\gamma^{-1}\,, \qquad x(t) - x(t_+^A) = v(t-t_+^A)\,,$$

so

$$c(t-t_+^A) = x(t) + d\gamma^{-1} - x(t_+^A)$$
$$= v(t-t_+^A) + d\gamma^{-1}\,,$$

and this implies that

$$(c-v)(t-t_+^A) = d\gamma^{-1}\,,$$

and finally,

$$t - t_+^A = \frac{d\gamma^{-1}}{c-v}\,.$$

In a similar way, one has

$$t - t_+^B = \frac{d\gamma^{-1}}{c+v}\,.$$

But a little algebra now shows that

$$\frac{1+\dfrac{v}{c}}{(t-t_+^A)^2\left(1-\dfrac{v}{c}\right)} = \frac{\gamma^2(c^2-v^2)}{d^2} = \frac{1-\dfrac{v}{c}}{(t-t_+^B)^2\left(1+\dfrac{v}{c}\right)}\,,$$

whence

$$\mathbf{F}^{\text{self}}(\text{constant velocity}) = 0\,.$$

This result is exact, since in the constant velocity case, one does not have the problem of having to make some special assumption about the varying length of the system.

7.4 Power Series Expansion of Self-Force

The time has come to examine the self-force in more detail. The first task is to analyse the retarded times $t_+^A = t_+^A(t,d)$ and $t_+^B = t_+^B(t,d)$, defined by (7.10) and (7.14) on p. 118, viz.,

$$\boxed{c(t - t_+^A) = x^B(t) - x^A(t_+^A), \qquad c(t - t_+^B) = x^B(t_+^B) - x^A(t)} \tag{7.25}$$

As mentioned there, these simple relations lie at the heart of the self-force effect, and in the more complex scenarios considered here, they are the source of all the difficulty in the calculation.

7.4.1 Expansions for Retarded Times and Retarded Points

Expansion of t_+^A

We seek an expansion of the form

$$t_+^A(t,d) = t_+^A(t,0) + \left.\frac{\partial t_+^A}{\partial d}\right|_{d=0} d + \frac{1}{2}\left.\frac{\partial^2 t_+^A}{\partial d^2}\right|_{d=0} d^2 + O(d^3). \tag{7.26}$$

We obviously have $t_+^A(t,0) = t$, i.e., the retarded time is just t itself when $d = 0$, because A and B lie at the same point in space. For notational purposes in what follows, we shall define

$$U := \left.\frac{\partial t_+^A}{\partial d}\right|_{d=0}, \qquad V := \left.\frac{\partial^2 t_+^A}{\partial d^2}\right|_{d=0},$$

so that the above expansion becomes

$$t_+(t,d) = t + Ud + \frac{1}{2}Vd^2 + O(d^3). \tag{7.27}$$

We need to satisfy

$$c(t - t_+^A) = x^B(t) - x^A(t_+^A),$$

where

$$x^A(t_+^A) = x(t_+^A), \qquad x^B(t) = x(t) + \frac{d}{\gamma}.$$

Hence,

$$c(t - t_+^A) = x(t) - x(t_+^A) + d\gamma^{-1}. \tag{7.28}$$

On the left-hand side, we can substitute the power series

$$c(t - t_+^A) = -cUd - \frac{1}{2}cVd^2 + O(d^3) \, . \tag{7.29}$$

We need an expansion for $x_+^A := x(t_+^A)$ considered as a function of t and d, viz.,

$$x(t_+^A) = x(t) + \frac{\partial x_+^A}{\partial d}\bigg|_{d=0} d + \frac{1}{2}\frac{\partial^2 x_+^A}{\partial d^2}\bigg|_{d=0} d^2 + O(d^3) \, ,$$

where

$$\frac{\partial x_+^A}{\partial d} = v_+^A \frac{\partial t_+^A}{\partial d} \, , \qquad \frac{\partial^2 x_+^A}{\partial d^2} = a_+^A \left(\frac{\partial t_+^A}{\partial d}\right)^2 + v_+^A \frac{\partial^2 t_+^A}{\partial d^2} \, ,$$

with $v_+^A := v(t_+^A)$ and $a_+^A := a(t_+^A)$. Hence,

$$\frac{\partial x_+^A}{\partial d}\bigg|_{d=0} = vU \, , \qquad \frac{\partial^2 x_+^A}{\partial d^2}\bigg|_{d=0} = aU^2 + vV \, ,$$

and we have the expansion for the retarded position x_+^A in the form

$$x_+^A = x + vUd + \frac{1}{2}(aU^2 + vV)d^2 + O(d^3) \, .$$

This in turn implies

$$x - x_+^A = -vUd - \frac{1}{2}(aU^2 + vV)d^2 + O(d^3) \, . \tag{7.30}$$

The defining relation (7.28) for t_+^A can now be written

$$-cUd - \frac{1}{2}cVd^2 + O(d^3) = -vUd - \frac{1}{2}(aU^2 + vV)d^2 + d\gamma^{-1} + O(d^3) \, ,$$

where we have simply substituted in (7.29) and (7.30). We now equate coefficients of powers of d on either side to obtain the two relations

$$\begin{cases} -Uc = \dfrac{1}{\gamma} - vU \, , \\ -\dfrac{1}{2}Vc = -\dfrac{1}{2}(aU^2 + vV) \, . \end{cases}$$

The solution is

$$\boxed{U = \frac{\partial t_+^A}{\partial d}\bigg|_{d=0} = -\frac{1}{\gamma(c - v)} = -\frac{1}{c}\sqrt{\frac{c + v}{c - v}}} \tag{7.31}$$

and

$$V = \frac{\partial^2 t_+^A}{\partial d^2}\bigg|_{d=0} = \frac{a}{\gamma^2 (c-v)^3} = \frac{a}{c^2} \frac{c+v}{(c-v)^2}$$
(7.32)

We now have the required expansion of t_+^A in the form

$$t - t_+^A = \frac{1}{\gamma(c-v)} d - \frac{a}{2\gamma^2 (c-v)^3} d^2 + O(d^3)$$
(7.33)

The power series expansion of $x - x_+^A$ is not needed for the self-force calculation.

Expansion of t_+^B

We seek an expansion of the form

$$t_+^B(t,d) = t_+^B(t,0) + \frac{\partial t_+^B}{\partial d}\bigg|_{d=0} d + \frac{1}{2} \frac{\partial^2 t_+^B}{\partial d^2}\bigg|_{d=0} d^2 + O(d^3).$$
(7.34)

As before, $t_+^B(t,0) = t$, i.e., the retarded time is just t itself when $d = 0$, because A and B lie at the same point in space. We define

$$U := \frac{\partial t_+^B}{\partial d}\bigg|_{d=0}, \qquad V := \frac{\partial^2 t_+^B}{\partial d^2}\bigg|_{d=0},$$

so that the above expansion becomes

$$t_+(t,d) = t + Ud + \frac{1}{2}Vd^2 + O(d^3),$$
(7.35)

bearing in mind that this U and V are not the same as those in the last section.

The equation to solve this time is

$$c(t - t_+^B) = x^B(t_+^B) - x^A(t),$$

where

$$x^A(t) = x(t), \qquad x^B(t_+^B) = x(t_+^B) + \frac{d}{\gamma(v(t_+^B))}.$$

So we have to expand both sides of

$$c(t - t_+^B) = x(t_+^B) + \frac{d}{\gamma(v(t_+^B))} - x(t).$$
(7.36)

On the left, we substitute the expansion

$$c(t - t_+^B) = -cUd - \frac{1}{2}cVd^2 + O(d^3) . \qquad (7.37)$$

We need an expansion for $x_+^B := x^B(t_+^B)$ considered as a function of t and d, viz.,

$$x^B(t_+^B) = x(t) + \left.\frac{\partial x_+^B}{\partial d}\right|_{d=0} d + \frac{1}{2}\left.\frac{\partial^2 x_+^B}{\partial d^2}\right|_{d=0} d^2 + O(d^3) ,$$

where we have used the fact that $x_+^B(t, 0) = x(t)$. Now

$$\frac{\partial x_+^B}{\partial d} = v(t_+^B)\frac{\partial t_+^B}{\partial d} + \frac{1}{\gamma(v(t_+^B))} - \frac{dv(t_+^B)a(t_+^B)/c^2}{[1 - v(t_+^B)^2/c^2]^{1/2}}\frac{\partial t_+^B}{\partial d} ,$$

and

$$\frac{\partial^2 x_+^B}{\partial d^2} = a(t_+^B)\left(\frac{\partial t_+^B}{\partial d}\right)^2 + v(t_+^B)\frac{\partial^2 t_+^B}{\partial d^2} - \frac{2v(t_+^B)a(t_+^B)/c^2}{[1 - v(t_+^B)^2/c^2]^{1/2}}\frac{\partial t_+^B}{\partial d}$$

$$- d\frac{\partial}{\partial d}\left\{\frac{v(t_+^B)a(t_+^B)/c^2}{[1 - v(t_+^B)^2/c^2]^{1/2}}\frac{\partial t_+^B}{\partial d}\right\} .$$

Hence,

$$\left.\frac{\partial x_+^B}{\partial d}\right|_{d=0} = vU + \frac{1}{\gamma} , \qquad \left.\frac{\partial^2 x_+^B}{\partial d^2}\right|_{d=0} = aU^2 + vV - \frac{2\gamma av}{c^2}U ,$$

and we have the expansion for the retarded position x_+^B in the form

$$x_+^B = x + \left(vU + \frac{1}{\gamma}\right)d + \frac{1}{2}\left(aU^2 + vV - \frac{2\gamma av}{c^2}U\right)d^2 + O(d^3) ,$$

or

$$x_+^B - x = \left(vU + \frac{1}{\gamma}\right)d + \frac{1}{2}\left(aU^2 + vV - \frac{2\gamma av}{c^2}U\right)d^2 + O(d^3) . \qquad (7.38)$$

Now (7.36) reads

$$c(t - t_+^B) = x_+^B - x ,$$

and substituting (7.37) and (7.38) into this, we have

$$-cUd - \frac{1}{2}cVd^2 + O(d^3) = \left(vU + \frac{1}{\gamma}\right)d + \frac{1}{2}\left(aU^2 + vV - \frac{2\gamma av}{c^2}U\right)d^2 + O(d^3) .$$

Equating coefficients of powers of d on either side, we obtain the two relations

$$\begin{cases} -Uc = vU + \dfrac{1}{\gamma}, \\[2mm] -\dfrac{1}{2}Vc = \dfrac{1}{2}\left(aU^2 + vV - \dfrac{2\gamma av}{c^2}U\right), \end{cases}$$

with solutions

$$U = \left.\frac{\partial t_+^{B}}{\partial d}\right|_{d=0} = -\frac{1}{\gamma(c+v)} = -\frac{1}{c}\sqrt{\frac{c-v}{c+v}} \tag{7.39}$$

and

$$V = \left.\frac{\partial^2 t_+^{B}}{\partial d^2}\right|_{d=0} = -\frac{a}{c^2(c+v)} \tag{7.40}$$

We now have the required expansion of t_+^{B} in the form

$$t - t_+^{B} = \frac{1}{\gamma(c+v)}d + \frac{a}{2c^2(c+v)}d^2 + O(d^3) \tag{7.41}$$

The power series expansion of $x - x_+^{B}$ is not needed for the self-force calculation.

7.4.2 Expansion for Self-Force

We now have to turn back to (7.24) on p. 121 for the self-force, viz.,

$$\mathbf{F}^{\text{self}} = \frac{e^2}{4c^2}\left[\frac{1 + \dfrac{v_+^{A}}{c}}{(t - t_+^{A})^2\left(1 - \dfrac{v_+^{A}}{c}\right)} - \frac{1 - \dfrac{v_+^{B}}{c}}{(t - t_+^{B})^2\left(1 + \dfrac{v_+^{B}}{c}\right)}\right]\begin{pmatrix}1 \\ 0 \\ 0\end{pmatrix} \tag{7.42}$$

This exercise can be broken down into parts.

Power Series Expansion of $1/(t - t_+^{A})^2$

We seek a series expansion of $1/(t - t_+^{A})^2$ to $O(d^{-1})$ in order to get $\mathbf{F}^{\text{self}}$ to this accuracy. Define

$$g(t,d) := \frac{d^2}{\left[t - t_+^{A}(t,d)\right]^2}.$$

The leading term in the expansion of g will be $O(d^0)$, and we obtain an expansion for $1/(t - t_+^A)^2$ from the expansion for $g(t,d)/d^2$. We use the results

$$t - t_+^A = \frac{1}{\gamma(c-v)}d - \frac{a}{2\gamma^2(c-v)^3}d^2 + O(d^3) , \tag{7.43}$$

$$t_+^A = t - \frac{1}{\gamma(c-v)}d + \frac{a}{2\gamma^2(c-v)^3}d^2 + O(d^3) , \tag{7.44}$$

and

$$\frac{\partial t_+^A}{\partial d} = -\frac{1}{\gamma(c-v)} + \frac{a}{\gamma^2(c-v)^3}d + O(d^2) . \tag{7.45}$$

First note that

$$g(t,0) = c^2\frac{c-v}{c+v} = \gamma^2(c-v)^2 ,$$

using (7.43). Further,

$$\frac{\partial g}{\partial d} = \frac{2d}{(t-t_+^A)^2} + \frac{2d^2}{(t-t_+^A)^3}\frac{\partial t_+^A}{\partial d}$$

$$= \frac{2d}{(t-t_+^A)^3}\left(t - t_+^A + d\frac{\partial t_+^A}{\partial d}\right)$$

$$= \frac{2d}{(t-t_+^A)^3}\left[\frac{a}{2\gamma^2(c-v)^3}d^2 + O(d^3)\right] ,$$

after substituting in the expansions (7.43) and (7.45). Now using (7.43) once more,

$$\left.\frac{\partial g}{\partial d}\right|_{d=0} = \lim_{d\to 0}\frac{2d}{(t-t_+^A)^3}\left[\frac{a}{2\gamma^2(c-v)^3}d^2 + O(d^3)\right] = a\gamma .$$

Finally, we have our expansion of g, viz.,

$$g(t,d) = \gamma^2(c-v)^2 + a\gamma d + O(d^2) ,$$

and the required expansion for $1/(t - t_+^A)^2$, viz.,

$$\boxed{\frac{1}{(t-t_+^A)^2} = \frac{\gamma^2(c-v)^2}{d^2} + \frac{a\gamma}{d} + O(d^0)} \tag{7.46}$$

Power Series Expansion of $1/(t - t_+^B)^2$

We seek a series expansion of $1/(t - t_+^B)^2$ to $O(d^{-1})$ in order to get $\mathbf{F}^{\text{self}}$ to this accuracy. Define

$$h(t,d) := \frac{d^2}{\left[t - t_+^B(t,d)\right]^2} .$$

The leading term in the expansion of h will be $O(d^0)$, and we obtain an expansion for $1/(t - t_+^B)^2$ from the expansion for $h(t,d)/d^2$. We use the results

$$t - t_+^B = \frac{1}{\gamma(c+v)}d + \frac{a}{2c^2(c+v)}d^2 + O(d^3) , \tag{7.47}$$

$$t_+^B = t - \frac{1}{\gamma(c+v)}d - \frac{a}{2c^2(c+v)}d^2 + O(d^3) , \tag{7.48}$$

and

$$\frac{\partial t_+^B}{\partial d} = -\frac{1}{\gamma(c+v)} - \frac{a}{c^2(c+v)}d + O(d^2) . \tag{7.49}$$

First note that

$$h(t,0) = \gamma^2(c+v)^2 ,$$

using (7.47). Further,

$$\begin{aligned}
\frac{\partial h}{\partial d} &= \frac{2d}{(t - t_+^B)^2} + \frac{2d^2}{(t - t_+^B)^3}\frac{\partial t_+^B}{\partial d} \\
&= \frac{2d}{(t - t_+^B)^3}\left(t - t_+^B + d\frac{\partial t_+^B}{\partial d}\right) \\
&= \frac{2d}{(t - t_+^B)^3}\left[-\frac{a}{2c^2(c+v)}d^2 + O(d^3)\right] \\
&= -\frac{d^3}{(t - t_+^B)^3}\left[\frac{a}{c^2(c+v)} + O(d)\right] ,
\end{aligned}$$

after substituting in the expansions (7.47) and (7.49). Now using (7.47) once more,

$$\frac{\partial h}{\partial d}\bigg|_{d=0} = -\lim_{d\to 0}\left\{\frac{d^3}{(t - t_+^B)^3}\left[\frac{a}{c^2(c+v)} + O(d)\right]\right\} = -\frac{ac^2}{\gamma(c-v)^2} .$$

Finally, we have our expansion of h, viz.,

$$h(t,d) = \gamma^2(c+v)^2 - \frac{ac^2}{\gamma(c-v)^2}d + O(d^2),$$

and the required expansion for $1/(t - t_+^B)^2$, viz.,

$$\boxed{\frac{1}{(t - t_+^B)^2} = \frac{\gamma^2(c+v)^2}{d^2} - \frac{ac^2}{\gamma(c-v)^2 d} + O(d^0)} \tag{7.50}$$

Power Series Expansion of $(c + v_+^A)/(c - v_+^A)$

Define

$$e(t,d) := \frac{c + v(t_+^A)}{c - v(t_+^A)},$$

recalling that $v_+^A := v(t_+^A)$. We require the series expansion up to $O(d)$. Obviously,

$$e(t,0) = \frac{c+v}{c-v}.$$

Now

$$\frac{\partial e}{\partial d} = \left\{ \frac{1}{c - v(t_+^A)} + \frac{c + v(t_+^A)}{[c - v(t_+^A)]^2} \right\} a(t_+^A) \frac{\partial t_+^A}{\partial d}$$

$$= \frac{2ca_+^A}{(c - v_+^A)^2} \frac{\partial t_+^A}{\partial d},$$

and this has to be evaluated at $d = 0$, using the result (7.31) on p. 124, viz.,

$$\left. \frac{\partial t_+^A}{\partial d} \right|_{d=0} = -\frac{1}{\gamma(c-v)}.$$

We soon find

$$\left. \frac{\partial e}{\partial d} \right|_{d=0} = -\frac{2ca}{\gamma(c-v)^3}.$$

Hence,

$$\boxed{e(t,d) = \frac{c + v(t_+^A)}{c - v(t_+^A)} = \frac{c+v}{c-v} - \frac{2ca}{\gamma(c-v)^3}d + O(d^2)} \tag{7.51}$$

Power Series Expansion of $(c - v_+^B)/(c + v_+^B)$

Define

$$f(t,d) := \frac{c - v_+^B}{c + v_+^B} \, .$$

We have to be careful here, because, recalling (7.2) on p. 116,

$$v_+^B := v^B(t_+^B) = v(t_+^B) - \frac{\gamma(v(t_+^B))v(t_+^B)a(t_+^B)d}{c^2} \, ,$$

which has an explicit dependence on d, as well as a more complicated dependence on d through t_+^B.

We require the series expansion up to $O(d)$. Obviously,

$$f(t,0) = \frac{c - v}{c + v} \, .$$

Now

$$\frac{\partial f}{\partial d} = \left[-\frac{1}{c + v_+^B} - \frac{c - v_+^B}{(c + v_+^B)^2} \right] \left[\frac{\partial v_+^B}{\partial t_+^B} \frac{\partial t_+^B}{\partial d} - \frac{\gamma(v(t_+^B))v(t_+^B)a(t_+^B)d}{c^2} \right]$$

$$= -\frac{2c}{(c + v_+^B)^2} \left[\frac{\partial v_+^B}{\partial t_+^B} \frac{\partial t_+^B}{\partial d} - \frac{\gamma(v(t_+^B))v(t_+^B)a(t_+^B)d}{c^2} \right] \, .$$

A short calculation shows that

$$\frac{dv^B}{dt} = \frac{d}{dt}\left[v(t) - \frac{\gamma(v(t))v(t)a(t)d}{c^2} \right]$$

$$= a - \frac{\gamma^3 a^2}{c^2}d - \frac{\gamma v \dot{a}}{c^2}d \, ,$$

whence

$$\frac{\partial v_+^B}{\partial t_+^B} = a(t_+^B) - \frac{\gamma(v(t_+^B))^3 a(t_+^B)^2}{c^2}d - \frac{\gamma(v(t_+^B))v(t_+^B)\dot{a}(t_+^B)}{c^2}d \, .$$

Note then that the more complicated terms here are irrelevant to us, because

$$\left.\frac{\partial v_+^B}{\partial t_+^B}\right|_{d=0} = a(t) \, .$$

Returning to the above expression for $\partial f/\partial d$ and inserting (7.39) on p. 127, viz.,

$$\frac{\partial t_+^B}{\partial d}\bigg|_{d=0} = -\frac{1}{\gamma(c+v)},$$

we have

$$\frac{\partial f}{\partial d} = -\frac{2c}{(c+v)^2}\left[-\frac{a}{\gamma(c+v)} - \frac{\gamma va}{c^2}\right].$$

Hence, after a little algebra,

$$\frac{\partial f}{\partial d}\bigg|_{d=0} = \frac{2a\gamma}{(c+v)^2},$$

and we have the required expansion

$$f(t,d) = \frac{c - v_+^B}{c + v_+^B} = \frac{c-v}{c+v} + \frac{2a\gamma}{(c+v)^2}d + O(d^2) \tag{7.52}$$

Power Series Expansion for $\mathbf{F}^{\text{self}}$

Returning to (7.42), we are now ready to expand the EM self-force to $O(d^{-1})$, using (7.46) and (7.51) to obtain

$$\frac{1}{(t-t_+^A)^2}\frac{c+v(t_+^A)}{c-v(t_+^A)} = \left[\frac{\gamma^2(c-v)^2}{d^2} + \frac{a\gamma}{d} + O(d^0)\right]\left[\frac{c+v}{c-v} - \frac{2ca}{\gamma(c-v)^3}d + O(d^2)\right]$$

$$= \frac{c^2}{d^2} - \frac{a\gamma}{d} + O(d^0),$$

and (7.50) and (7.52) to obtain

$$\frac{1}{(t-t_+^B)^2}\frac{c-v_+^B}{c+v_+^B} = \left[\frac{\gamma^2(c+v)^2}{d^2} - \frac{ac^2}{\gamma(c-v)^2 d} + O(d^0)\right]\left[\frac{c-v}{c+v} + \frac{2a\gamma}{(c+v)^2}d + O(d^2)\right]$$

$$= \frac{c^2}{d^2} + a\gamma(2\gamma^2 - 1)\frac{1}{d} + O(d^0),$$

whence finally,

$$\mathbf{F}^{\text{self}} = -\frac{e^2}{2c^2 d}a\gamma^3\begin{pmatrix} 1 \\ 0 \\ 0 \end{pmatrix} + O(d^0) \tag{7.53}$$

7.5 Interpreting the Leading Order Term in the Self-Force

As for the calculation in Chap. 6, the first observation here is that there is no term going as d^{-2}. Once again the leading order goes as d^{-1} so the self-force diverges if we let $d \to 0$, i.e., if we let the system tend to a point particle. In the present case, we have not calculated the $O(d^0)$ term in the self-force, which would correspond to the radiation reaction force, independent of the system dimensions. This is because the calculation would be much more involved. In fact, one then requires the expansions of the retarded times t_+^A and t_+^B to $O(d^3)$.

There is no point repeating everything that was said in Sect. 6.6.1. Let us just observe that the mass renormalisation is slightly different here, viz.,

$$m_{\text{ren}}^{\text{longitudinal}} := m_0 + \frac{e^2}{2c^2 d} \, , \tag{7.54}$$

to be compared with (6.51) on p. 110, viz.,

$$m_{\text{ren}}^{\text{transverse}} := m_0 + \frac{e^2}{4c^2 d} \, , \tag{7.55}$$

So here is a very interesting feature of the electromagnetic mass, as derived from the self-force, namely, it may depend which way the object is moving relative to its own geometry. This is not something one normally expects of the inertial mass, and yet it is a fact that, in bound systems that do not have spherical symmetry, the inertia of the system will depend which way it moves relative to its own geometry. This is what we have just shown for an electromagnetic binding force.

We should also mention the miraculous factor of γ^3 in (7.53), which allows one to say that the contribution $e^2/2c^2 d$ to the total inertia of the system will increase as $\gamma(v)$ with the speed v of the system, in just the way mass is supposed to increase with speed in special relativity (see the discussion in Sect. 6.6.1). But there is a very important proviso in this case, because it is not really clear what the speed of the system is when it moves longitudinally. The point is that A and B generally have different speeds, and we have calculated γ for the speed of A. Indeed, since different γ factors are associated with A and B, it is not obvious that we can just add up a force on A and a force on B.

In pre-relativistic physics, we had a whole theory of rigid bodies, and it made sense to add up forces acting on different parts of a body, because one could deduce something about the center of mass of the body. What we lack here is a proper theory of rigid bodies in relativity theory, something which will be discussed further in Chap. 12. For the moment though, let us just say that it is still satisfying to see that we get the γ factor at a certain level of approximation.

Chapter 8
Self-Force for Transverse Rotational Motion

8.1 Setting the Scene

We consider the dumbbell charge system rotating about a fixed center, perpendicular
to its axis, in such a way that it always lies along a radial line from the center of
rotation, as shown in Fig. 8.1. So as usual we have two like charges $q_e/2$, and they
are assumed to be separated by a constant distance d. As we shall see, if this distance
is to be the rest frame length of the system, this is something that has to be cleverly
engineered by the binding force in the system. At least in this case, both A and B
have constant speeds, always instantaneously perpendicular to the axis joining them,
so we do not have to worry about FitzGerald contraction.

The beauty of this scenario, as compared with those discussed in Chaps. 6 and
7, is that the acceleration of either A or B is now along the system axis (the line
joining A and B), while the velocity is always instantaneously perpendicular to the
acceleration. A priori, it is not at all obvious that the electromagnetic self-force will
lie along the system axis, i.e., parallel to the acceleration.

Let us formulate this. We shall take the position vectors of A and B at time t to
be

$$\mathbf{r}^A(t) = \begin{pmatrix} R\cos\omega t \\ R\sin\omega t \\ 0 \end{pmatrix} , \qquad \mathbf{r}^B(t) = \begin{pmatrix} (R+d)\cos\omega t \\ (R+d)\sin\omega t \\ 0 \end{pmatrix} , \qquad (8.1)$$

where R is thus the radial distance of A from the center of rotation, d is the length
of the system (the separation of A and B), which is constant in this scenario, and ω
is the angular velocity of either A or B. The velocities are

$$\mathbf{v}^A(t) = \omega R \begin{pmatrix} -\sin\omega t \\ \cos\omega t \\ 0 \end{pmatrix} , \qquad \mathbf{v}^B(t) = \omega(R+d) \begin{pmatrix} -\sin\omega t \\ \cos\omega t \\ 0 \end{pmatrix} , \qquad (8.2)$$

Lyle, S.N.: *Self-Force for Transverse Rotational Motion*. Lect. Notes Phys. **796**, 135–155 (2010)
DOI 10.1007/978-3-642-04785-5_8   © Springer-Verlag Berlin Heidelberg 2010

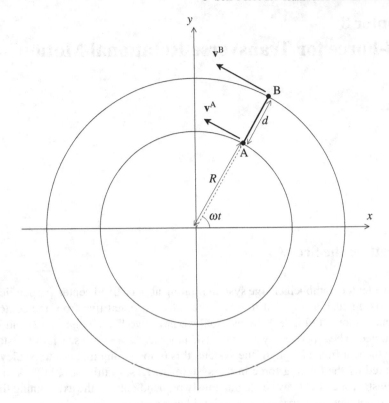

Fig. 8.1 Dumbbell charge system under rotational motion perpendicular to its axis

whence A and B move at different speeds ωR and $\omega(R+d)$, respectively. They also have different accelerations

$$
\mathbf{a}^A(t) = -\omega^2 R \begin{pmatrix} \cos\omega t \\ \sin\omega t \\ 0 \end{pmatrix}, \qquad
\mathbf{a}^B(t) = -\omega^2 (R+d) \begin{pmatrix} \cos\omega t \\ \sin\omega t \\ 0 \end{pmatrix}, \qquad (8.3)
$$

but both directed toward the center of rotation.

8.2 Retarded Times and Retarded Displacement Vectors

We know that the first step in calculating the electric and magnetic fields produced by A at B, and by B at A, is to find the retarded times t_+^A and t_+^B, treating them as functions of both t and d and expanding them to second order in a power series in the small quantity d.

When we consider B at time t, the condition on the relevant retarded time t_+^A for A, i.e., the time $t_+^A < t$ at which A generated the fields that now concern B, is

$$\left| \mathbf{r}^B(t) - \mathbf{r}^A(t_+^A) \right|^2 = c^2(t - t_+^A)^2 . \tag{8.4}$$

This leads to

$$\boxed{(R+d)^2 + R^2 - 2R(R+d)\cos\omega(t - t_+^A) = c^2(t - t_+^A)^2} \tag{8.5}$$

having used the identity

$$\cos\omega t \cos\omega t_+^A + \sin\omega t \sin\omega t_+^A = \cos\omega(t - t_+^A) .$$

Actually, we shall define the retarded displacement vector from A at the appropriate retarded time t_+^A to B at the present time t, viz.,

$$\mathbf{r}_+^{AB}(t) := \mathbf{r}^B(t) - \mathbf{r}^A(t_+^A) = \begin{pmatrix} (R+d)\cos\omega t - R\cos\omega t_+^A \\ (R+d)\sin\omega t - R\sin\omega t_+^A \\ 0 \end{pmatrix}, \tag{8.6}$$

since this is a vector we must also eventually expand in a power series in d.

In a like manner, when we consider A at time t, the condition on the relevant retarded time t_+^B for B, i.e., the time $t_+^B < t$ at which B generated the fields that now concern A, is

$$\left| \mathbf{r}^A(t) - \mathbf{r}^B(t_+^B) \right|^2 = c^2(t - t_+^B)^2 . \tag{8.7}$$

This leads to

$$\boxed{(R+d)^2 + R^2 - 2R(R+d)\cos\omega(t - t_+^B) = c^2(t - t_+^B)^2} \tag{8.8}$$

which is exactly the same as the condition on t_+^A. This means that we will only need to work out one expansion, say for t_+^A, and then the other will be identical. Once again, we shall define the retarded displacement vector from B at the appropriate retarded time t_+^B to A at the present time t, viz.,

$$\mathbf{r}_+^{BA}(t) := \mathbf{r}^A(t) - \mathbf{r}^B(t_+^B) = \begin{pmatrix} R\cos\omega t - (R+d)\cos\omega t_+^B \\ R\sin\omega t - (R+d)\sin\omega t_+^B \\ 0 \end{pmatrix}, \tag{8.9}$$

since we shall also eventually expand this in a power series in d.

8.3 Power Series Expansion of Retarded Times

As mentioned above, we shall only need to do this for t_+^A, whence the other retarded time will have the same expansion. As usual, we treat t_+^A as a function of t and d, where t is the time at which we are considering B, and seek an expansion of the form

$$t_+^A(t,d) = t_+^A(t,0) + \left.\frac{\partial t_+^A}{\partial d}\right|_{d=0} d + \frac{1}{2}\left.\frac{\partial^2 t_+^A}{\partial d^2}\right|_{d=0} d^2 + O(d^3) . \qquad (8.10)$$

As usual, $t_+^A(t,0) = t$. Now one way to obtain $\partial t_+^A/\partial d$ is just to differentiate both sides of the relation (8.5) with respect to d, to obtain

$$\frac{\partial t_+^A}{\partial d} = \frac{R + d - R\cos\omega(t - t_+^A)}{\omega R(R+d)\sin\omega(t - t_+^A) - c^2(t - t_+^A)} ,$$

then take the limit of the right-hand side as $d \to 0$. This is reasonably effective, but the calculation for $\partial^2 t_+^A/\partial d^2$ is not. A better approach is to define functions

$$f(d) := c^2(t - t_+^A)^2 , \qquad g(d) := (R+d)^2 + R^2 - 2R(R+d)\cos\omega(t - t_+^A) ,$$

expand each to $O(d^3)$, and equate coefficients of equal powers of d on either side of the relation $f = g$ [which is just (8.5)].

First define the notation

$$U := \left.\frac{\partial t_+^A}{\partial d}\right|_{d=0} , \qquad V := \left.\frac{\partial^2 t_+^A}{\partial d^2}\right|_{d=0} ,$$

so that (8.10) becomes

$$t_+^A = t + Ud + \frac{1}{2}Vd^2 + O(d^3) , \qquad t - t_+^A = -Ud - \frac{1}{2}Vd^2 + O(d^3) . \qquad (8.11)$$

Further

$$\frac{\partial t_+^A}{\partial d} = U + Vd + O(d^2) , \qquad \frac{\partial^2 t_+^A}{\partial d^2} = V + O(d) . \qquad (8.12)$$

Now

$$f(d) = f(0) + f^{(1)}(0)d + \frac{1}{2}f^{(2)}(0)d^2 + \frac{1}{6}f^{(3)}(0)d^3 + O(d^4) ,$$

where $f(0) = 0$, and

$$f^{(1)}(d) := \frac{\mathrm{d}f}{\mathrm{d}d} = -2c^2(t - t_+^A)\frac{\partial t_+^A}{\partial d} ,$$

$$f^{(2)}(d) := \frac{d^2 f}{dd^2} = 2c^2 \left(\frac{\partial t_+^A}{\partial d}\right)^2 - 2c^2(t - t_+^A)\frac{\partial^2 t_+^A}{\partial d^2},$$

and

$$f^{(3)}(d) := \frac{d^3 f}{dd^3} = 6c^2 \frac{\partial t_+^A}{\partial d}\frac{\partial^2 t_+^A}{\partial d^2} - 2c^2(t - t_+^A)\frac{\partial^3 t_+^A}{\partial d^3}.$$

Evaluating these at $d = 0$,

$$f^{(1)}(0) = 0, \qquad f^{(2)}(0) = 2c^2 U^2, \qquad f^{(3)}(0) = 6c^2 UV,$$

whence

$$\boxed{c^2(t - t_+^A)^2 = c^2 U^2 d^2 + c^2 UV d^3 + O(d^4)} \qquad (8.13)$$

Likewise for g, we have

$$g(d) = g(0) + g^{(1)}(0)d + \frac{1}{2}g^{(2)}(0)d^2 + \frac{1}{6}g^{(3)}(0)d^3 + O(d^4),$$

where $g(0) = 0$, and

$$g^{(1)}(d) = 2(R + d) - 2R\cos\omega(t - t_+^A) - 2\omega R(R + d)\sin\omega(t - t_+^A)\frac{\partial t_+^A}{\partial d},$$

$$g^{(2)}(d) = 2 - 4\omega R\sin\omega(t - t_+^A)\frac{\partial t_+^A}{\partial d} + 2\omega^2 R(R + d)\cos\omega(t - t_+^A)\left(\frac{\partial t_+^A}{\partial d}\right)^2$$
$$- 2\omega R(R + d)\sin\omega(t - t_+^A)\frac{\partial^2 t_+^A}{\partial d^2},$$

and

$$g^{(3)}(d) = 6\omega^2 R\cos\omega(t - t_+^A)\left(\frac{\partial t_+^A}{\partial d}\right)^2 - 6\omega R\sin\omega(t - t_+^A)\frac{\partial^2 t_+^A}{\partial d^2}$$
$$- 2\omega^3 R(R + d)\sin\omega(t - t_+^A)\left(\frac{\partial t_+^A}{\partial d}\right)^3 - 2\omega R(R + d)\sin\omega(t - t_+^A)\frac{\partial^3 t_+^A}{\partial d^3}$$
$$+ 6\omega^2 R(R + d)\cos\omega(t - t_+^A)\frac{\partial t_+^A}{\partial d}\frac{\partial^2 t_+^A}{\partial d^2}.$$

Evaluating these at $d = 0$,

$$g^{(1)}(0) = 0, \qquad g^{(2)}(0) = 2(1 + \omega^2 R^2 U^2), \qquad g^{(3)}(0) = 6\omega^2 RU(U + RV),$$

whence

$$(R+d)^2 + R^2 - 2R(R+d)\cos\omega(t - t_+^A)$$
$$= (1 + \omega^2 R^2 U^2)d^2 + \omega^2 RU(U + RV)d^3 + O(d^4) \qquad (8.14)$$

Now equating the coefficients of equal powers of d on the right-hand sides of (8.13) and (8.14), we obtain the relations

$$\begin{cases} c^2 U^2 = 1 + \omega^2 R^2 U^2 , \\ c^2 UV = \omega^2 RU(U + RV) . \end{cases}$$

The solution is

$$U = -\frac{\gamma}{c} , \qquad V = -\frac{\omega^2 R \gamma^3}{c^3} ,$$

where we have defined

$$\gamma := \left(1 - \frac{\omega^2 R^2}{c^2}\right)^{-1/2} ,$$

the notation being justified by the fact that ωR is the speed of A, so that this is the usual relativistic γ factor for A. (But note that it is not the γ factor for B.)

With these results, we can now write down the expansions for t_+^A, but also for t_+^B, from the observation that the determining relation (8.8) for the latter is precisely the same as the determining relation (8.5) for t_+^A. All the results needed for the following calculation are summarised here.

Summary of Results

Regarding t_+^A, we now have

$$U := \left.\frac{\partial t_+^A}{\partial d}\right|_{d=0} = -\frac{\gamma}{c} , \qquad V := \left.\frac{\partial^2 t_+^A}{\partial d^2}\right|_{d=0} = -\frac{\omega^2 R \gamma^3}{c^3} \qquad (8.15)$$

$$t_+^A(t,d) = t - \frac{\gamma}{c}d - \frac{\omega^2 R \gamma^3}{2c^3}d^2 + O(d^3) \qquad (8.16)$$

$$t - t_+^A(t,d) = \frac{\gamma}{c}d + \frac{\omega^2 R \gamma^3}{2c^3}d^2 + O(d^3) \qquad (8.17)$$

$$\frac{\partial t_+^A}{\partial d} = -\frac{\gamma}{c} - \frac{\omega^2 R \gamma^3}{c^3} d + O(d^3) , \qquad \frac{\partial^2 t_+^A}{\partial d^2} = -\frac{\omega^2 R \gamma^3}{c^3} + O(d) \qquad (8.18)$$

Regarding t_+^B,

$$\left.\frac{\partial t_+^B}{\partial d}\right|_{d=0} = -\frac{\gamma}{c} , \qquad \left.\frac{\partial^2 t_+^B}{\partial d^2}\right|_{d=0} = -\frac{\omega^2 R \gamma^3}{c^3} \qquad (8.19)$$

$$t_+^B(t,d) = t - \frac{\gamma}{c} d - \frac{\omega^2 R \gamma^3}{2c^3} d^2 + O(d^3) \qquad (8.20)$$

$$t - t_+^B(t,d) = \frac{\gamma}{c} d + \frac{\omega^2 R \gamma^3}{2c^3} d^2 + O(d^3) \qquad (8.21)$$

$$\frac{\partial t_+^B}{\partial d} = -\frac{\gamma}{c} - \frac{\omega^2 R \gamma^3}{c^3} d + O(d^3) , \qquad \frac{\partial^2 t_+^B}{\partial d^2} = -\frac{\omega^2 R \gamma^3}{c^3} + O(d) \qquad (8.22)$$

One thing to note here is that $t - t_+^A$ and $t - t_+^B$ are constant in time to this order. Of course, they must be constant in time to all orders, by the symmetry of the scenario we are considering.

8.4 Power Series Expansion of Retarded Displacement Vectors

The next task to prepare for calculating the self-force is to expand the retarded displacement vectors $\mathbf{r}_+^{AB}(t)$ and $\mathbf{r}_+^{BA}(t)$ defined by

$$\mathbf{r}_+^{AB}(t) = \begin{pmatrix} (R+d)\cos \omega t - R\cos \omega t_+^A \\ (R+d)\sin \omega t - R\sin \omega t_+^A \\ 0 \end{pmatrix} , \qquad \mathbf{r}_+^{BA}(t) = \begin{pmatrix} R\cos \omega t - (R+d)\cos \omega t_+^B \\ R\sin \omega t - (R+d)\sin \omega t_+^B \\ 0 \end{pmatrix} .$$

$$(8.23)$$

For this purpose, we shall require expansions of $\cos \omega t_+^A$, $\sin \omega t_+^A$, $\cos \omega t_+^B$, and $\sin \omega t_+^B$ to $O(d^2)$. The latter two will be immediate once we have the first two. [But note that the expansions of $\mathbf{r}_+^{AB}(t)$ and $\mathbf{r}_+^{BA}(t)$ in powers of d will not be the same.]

Define $h(d) := \cos \omega t_+^A$. Then $h(0) = \cos \omega t$ and

$$h^{(1)}(d) = -\omega \sin \omega t_+^A \frac{\partial t_+^A}{\partial d} , \qquad h^{(2)}(d) = -\omega^2 \cos \omega t_+^A \left(\frac{\partial t_+^A}{\partial d}\right)^2 - \omega \sin \omega t_+^A \frac{\partial^2 t_+^A}{\partial d^2} ,$$

whence

$$h^{(1)}(0) = \frac{\gamma \omega}{c} \sin \omega t , \quad h^{(2)}(0) = -\frac{\omega^2 \gamma^2}{c^2} \cos \omega t + \frac{\omega^3 \gamma^3 R}{c^3} \sin \omega t ,$$

using (8.15). Since

$$h(d) = h(0) + h^{(1)}(0)d + \frac{1}{2}h^{(2)}(0)d^2 + O(d^3) ,$$

we thus find

$$\boxed{\cos \omega t_+^A = \cos \omega t + \frac{\gamma \omega}{c}d \sin \omega t + \frac{\omega^2 \gamma^2}{2c^2}d^2 \left(\frac{\omega \gamma R}{c} \sin \omega t - \cos \omega t \right) + O(d^3)}$$

(8.24)

Now define $i(d) := \sin \omega t_+^A$, so that $i(0) = \sin \omega t$ and

$$i^{(1)}(d) = \omega \cos \omega t_+^A \frac{\partial t_+^A}{\partial d} , \quad i^{(2)}(d) = -\omega^2 \sin \omega t_+^A \left(\frac{\partial t_+^A}{\partial d} \right)^2 + \omega \cos \omega t_+^A \frac{\partial^2 t_+^A}{\partial d^2} ,$$

whence

$$i^{(1)}(0) = -\frac{\gamma \omega}{c} \cos \omega t , \quad i^{(2)}(0) = -\frac{\omega^2 \gamma^2}{c^2} \sin \omega t - \frac{\omega^3 \gamma^3 R}{c^3} \cos \omega t ,$$

using (8.15). Since

$$i(d) = i(0) + i^{(1)}(0)d + \frac{1}{2}i^{(2)}(0)d^2 + O(d^3) ,$$

we thus find

$$\boxed{\sin \omega t_+^A = \sin \omega t - \frac{\gamma \omega}{c}d \cos \omega t - \frac{\omega^2 \gamma^2}{2c^2}d^2 \left(\frac{\omega \gamma R}{c} \cos \omega t + \sin \omega t \right) + O(d^3)}$$

(8.25)

Hence we have immediately the corresponding results for t_+^B, viz.,

$$\boxed{\cos \omega t_+^B = \cos \omega t + \frac{\gamma \omega}{c}d \sin \omega t + \frac{\omega^2 \gamma^2}{2c^2}d^2 \left(\frac{\omega \gamma R}{c} \sin \omega t - \cos \omega t \right) + O(d^3)}$$

(8.26)

and

$$\sin \omega t_+^B = \sin \omega t - \frac{\gamma \omega}{c} d \cos \omega t - \frac{\omega^2 \gamma^2}{2c^2} d^2 \left(\frac{\omega \gamma R}{c} \cos \omega t + \sin \omega t \right) + O(d^3)$$

$$(8.27)$$

These can be inserted in (8.23) to obtain expansions of the retarded displacement vectors:

$$\mathbf{r}_+^{AB}(t) = \left(d + \frac{\gamma^2 \omega^2 R}{2c^2} d^2 \right) \begin{pmatrix} \cos \omega t - \dfrac{\gamma \omega R}{c} \sin \omega t \\ \sin \omega t + \dfrac{\gamma \omega R}{c} \cos \omega t \\ 0 \end{pmatrix} + O(d^3) \qquad (8.28)$$

and

$$\mathbf{r}_+^{BA}(t) = -d \begin{pmatrix} \cos \omega t + \dfrac{\gamma \omega R}{c} \sin \omega t \\ \sin \omega t - \dfrac{\gamma \omega R}{c} \cos \omega t \\ 0 \end{pmatrix}$$

$$+ d^2 \begin{pmatrix} -\dfrac{\gamma \omega}{c} \sin \omega t - \dfrac{\gamma^2 \omega^2 R}{2c^2} \left[\dfrac{\gamma \omega R}{c} \sin \omega t - \cos \omega t \right] \\ \dfrac{\gamma \omega}{c} \cos \omega t + \dfrac{\gamma^2 \omega^2 R}{2c^2} \left[\dfrac{\gamma \omega R}{c} \cos \omega t + \sin \omega t \right] \\ 0 \end{pmatrix} + O(d^3)$$

$$(8.29)$$

Now let us estimate the lengths of these two vectors to $O(d^2)$ and check that we do indeed obtain $|\mathbf{r}_+^{AB}| = r_+^{AB}$ as given by (8.17), viz.,

$$r_+^{AB} = c(t - t_+^A) = \gamma d + \frac{\gamma^3 \omega^2 R}{2c^2} d^2 + O(d^3), \qquad (8.30)$$

and $|\mathbf{r}_+^{BA}| = r_+^{BA}$ as given by (8.21), viz.,

$$r_+^{BA} = c(t - t_+^B) = \gamma d + \frac{\gamma^3 \omega^2 R}{2c^2} d^2 + O(d^3). \qquad (8.31)$$

This will just be a confirmation that no mistakes have been made. Both $\mathbf{r}_+^{AB}$ and $\mathbf{r}_+^{BA}$ have the form

$$\mathbf{r} = \mathbf{p} d + \mathbf{q} d^2 + O(d^3),$$

and

$$|\mathbf{r}|^2 = \mathbf{p}^2 d^2 + 2\mathbf{p} \cdot \mathbf{q} d^3 + O(d^4) = \mathbf{p}^2 d^2 \left[1 + \frac{2\mathbf{p} \cdot \mathbf{q}}{\mathbf{p}^2} d + O(d^2) \right] ,$$

whence

$$|\mathbf{r}| = |\mathbf{p}| d \left[1 + \frac{\mathbf{p} \cdot \mathbf{q}}{\mathbf{p}^2} d + O(d^2) \right] = |\mathbf{p}| d \left(1 + \frac{\mathbf{p} \cdot \mathbf{q}}{\mathbf{p}^2} d \right) + O(d^3) .$$

So, for example, for $\mathbf{r}_+^{AB}$ in (8.28),

$$\mathbf{p} = \begin{pmatrix} \cos \omega t - \dfrac{\gamma \omega R}{c} \sin \omega t \\ \sin \omega t + \dfrac{\gamma \omega R}{c} \cos \omega t \\ 0 \end{pmatrix} , \qquad \mathbf{q} = \frac{\gamma^2 \omega^2 R}{2c^2} \begin{pmatrix} \cos \omega t - \dfrac{\gamma \omega R}{c} \sin \omega t \\ \sin \omega t + \dfrac{\gamma \omega R}{c} \cos \omega t \\ 0 \end{pmatrix} , \tag{8.32}$$

and we soon find

$$|\mathbf{p}| = \gamma , \qquad \mathbf{p} \cdot \mathbf{q} = \frac{\gamma^4 \omega^2 R}{2c^2} ,$$

whence

$$|\mathbf{r}_+^{AB}| = \gamma d + \frac{\gamma^3 \omega^2 R}{2c^2} d^2 + O(d^3) ,$$

which confirms (8.30).

Despite the greater complexity of the expression (8.29), with

$$\mathbf{p} = - \begin{pmatrix} \cos \omega t + \dfrac{\gamma \omega R}{c} \sin \omega t \\ \sin \omega t - \dfrac{\gamma \omega R}{c} \cos \omega t \\ 0 \end{pmatrix} \tag{8.33}$$

and

$$\mathbf{q} = \begin{pmatrix} -\dfrac{\gamma \omega}{c} \sin \omega t - \dfrac{\gamma^2 \omega^2 R}{2c^2} \left[\dfrac{\gamma \omega R}{c} \sin \omega t - \cos \omega t \right] \\ \dfrac{\gamma \omega}{c} \cos \omega t + \dfrac{\gamma^2 \omega^2 R}{2c^2} \left[\dfrac{\gamma \omega R}{c} \cos \omega t + \sin \omega t \right] \\ 0 \end{pmatrix} , \tag{8.34}$$

we still find that

$$|\mathbf{p}| = \gamma , \qquad \mathbf{p} \cdot \mathbf{q} = \frac{\gamma^4 \omega^2 R}{2c^2} ,$$

whence

$$|\mathbf{r}_+^{BA}| = \gamma d + \frac{\gamma^3 \omega^2 R}{2c^2} d^2 + O(d^3) \,,$$

which confirms (8.31).

Before ending this section and beginning on the calculation of the self-force proper, one thing we shall be needing eventually, to calculate the magnetic fields generated by A and B, is the unit retarded displacement vectors $\mathbf{r}_+^{AB}/r_+^{AB}$ and $\mathbf{r}_+^{BA}/r_+^{BA}$. Returning to the general case of a vector of the form $\mathbf{r} = \mathbf{p}d + \mathbf{q}d^2 + O(d^3)$, we find

$$\frac{\mathbf{r}}{|\mathbf{r}|} = \frac{\mathbf{p} + \mathbf{q}d + O(d^2)}{|\mathbf{p}| \left[1 + \frac{\mathbf{p} \cdot \mathbf{q}}{\mathbf{p}^2} d + O(d^2) \right]}$$

$$= \frac{1}{|\mathbf{p}|} [\mathbf{p} + \mathbf{q}d + O(d^2)] \left[1 - \frac{\mathbf{p} \cdot \mathbf{q}}{\mathbf{p}^2} d + O(d^2) \right]$$

$$= \frac{\mathbf{p}}{|\mathbf{p}|} + \frac{1}{|\mathbf{p}|} \left(\mathbf{q} - \frac{\mathbf{p} \cdot \mathbf{q}}{\mathbf{p}^2} \mathbf{p} \right) d + O(d^2) \,. \tag{8.35}$$

Applying this to the case $\mathbf{r} = \mathbf{r}_+^{AB}$, where $\mathbf{p}$ and $\mathbf{q}$ are given in (8.32), we find that $\mathbf{q} = \mathbf{p} \cdot \mathbf{q}/\mathbf{p}^2$, whence the unit vector has the simple form

$$\frac{\mathbf{r}_+^{AB}}{r_+^{AB}} = \frac{1}{\gamma} \begin{pmatrix} \cos \omega t - \dfrac{\gamma \omega R}{c} \sin \omega t \\ \sin \omega t + \dfrac{\gamma \omega R}{c} \cos \omega t \\ 0 \end{pmatrix} + O(d^2) \tag{8.36}$$

with no term of $O(d)$. It turns out that this level of accuracy will be sufficient later for calculating the magnetic field at B due to A.

Regarding the other unit retarded displacement vector $\mathbf{r}_+^{BA}/r_+^{BA}$, application of the above rule for $\mathbf{p}$ and $\mathbf{q}$ as given in (8.33) and (8.34) leads to

$$\frac{\mathbf{r}_+^{BA}}{r_+^{BA}} = -\frac{1}{\gamma} \begin{pmatrix} \cos \omega t + \dfrac{\gamma \omega R}{c} \sin \omega t \\ \sin \omega t - \dfrac{\gamma \omega R}{c} \cos \omega t \\ 0 \end{pmatrix} + \frac{\omega}{c} \begin{pmatrix} \dfrac{\gamma \omega R}{c} \cos \omega t - \sin \omega t \\ \dfrac{\gamma \omega R}{c} \sin \omega t + \cos \omega t \\ 0 \end{pmatrix} d + O(d^2) \tag{8.37}$$

8.5 Power Series Expansion of Electric Self-Force

We come back now to the expression (2.61) on p. 21 for the electric field generated
by a point charge in motion, viz.,

$$
\mathbf{E} = \frac{q_e}{8\pi\varepsilon_0} \frac{\left(\mathbf{r}_{01} - \frac{r_{01}\mathbf{v}}{c}\right)\left(1 - \frac{v^2}{c^2}\right) + \frac{\mathbf{r}_{01}}{c^2} \times \left[\left(\mathbf{r}_{01} - \frac{r_{01}\mathbf{v}}{c}\right) \times \frac{d\mathbf{v}}{dt}\right]}{(r_{01} - \mathbf{r}_{01}\cdot\mathbf{v}/c)^3} , \tag{8.38}
$$

as presented in Chap. 2, and as already applied in Chaps. 6 and 7.

8.5.1 Electric Force of A on B

We shall find the electric field $\mathbf{E}^A(\text{at } B)$ at B due to A as given by

$$
\mathbf{E}^A(\text{at } B) \tag{8.39}
$$

$$
= \frac{q_e}{8\pi\varepsilon_0} \frac{\left(\mathbf{r}_+^{AB} - \frac{r_+^{AB}\mathbf{v}_+^A}{c}\right)\left[1 - \frac{(\mathbf{v}_+^A)^2}{c^2}\right] + \frac{\mathbf{r}_+^{AB}}{c^2} \times \left[\left(\mathbf{r}_+^{AB} - \frac{r_+^{AB}\mathbf{v}_+^A}{c}\right) \times \frac{d\mathbf{v}^A}{dt}\Big|_{t=t_+^A}\right]}{\left(r_+^{AB} - \mathbf{r}_+^{AB}\cdot\mathbf{v}_+^A/c\right)^3} ,
$$

then apply the usual rule to get the electric force

$$
\mathbf{F}_{\text{elec}}^A(\text{on } B) = \frac{q_e}{2}\mathbf{E}^A(\text{at } B) . \tag{8.40}
$$

Let us begin with the expansion of $\mathbf{v}_+^A = \mathbf{v}^A(t_+^A)$, recalling from (8.2) on p. 135 that

$$
\mathbf{v}^A(t_+^A) = \omega R \begin{pmatrix} -\sin \omega t_+^A \\ \cos \omega t_+^A \\ 0 \end{pmatrix} ,
$$

and substituting in the expansions (8.24) and (8.25) for $\cos \omega t_+^A$ and $\sin \omega t_+^A$ to obtain

$$
\mathbf{v}_+^A = \omega R \begin{pmatrix} -\sin \omega t \\ \cos \omega t \\ 0 \end{pmatrix} + \frac{\omega^2 R \gamma}{c} \begin{pmatrix} \cos \omega t \\ \sin \omega t \\ 0 \end{pmatrix} d
$$

$$
+ \frac{\gamma^2 \omega^3 R}{2c^2} \begin{pmatrix} \frac{\gamma \omega R}{c}\cos \omega t + \sin \omega t \\ \frac{\gamma \omega R}{c}\sin \omega t - \cos \omega t \\ 0 \end{pmatrix} d^2 + O(d^3)
$$

$$
\tag{8.41}
$$

A short calculation now finds

$$(v_+^A)^2 = \omega^2 R^2 + O(d^3),$$

whence

$$1 - \frac{(v_+^A)^2}{c^2} = \gamma^{-2} + O(d^3) \tag{8.42}$$

Further efforts using (8.28) for $\mathbf{r}_+^{AB}$, (8.30) for r_+^{AB}, and (8.41) for v_+^A lead to

$$r_+^{AB} - \frac{\mathbf{r}_+^{AB} \cdot \mathbf{v}_+^A}{c} = \frac{d}{\gamma} - \frac{\omega^2 R \gamma}{2c^2} d^2 + O(d^3) \tag{8.43}$$

and

$$\mathbf{r}_+^{AB} - \frac{r_+^{AB} \mathbf{v}_+^A}{c} = \left(d - \frac{\gamma^2 \omega^2 R}{2c^2} d^2 \right) \begin{pmatrix} \cos \omega t \\ \sin \omega t \\ 0 \end{pmatrix} + O(d^3) \tag{8.44}$$

We now have

$$\left(\mathbf{r}_+^{AB} - \frac{r_+^{AB} \mathbf{v}_+^A}{c} \right) \left[1 - \frac{(v_+^A)^2}{c^2} \right] = \left(\frac{d}{\gamma^2} - \frac{\omega^2 R}{2c^2} d^2 \right) \begin{pmatrix} \cos \omega t \\ \sin \omega t \\ 0 \end{pmatrix} + O(d^3) \tag{8.45}$$

Furthermore, the other term in the numerator of (8.39) is zero to $O(d^2)$, i.e.,

$$\frac{\mathbf{r}_+^{AB}}{c^2} \times \left[\left(\mathbf{r}_+^{AB} - \frac{r_+^{AB} \mathbf{v}_+^A}{c} \right) \times \frac{d\mathbf{v}^A}{dt} \Big|_{t=t_+^A} \right] = O(d^3) \tag{8.46}$$

for the following reason. By (8.3),

$$\frac{d\mathbf{v}^A}{dt} \Big|_{t=t_+^A} = -\omega^2 R \begin{pmatrix} \cos \omega t_+^A \\ \sin \omega t_+^A \\ 0 \end{pmatrix} = -\omega^2 R \begin{pmatrix} \cos \omega t \\ \sin \omega t \\ 0 \end{pmatrix} + O(d),$$

whence

$$\left(\mathbf{r}_+^{AB} - \frac{r_+^{AB}\mathbf{v}_+^A}{c}\right) \times \frac{d\mathbf{v}^A}{dt}\bigg|_{t=t_+^A} = -\omega^2 R\left(d - \frac{\gamma^2\omega^2 R}{2c^2}d^2\right)\begin{pmatrix}\cos\omega t\\ \sin\omega t\\ 0\end{pmatrix} \times \begin{pmatrix}\cos\omega t\\ \sin\omega t\\ 0\end{pmatrix} + O(d^2)$$

$$= O(d^2).$$

But $\mathbf{r}_+^{AB} = O(d)$, so (8.46) holds.

Referring back to (8.39), we now have

$$\mathbf{E}^A(\text{at } B) = \frac{q_e}{8\pi\varepsilon_0}\frac{\left(\dfrac{d}{\gamma^2} - \dfrac{\omega^2 R}{2c^2}d^2\right)\begin{pmatrix}\cos\omega t\\ \sin\omega t\\ 0\end{pmatrix} + O(d^3)}{\left[\dfrac{d}{\gamma} - \dfrac{\omega^2 R\gamma}{2c^2}d^2 + O(d^3)\right]^3},$$

whence finally,

$$\boxed{\mathbf{E}^A(\text{at } B) = \frac{q_e\gamma}{8\pi\varepsilon_0 d^2}\begin{pmatrix}\cos\omega t\\ \sin\omega t\\ 0\end{pmatrix} + \frac{q_e}{8\pi\varepsilon_0}\frac{\gamma^3\omega^2 R}{c^2 d}\begin{pmatrix}\cos\omega t\\ \sin\omega t\\ 0\end{pmatrix} + O(d^0)} \qquad (8.47)$$

Then by (8.40), and defining e by $e^2 = q_e^2/4\pi\varepsilon_0$, we have the electric force of A on B in the form

$$\boxed{\mathbf{F}_{\text{elec}}^A(\text{on } B) = \frac{e^2\gamma}{4d^2}\begin{pmatrix}\cos\omega t\\ \sin\omega t\\ 0\end{pmatrix} + \frac{e^2}{4c^2 d}\omega^2 R\gamma^3\begin{pmatrix}\cos\omega t\\ \sin\omega t\\ 0\end{pmatrix} + O(d^0)} \qquad (8.48)$$

which we may note immediately to be radially outward.

8.5.2 Electric Force of B on A

We shall find the electric field $\mathbf{E}^B(\text{at } A)$ at A due to B as given by

$$\mathbf{E}^B(\text{at } A) \qquad\qquad\qquad\qquad\qquad\qquad\qquad\qquad\qquad\qquad\qquad (8.49)$$

$$= \frac{q_e}{8\pi\varepsilon_0}\frac{\left(\mathbf{r}_+^{BA} - \dfrac{r_+^{BA}\mathbf{v}_+^B}{c}\right)\left[1 - \dfrac{(v_+^B)^2}{c^2}\right] + \dfrac{r_+^{BA}}{c^2}\times\left[\left(\mathbf{r}_+^{BA} - \dfrac{r_+^{BA}\mathbf{v}_+^B}{c}\right)\times\dfrac{d\mathbf{v}^B}{dt}\bigg|_{t=t_+^B}\right]}{\left(r_+^{BA} - \mathbf{r}_+^{BA}\cdot\mathbf{v}_+^B/c\right)^3},$$

then apply the usual rule to get the electric force

$$\mathbf{F}^B_{\text{elec}}(\text{on A}) = \frac{q_e}{2}\mathbf{E}^B(\text{at A}) . \tag{8.50}$$

Let us begin with the expansion of $\mathbf{v}^B_+ = \mathbf{v}^B(t^B_+)$, recalling from (8.2) on p. 135 that

$$\mathbf{v}^B(t^B_+) = \omega(R+d)\begin{pmatrix} -\sin \omega t^B_+ \\ \cos \omega t^B_+ \\ 0 \end{pmatrix},$$

and substituting in the expansions (8.26) and (8.27) for $\cos \omega t^B_+$ and $\sin \omega t^B_+$ to obtain

$$\mathbf{v}^B_+ = \omega R \begin{pmatrix} -\sin \omega t \\ \cos \omega t \\ 0 \end{pmatrix} + \begin{pmatrix} \dfrac{\omega^2 R \gamma}{c}\cos \omega t - \omega \sin \omega t \\ \dfrac{\omega^2 R \gamma}{c}\sin \omega t + \omega \cos \omega t \\ 0 \end{pmatrix} d + O(d^2) \tag{8.51}$$

keeping only terms to $O(d)$. (Note that only this order was ever used in the expression for $\mathbf{v}^A_+$ too.) A short calculation now finds

$$\left(\mathbf{v}^B_+\right)^2 = \omega^2 R^2 + 2\omega^2 R d + O(d^2) ,$$

whence

$$1 - \frac{(\mathbf{v}^B_+)^2}{c^2} = \gamma^{-2} - \frac{2\omega^2 R}{c^2}d + O(d^2) \tag{8.52}$$

Further efforts using (8.29) for $\mathbf{r}^{BA}_+$, (8.31) for r^{BA}_+, and (8.51) for $\mathbf{v}^B_+$ lead to

$$r^{BA}_+ - \frac{\mathbf{r}^{BA}_+ \cdot \mathbf{v}^B_+}{c} = \frac{d}{\gamma} - \frac{\omega^2 R \gamma}{2c^2}d^2 + O(d^3) \tag{8.53}$$

Interestingly, this is exactly the same as the expression (8.43) for $r^{AB}_+ - \mathbf{r}^{AB}_+ \cdot \mathbf{v}^A_+/c$. We also find

$$\mathbf{r}^{BA}_+ - \frac{r^{BA}_+ \mathbf{v}^B_+}{c} = -\left(d + \frac{\gamma^2 \omega^2 R}{2c^2}d^2\right)\begin{pmatrix} \cos \omega t \\ \sin \omega t \\ 0 \end{pmatrix} + O(d^3) \tag{8.54}$$

We now have

$$\left(\mathbf{r}_+^{BA} - \frac{r_+^{BA}\mathbf{v}_+^B}{c}\right)\left[1 - \frac{(\mathbf{v}_+^B)^2}{c^2}\right] = \left(-\frac{d}{\gamma^2} + \frac{3\omega^2 R}{2c^2}d^2\right)\begin{pmatrix} \cos \omega t \\ \sin \omega t \\ 0 \end{pmatrix} + O(d^3) \quad (8.55)$$

Furthermore, the other term in the numerator of (8.49) is zero to $O(d^2)$, i.e.,

$$\frac{\mathbf{r}_+^{BA}}{c^2} \times \left[\left(\mathbf{r}_+^{BA} - \frac{r_+^{BA}\mathbf{v}_+^B}{c}\right) \times \frac{d\mathbf{v}^B}{dt}\Big|_{t=t_+^B}\right] = O(d^3) \quad (8.56)$$

for the following reason. By (8.3),

$$\frac{d\mathbf{v}^B}{dt}\Big|_{t=t_+^B} = -\omega^2(R+d)\begin{pmatrix} \cos \omega t_+^A \\ \sin \omega t_+^A \\ 0 \end{pmatrix} = -\omega^2 R\begin{pmatrix} \cos \omega t \\ \sin \omega t \\ 0 \end{pmatrix} + O(d) ,$$

whence

$$\left(\mathbf{r}_+^{BA} - \frac{r_+^{BA}\mathbf{v}_+^B}{c}\right) \times \frac{d\mathbf{v}^B}{dt}\Big|_{t=t_+^B} = \omega^2 R\left(d + \frac{\gamma^2\omega^2 R}{2c^2}d^2\right)\begin{pmatrix} \cos \omega t \\ \sin \omega t \\ 0 \end{pmatrix} \times \begin{pmatrix} \cos \omega t \\ \sin \omega t \\ 0 \end{pmatrix} + O(d^2)$$

$$= O(d^2) .$$

But $\mathbf{r}_+^{BA} = O(d)$, so (8.56) holds.

Referring back to (8.49), we now have

$$\mathbf{E}^B(\text{at } A) = \frac{q_e}{8\pi\varepsilon_0} \frac{\left(-\dfrac{d}{\gamma^2} + \dfrac{3\omega^2 R}{2c^2}d^2\right)\begin{pmatrix} \cos \omega t \\ \sin \omega t \\ 0 \end{pmatrix} + O(d^3)}{\left[\dfrac{d}{\gamma} - \dfrac{\omega^2 R\gamma}{2c^2}d^2 + O(d^3)\right]^3} ,$$

whence finally,

$$\mathbf{E}^B(\text{at } A) = -\frac{q_e\gamma}{8\pi\varepsilon_0 d^2}\begin{pmatrix} \cos \omega t \\ \sin \omega t \\ 0 \end{pmatrix} + O(d^0) \quad (8.57)$$

with no term $O(d^{-1})$. Then by (8.50), and defining e by $e^2 = q_e^2/4\pi\varepsilon_0$, we have the electric force of B on A in the form

$$\mathbf{F}^{B}_{elec}(\text{on A}) = -\frac{e^2\gamma}{4d^2}\begin{pmatrix} \cos\omega t \\ \sin\omega t \\ 0 \end{pmatrix} + O(d^0) \tag{8.58}$$

which we may note immediately to be radially outward.

8.5.3 Electric Self-Force

We now simply add together the results in (8.48) and (8.58) to obtain

$$\mathbf{F}^{self}_{elec} = \frac{e^2}{4c^2d}\omega^2 R\gamma^3\begin{pmatrix} \cos\omega t \\ \sin\omega t \\ 0 \end{pmatrix} + O(d^0) \tag{8.59}$$

The first thing to note is that the terms $O(d^{-2})$, which are basically Coulomb terms, cancel one another. Secondly, it is worth comparing this with the $O(d^{-1})$ term of (6.48) on p. 109, for the scenario in which the system is accelerating along a straight line perpendicular to its axis. Note that $\omega^2 R$ is the magnitude of the acceleration, so we have exactly the same magnitude, and $\mathbf{F}^{self}_{elec}$ is directed radially outward, i.e., in the direction opposite to the acceleration. It also contains a γ^3 factor, but note that the comments in Sect. 7.5 apply, because different γ factors are associated with A and B, and the γ appearing in (8.59) is just the one associated with A.

So here we have a different scenario to the one in Chap. 6, since the acceleration is along the system axis and perpendicular to the velocity, whereas it was perpendicular to the system axis and parallel to the velocity in Chap. 6. But the mass renormalisation due to the electric self-force is the same in both cases, viz.,

$$m_{ren} := m_0 + \frac{e^2}{4c^2d}. \tag{8.60}$$

However, there are magnetic fields and we need to consider them.

8.6 Power Series Expansion of Magnetic Self-Force

According to the formula (2.62) given on p. 21 of Chap. 2, the magnetic field at B due to A is given by

$$\mathbf{B}^A(\text{at B}) = \frac{\mathbf{r}^{AB}_+ \times \mathbf{E}^A(\text{at B})}{cr^{AB}_+}, \tag{8.61}$$

where $\mathbf{E}^A$(at B) is given by (8.47), and the magnetic force of A on B is then

$$\mathbf{F}^A_{\text{mag}}(\text{on B}) = \frac{q_e}{2}\mathbf{v}^B(t) \times \mathbf{B}^A(\text{at B}) . \tag{8.62}$$

We have already calculated the unit retarded displacement vector $\mathbf{r}^{AB}_+/r^{AB}_+$ in (8.36) on p. 145, and a short calculation gives

$$\mathbf{B}^A(\text{at B}) = -\frac{q_e}{8\pi\varepsilon_0 d}\left(\frac{1}{d} + \frac{\omega^2 R\gamma^2}{c^2}\right)\frac{\gamma\omega R}{c^2}\begin{pmatrix} 0 \\ 0 \\ 1 \end{pmatrix} + O(d^0) .$$

As viewed in Fig. 8.1, this is perpendicular to the page. Now according to (8.2),

$$\mathbf{v}^B(t) = \omega(R+d)\begin{pmatrix} -\sin\omega t \\ \cos\omega t \\ 0 \end{pmatrix} ,$$

and we soon find

$$\mathbf{F}^A_{\text{mag}}(\text{on B}) = -\frac{e^2}{4d^2}\frac{\gamma\omega^2 R^2}{c^2}\begin{pmatrix} \cos\omega t \\ \sin\omega t \\ 0 \end{pmatrix} - \frac{e^2}{4c^2 d}\omega^2 R\gamma^3\begin{pmatrix} \cos\omega t \\ \sin\omega t \\ 0 \end{pmatrix} + O(d^0) \tag{8.63}$$

having made the usual definition $e^2 := q_e/4\pi\varepsilon_0$.

The magnetic field at A due to B is given by

$$\mathbf{B}^B(\text{at A}) = \frac{\mathbf{r}^{BA}_+ \times \mathbf{E}^B(\text{at A})}{cr^{BA}_+} , \tag{8.64}$$

where $\mathbf{E}^B$(at A) is given by (8.57), and the magnetic force of B on A is then

$$\mathbf{F}^B_{\text{mag}}(\text{on A}) = \frac{q_e}{2}\mathbf{v}^A(t) \times \mathbf{B}^B(\text{at A}) . \tag{8.65}$$

We have already calculated the unit retarded displacement vector $\mathbf{r}^{BA}_+/r^{BA}_+$ in (8.37) on p. 145, and a short calculation gives

$$\mathbf{B}^B(\text{at A}) = \frac{q_e\omega\gamma}{8\pi\varepsilon_0 dc^2}\left(\frac{R}{d} + 1\right)\begin{pmatrix} 0 \\ 0 \\ 1 \end{pmatrix} + O(d^0) .$$

Once again, as viewed in Fig. 8.1, this is perpendicular to the page. Now according to (8.2),

$$\mathbf{v}^A(t) = \omega R \begin{pmatrix} -\sin \omega t \\ \cos \omega t \\ 0 \end{pmatrix},$$

and we soon find

$$\mathbf{F}^B_{mag}(\text{on } A) = \frac{e^2}{4d^2} \frac{\gamma \omega^2 R^2}{c^2} \begin{pmatrix} \cos \omega t \\ \sin \omega t \\ 0 \end{pmatrix} + \frac{e^2}{4c^2 d} \omega^2 R \gamma \begin{pmatrix} \cos \omega t \\ \sin \omega t \\ 0 \end{pmatrix} + O(d^0) \quad (8.66)$$

We can now write down the magnetic self-force, simply taking the sum of (8.63) and (8.66) to obtain

$$\mathbf{F}^{self}_{mag} = \frac{e^2}{4c^2 d} \omega^2 R \gamma (1 - \gamma^2) \begin{pmatrix} \cos \omega t \\ \sin \omega t \\ 0 \end{pmatrix} + O(d^0) \quad (8.67)$$

Note that the $O(d^{-2})$ terms cancel. The fact that the $O(d^{-1})$ terms do not cancel because of the different powers of the γ factors looks odd, and more will be said about that below.

8.7 Power Series Expansion of Total Self-Force and Interpretation

We now obtain the power series expansion of the total self-force by simply adding (8.59) and (8.67) to obtain

$$\mathbf{F}^{self} = \frac{e^2}{4c^2 d} \omega^2 R \gamma \begin{pmatrix} \cos \omega t \\ \sin \omega t \\ 0 \end{pmatrix} + O(d^0) \quad (8.68)$$

If one is interested in renormalising the inertial mass of the system by absorbing this leading order term into whatever else constitutes its inertia, there are several features about this that look helpful. The first is that it is radially outward, directly opposing the acceleration of either A or B. Secondly, it is proportional to the acceleration $\omega^2 R$ of A. And finally, we recover the same factor of $e^2/4c^2 d$ that seems to tag along with this system when its velocity is perpendicular to its axis [compare with the result (6.50) on p. 110].

Another crucial point about (8.68) is that the factor of γ is just right to be able to absorb this contribution into the inertial mass. The point is that, when the velocity is perpendicular to the acceleration, as in this case, we have

$$\frac{d}{dt}(m_0\gamma v) = m_0\gamma\frac{dv}{dt} \ .$$

Finally, it is intriguing to note that the total force of A on B contains no $O(d^{-1})$ term:

$$\mathbf{F}^A_{total}(\text{on B}) = \frac{e^2}{4d^2\gamma}\begin{pmatrix}\cos\omega t\\ \sin\omega t\\ 0\end{pmatrix} + O(d^0) \tag{8.69}$$

while the total force of B on A is

$$\mathbf{F}^B_{total}(\text{on A}) = -\frac{e^2}{4d^2\gamma}\begin{pmatrix}\cos\omega t\\ \sin\omega t\\ 0\end{pmatrix} + \frac{e^2}{4c^2d}\omega^2R\gamma\begin{pmatrix}\cos\omega t\\ \sin\omega t\\ 0\end{pmatrix} + O(d^0) \tag{8.70}$$

We come back here to the problem discussed in Sect. 7.5, namely that A and B have different motions. The acceleration ω^2R appearing in (8.68) is the acceleration of A, while B has acceleration $\omega^2(R+d)$, and the γ factor appearing in all the above formulas is $\gamma_A := (1 - v_A^2/c^2)^{-1/2}$, not $\gamma_B := (1 - v_B^2/c^2)^{-1/2}$. These things are not very democratic! This issue also highlights the fact that we simply add up 3-forces acting at different points of the system, with different motions. It highlights the fact that we have not properly addressed the problem of the dynamics of spatially extended objects in relativity theory.

To give some idea of the difference that can be made by the different γ factors, let us expand γ_B in terms of γ_A and d. We have

$$\gamma_B(d) = \left[1 - \frac{\omega^2(R+d)^2}{c^2}\right]^{-1/2} , \qquad \gamma_A = \gamma_B(0) \ .$$

Now

$$\gamma_B^{(1)}(d) := \frac{d\gamma_B}{dd} = \frac{\omega^2}{c^2}(R+d)\gamma_B^3 ,$$

whence

$$\gamma_B^{(1)}(0) = \frac{\omega^2R}{c^2}\gamma_A^3 ,$$

and we obtain the first order expansion

$$\gamma_B = \gamma_A + \frac{\omega^2R}{c^2}\gamma_A^3 d \ . \tag{8.71}$$

Now the electric force of A on B in (8.48) can be rewritten

$$\mathbf{F}^A_{elec}(\text{on B}) = \frac{e^2 \gamma_B}{4d^2} \begin{pmatrix} \cos \omega t \\ \sin \omega t \\ 0 \end{pmatrix} + O(d^0) \,,$$

while the electric force of B on A in (8.58) has the form

$$\mathbf{F}^B_{elec}(\text{on A}) = -\frac{e^2 \gamma_A}{4d^2} \begin{pmatrix} \cos \omega t \\ \sin \omega t \\ 0 \end{pmatrix} + O(d^0) \,.$$

The first force applies at B and the second at A.

One final point to remember is that all the four forces we have evaluated act to $O(d^{-2})$ and $O(d^{-1})$ along the axis of the system. This means that they will have stretching or compressing effects, and these have to be cleverly balanced by the binding force so that the length remains the same.

We shall return to the issue of the dynamics of spatially extended bodies in relativity theory, but let us first examine the last of the four scenarios, in which the system rotates about a center, but with velocity always instantaneously along its axis, to a good approximation. The γ factors and accelerations of A and B are equal in that case, so one hopes for a similarly neat result.

Chapter 9
Self-Force for Longitudinal Rotational Motion

9.1 Setting the Scene

We consider the dumbbell charge system, consisting of two like charges $q_e/2$ rotating about a fixed center, in such a way that its axis always lies perpendicular to a radial line from the center of rotation passing through its midpoint, as shown in Fig. 9.1. As in the last chapter, both A and B have constant speeds. Although the velocities of A and B are not always exactly parallel to the axis joining them, we do not have to worry about changing FitzGerald contractions during the motion, as we did in Chap. 7 (where the whole system was changing length all the time), because each of A and B always has the same speed. So we can assume a fixed length for the system, but there is a question about what the fixed length should be, because we expect it to be contracted in some way as compared with its length when stationary relative to some inertial frame. We shall nevertheless denote the length by d for the moment, and return to the question of its value at the end of the calculation.

Once again the beauty of this scenario, as compared with those discussed in Chaps. 6, 7, and 8, is that the acceleration of either A or B is now almost perpendicular to the system axis (the line joining A and B), while the velocity is always perpendicular to the acceleration for each particle. A priori, it is not at all obvious that the electromagnetic self-force will lie perpendicular to the system axis, i.e., parallel to the acceleration.

We now formulate this final scenario. We take the position vectors of A and B at time t to be

$$\mathbf{r}^A(t) = \begin{pmatrix} R\cos\omega t \\ R\sin\omega t \\ 0 \end{pmatrix}, \qquad \mathbf{r}^B(t) = \begin{pmatrix} R\cos(\omega t + \phi) \\ R\sin(\omega t + \phi) \\ 0 \end{pmatrix}, \qquad (9.1)$$

where R is thus the radial distance of A or B from the center of rotation, d is the length of the system (the separation of A and B), which is constant in this scenario, ω is the angular velocity of either A or B, and ϕ is a simple function of d satisfying

Lyle, S.N.: *Self-Force for Longitudinal Rotational Motion*. Lect. Notes Phys. **796**, 157–177 (2010)
DOI 10.1007/978-3-642-04785-5_9 © Springer-Verlag Berlin Heidelberg 2010

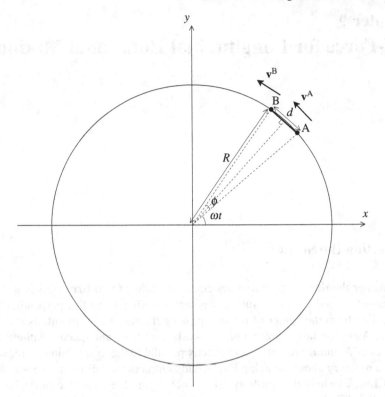

Fig. 9.1 Dumbbell charge system under rotational motion along its axis

$$d = 2R\sin\frac{\phi(d)}{2} \ . \tag{9.2}$$

The velocities are

$$\mathbf{v}^A(t) = \omega R \begin{pmatrix} -\sin\omega t \\ \cos\omega t \\ 0 \end{pmatrix} \ , \qquad \mathbf{v}^B(t) = \omega R \begin{pmatrix} -\sin(\omega t + \phi) \\ \cos(\omega t + \phi) \\ 0 \end{pmatrix} \ , \tag{9.3}$$

whence A and B both move at speed ωR. They also have different accelerations

$$\mathbf{a}^A(t) = -\omega^2 R \begin{pmatrix} \cos\omega t \\ \sin\omega t \\ 0 \end{pmatrix} \ , \qquad \mathbf{a}^B(t) = -\omega^2 R \begin{pmatrix} \cos(\omega t + \phi) \\ \sin(\omega t + \phi) \\ 0 \end{pmatrix} \ , \tag{9.4}$$

with the same magnitude, and both directed toward the center of rotation.

We shall need an expansion of $\phi(d)$ in powers of d. Differentiating both sides of (9.2) with respect to d, we obtain

$$1 = R\cos\frac{\phi}{2}\frac{d\phi}{dd},$$

whence

$$\frac{d\phi}{dd} = \frac{2}{(4R^2 - d^2)^{1/2}}. \tag{9.5}$$

Then

$$\frac{d^2\phi}{dd^2} = \frac{2d}{(4R^2 - d^2)^{3/2}}. \tag{9.6}$$

Hence we have the expansion

$$\boxed{\phi(d) = \frac{d}{R} + O(d^3)} \tag{9.7}$$

9.2 Retarded Times and Retarded Displacement Vectors

As usual, we define retarded displacement vectors

$$\mathbf{r}_+^{AB}(t) := \mathbf{r}^B(t) - \mathbf{r}^A(t_+^A) = \begin{pmatrix} R\cos(\omega t + \phi) - R\cos\omega t_+^A \\ R\sin(\omega t + \phi) - R\sin\omega t_+^A \\ 0 \end{pmatrix} \tag{9.8}$$

and

$$\mathbf{r}_+^{BA}(t) := \mathbf{r}^A(t) - \mathbf{r}^B(t_+^B) = \begin{pmatrix} R\cos\omega t - R\cos(\omega t_+^B + \phi) \\ R\sin\omega t - R\sin(\omega t_+^B + \phi) \\ 0 \end{pmatrix}, \tag{9.9}$$

in which the retarded times t_+^A and t_+^B are defined by

$$\mathbf{r}_+^{AB}(t)\cdot\mathbf{r}_+^{AB}(t) = c^2(t - t_+^A)^2, \quad \mathbf{r}_+^{BA}(t)\cdot\mathbf{r}_+^{BA}(t) = c^2(t - t_+^B)^2.$$

These just say that the lengths of $\mathbf{r}_+^{AB}(t)$ and $\mathbf{r}_+^{BA}(t)$ are the appropriate light travel distances. A little algebra gets these relations into the form

$$\boxed{c^2(t - t_+^A)^2 = 2R^2\left\{1 - \cos\left[\omega(t - t_+^A) + \phi\right]\right\}} \tag{9.10}$$

and

$$\boxed{c^2(t - t_+^B)^2 = 2R^2\left\{1 - \cos\left[\omega(t - t_+^B) - \phi\right]\right\}} \tag{9.11}$$

9.3 Power Series Expansion of Retarded Times

As usual, we treat t_+^A and t_+^B as functions of t and d, where t is the time at which we are considering B or A, respectively, and seek expansions of the form

$$t_+^A(t,d) = t_+^A(t,0) + \left.\frac{\partial t_+^A}{\partial d}\right|_{d=0} d + \frac{1}{2}\left.\frac{\partial^2 t_+^A}{\partial d^2}\right|_{d=0} d^2 + O(d^3) \qquad (9.12)$$

and

$$t_+^B(t,d) = t_+^B(t,0) + \left.\frac{\partial t_+^B}{\partial d}\right|_{d=0} d + \frac{1}{2}\left.\frac{\partial^2 t_+^B}{\partial d^2}\right|_{d=0} d^2 + O(d^3) . \qquad (9.13)$$

Once again, $t_+^A(t,0) = t$ and $t_+^B(t,0) = t$. We use the same method as in Sect. 8.3. In contrast to the situation in Chap. 8, however, we shall need to expand both t_+^A and t_+^B, since the defining relations (9.10) and (9.11) are different here.

9.3.1 Expansion for t_+^A

We thus define functions

$$f(d) := c^2(t - t_+^A)^2 , \qquad g(d) := 2R^2\left\{1 - \cos\left[\omega(t - t_+^A) + \phi\right]\right\} ,$$

expand each to $O(d^3)$, and equate coefficients of equal powers of d on either side of the relation $f = g$ [which is just (9.10)].

First define the notation

$$U := \left.\frac{\partial t_+^A}{\partial d}\right|_{d=0} , \qquad V := \left.\frac{\partial^2 t_+^A}{\partial d^2}\right|_{d=0} ,$$

so that (9.12) becomes

$$t_+^A = t + Ud + \frac{1}{2}Vd^2 + O(d^3) , \qquad t - t_+^A = -Ud - \frac{1}{2}Vd^2 + O(d^3) . \qquad (9.14)$$

Further

$$\frac{\partial t_+^A}{\partial d} = U + Vd + O(d^2) , \qquad \frac{\partial^2 t_+^A}{\partial d^2} = V + O(d) . \qquad (9.15)$$

Now

$$f(d) = f(0) + f^{(1)}(0)d + \frac{1}{2}f^{(2)}(0)d^2 + \frac{1}{6}f^{(3)}(0)d^3 + O(d^4) ,$$

where $f(0) = 0$, and

$$f^{(1)}(d) := \frac{df}{dd} = -2c^2(t - t_+^A)\frac{\partial t_+^A}{\partial d} \, ,$$

$$f^{(2)}(d) := \frac{d^2 f}{dd^2} = 2c^2\left(\frac{\partial t_+^A}{\partial d}\right)^2 - 2c^2(t - t_+^A)\frac{\partial^2 t_+^A}{\partial d^2} \, ,$$

and

$$f^{(3)}(d) := \frac{d^3 f}{dd^3} = 6c^2\frac{\partial t_+^A}{\partial d}\frac{\partial^2 t_+^A}{\partial d^2} - 2c^2(t - t_+^A)\frac{\partial^3 t_+^A}{\partial d^3} \, .$$

Evaluating these at $d = 0$,

$$f^{(1)}(0) = 0 \, , \qquad f^{(2)}(0) = 2c^2 U^2 \, , \qquad f^{(3)}(0) = 6c^2 UV \, ,$$

whence

$$\boxed{c^2(t - t_+^A)^2 = c^2 U^2 d^2 + c^2 UV d^3 + O(d^4)} \tag{9.16}$$

Likewise for g, we have

$$g(d) = g(0) + g^{(1)}(0)d + \frac{1}{2}g^{(2)}(0)d^2 + \frac{1}{6}g^{(3)}(0)d^3 + O(d^4) \, ,$$

where $g(0) = 0$. Setting $\Delta := t - t_+^A$, considered as a function of d, we have

$$g^{(1)}(d) = 2R^2 \sin(\omega\Delta + \phi)\left[\omega\Delta^{(1)} + \phi^{(1)}\right] \, ,$$

$$g^{(2)}(d) = 2R^2 \sin(\omega\Delta + \phi)\left[\omega\Delta^{(2)} + \phi^{(2)}\right] + 2R^2 \cos(\omega\Delta + \phi)\left[\omega\Delta^{(1)} + \phi^{(1)}\right]^2 \, ,$$

and

$$g^{(3)}(d) = 2R^2 \sin(\omega\Delta + \phi)\left[\omega\Delta^{(3)} + \phi^{(3)}\right] - 2R^2 \sin(\omega\Delta + \phi)\left[\omega\Delta^{(1)} + \phi^{(1)}\right]^3$$
$$+ 6R^2 \cos(\omega\Delta + \phi)\left[\omega\Delta^{(1)} + \phi^{(1)}\right]\left[\omega\Delta^{(2)} + \phi^{(2)}\right] \, ,$$

where $\phi^{(1)}$ is given by (9.5), $\phi^{(2)}$ by (9.6), and

$$\Delta^{(1)} = -\frac{\partial t_+^A}{\partial t} \, , \qquad \Delta^{(2)} = -\frac{\partial^2 t_+^A}{\partial t^2} \, .$$

Evaluating everything at $d = 0$, we have

$$g^{(1)}(0) = 0 \, , \quad g^{(2)}(0) = 2(1 - \omega RU)^2 \, , \quad g^{(3)}(0) = -6\omega VR(1 - \omega RU) \, .$$

We thus have the expansion

$$2R^2\left\{1 - \cos\left[\omega(t - t_+^A) + \phi\right]\right\} = (1 - \omega RU)^2 d^2 - \omega VR(1 - \omega RU)d^3 + O(d^4)$$

(9.17)

Equating coefficients of equal powers of d in (9.16) and (9.17), we obtain

$$\begin{cases} c^2 U^2 = (1 - \omega RU)^2, \\ c^2 UV = -\omega VR(1 - \omega RU), \end{cases}$$

with solution

$$U := \left.\frac{\partial t_+^A}{\partial d}\right|_{d=0} = -\frac{1}{c - \omega R}, \quad V := \left.\frac{\partial^2 t_+^A}{\partial d^2}\right|_{d=0} = 0 \qquad (9.18)$$

We thus have the expansions

$$t_+^A = t - \frac{1}{c - \omega R}d + O(d^3) \qquad (9.19)$$

and

$$t - t_+^A = \frac{1}{c - \omega R}d + O(d^3) \qquad (9.20)$$

As expected, this is constant in time to this order, and should be to all orders, from the symmetry of the situation.

9.3.2 Expansion for t_+^B

Likewise we define functions

$$h(d) := c^2(t - t_+^B)^2, \qquad i(d) := 2R^2\left\{1 - \cos\left[\omega(t - t_+^B) - \phi\right]\right\},$$

expand each to $O(d^3)$, and equate coefficients of equal powers of d on either side of the relation $h = i$ [which is just (9.11)].

As before, we define the notation

$$U := \left.\frac{\partial t_+^B}{\partial d}\right|_{d=0}, \qquad V := \left.\frac{\partial^2 t_+^B}{\partial d^2}\right|_{d=0},$$

bearing in mind that this U and V are not the same as in the last section. Then (9.13) becomes

$$t^B_+ = t + Ud + \frac{1}{2}Vd^2 + O(d^3), \qquad t - t^B_+ = -Ud - \frac{1}{2}Vd^2 + O(d^3). \qquad (9.21)$$

Further

$$\frac{\partial t^B_+}{\partial d} = U + Vd + O(d^2), \qquad \frac{\partial^2 t^B_+}{\partial d^2} = V + O(d). \qquad (9.22)$$

Now the calculation for

$$h(d) = h(0) + h^{(1)}(0)d + \frac{1}{2}h^{(2)}(0)d^2 + \frac{1}{6}h^{(3)}(0)d^3 + O(d^4),$$

is exactly the same as the calculation for f in the last section, with the new symbols U and V, and so

$$\boxed{c^2(t - t^B_+)^2 = c^2U^2d^2 + c^2UVd^3 + O(d^4)} \qquad (9.23)$$

For i, we have

$$i(d) = i(0) + i^{(1)}(0)d + \frac{1}{2}i^{(2)}(0)d^2 + \frac{1}{6}i^{(3)}(0)d^3 + O(d^4),$$

where $i(0) = 0$. Setting $\Delta := t - t^B_+$, considered as a function of d (again this differs from Δ in the last section), we have

$$i^{(1)}(d) = 2R^2 \sin(\omega\Delta - \phi)\left[\omega\Delta^{(1)} - \phi^{(1)}\right],$$

$$i^{(2)}(d) = 2R^2 \sin(\omega\Delta - \phi)\left[\omega\Delta^{(2)} - \phi^{(2)}\right] + 2R^2 \cos(\omega\Delta - \phi)\left[\omega\Delta^{(1)} - \phi^{(1)}\right]^2,$$

and

$$i^{(3)}(d) = 2R^2 \sin(\omega\Delta - \phi)\left[\omega\Delta^{(3)} - \phi^{(3)}\right] - 2R^2 \sin(\omega\Delta - \phi)\left[\omega\Delta^{(1)} - \phi^{(1)}\right]^3$$

$$+ 6R^2 \cos(\omega\Delta - \phi)\left[\omega\Delta^{(1)} - \phi^{(1)}\right]\left[\omega\Delta^{(2)} - \phi^{(2)}\right],$$

where $\phi^{(1)}$ is given by (9.5), $\phi^{(2)}$ by (9.6), and

$$\Delta^{(1)} = -\frac{\partial t^B_+}{\partial t}, \qquad \Delta^{(2)} = -\frac{\partial^2 t^B_+}{\partial t^2}.$$

Evaluating everything at $d = 0$, we have

$$i^{(1)}(0) = 0, \quad i^{(2)}(0) = 2(1 + \omega RU)^2, \quad i^{(3)}(0) = 6\omega VR(1 + \omega RU).$$

We thus have the expansion

$$2R^2\left\{1-\cos\left[\omega(t-t_+^B)+\phi\right]\right\}=(1+\omega RU)^2d^2+\omega VR(1+\omega RU)d^3+O(d^4)$$

$$(9.24)$$

Equating coefficients of equal powers of d in (9.23) and (9.24), we obtain

$$\begin{cases} c^2U^2=(1+\omega RU)^2\,, \\ c^2UV=\omega VR(1+\omega RU)\,, \end{cases}$$

with solution

$$U:=\left.\frac{\partial t_+^B}{\partial d}\right|_{d=0}=-\frac{1}{c+\omega R}\,,\quad V:=\left.\frac{\partial^2 t_+^B}{\partial d^2}\right|_{d=0}=0 \qquad (9.25)$$

We thus have the expansions

$$t_+^B=t-\frac{1}{c+\omega R}d+O(d^3) \qquad (9.26)$$

and

$$t-t_+^B=\frac{1}{c+\omega R}d+O(d^3) \qquad (9.27)$$

As expected, this is constant in time to this order, and should be to all orders, from the symmetry of the situation.

9.4 Power Series Expansion of Retarded Displacement Vectors

The next task to prepare for calculating the self-force is to expand the retarded displacement vectors $\mathbf{r}_+^{AB}(t)$ and $\mathbf{r}_+^{BA}(t)$ defined by

$$\mathbf{r}_+^{AB}(t)=\begin{pmatrix} R\cos(\omega t+\phi)-R\cos\omega t_+^A \\ R\sin(\omega t+\phi)-R\sin\omega t_+^A \\ 0 \end{pmatrix}\,,\quad \mathbf{r}_+^{BA}(t)=\begin{pmatrix} R\cos\omega t-R\cos(\omega t_+^B+\phi) \\ R\sin\omega t-R\sin(\omega t_+^B+\phi) \\ 0 \end{pmatrix}\,.$$

$$(9.28)$$

For this purpose, we shall require expansions of $\cos\omega t_+^A$, $\sin\omega t_+^A$, $\cos(\omega t+\phi)$, $\sin(\omega t+\phi)$, $\cos(\omega t_+^B+\phi)$, and $\sin(\omega t_+^B+\phi)$ to $O(d^2)$.

Since $\phi=d/R+O(d^3)$, we have

$$\cos(\omega t+\phi)=\cos\omega t-\phi\sin\omega t-\frac{1}{2}\phi^2\cos\omega t+O(d^3)\,,$$

and

$$\sin(\omega t + \phi) = \sin \omega t + \phi \cos \omega t - \frac{1}{2} \phi^2 \cos \omega t + O(d^3) \,,$$

whence

$$\cos(\omega t + \phi) = \cos \omega t - \frac{d}{R} \sin \omega t - \frac{d^2}{2R^2} \cos \omega t + O(d^3) \tag{9.29}$$

and

$$\sin(\omega t + \phi) = \sin \omega t + \frac{d}{R} \cos \omega t - \frac{d^2}{2R^2} \sin \omega t + O(d^3) \tag{9.30}$$

Now define $j(d) := \cos \omega t_+^A$, so that $j(0) = \cos \omega t$ and

$$j^{(1)}(d) = -\omega \sin \omega t_+^A \frac{\partial t_+^A}{\partial d} \,, \quad j^{(2)}(d) = -\omega^2 \cos \omega t_+^A \left(\frac{\partial t_+^A}{\partial d} \right)^2 - \omega \sin \omega t_+^A \frac{\partial^2 t_+^A}{\partial d^2} \,,$$

whence

$$j^{(1)}(0) = \frac{\omega}{c - \omega R} \sin \omega t \,, \quad j^{(2)}(0) = -\frac{\omega^2}{(c - \omega R)^2} \cos \omega t \,,$$

using (9.18). Since

$$j(d) = j(0) + j^{(1)}(0)d + \frac{1}{2} j^{(2)}(0)d^2 + O(d^3) \,,$$

we thus find

$$\cos \omega t_+^A = \cos \omega t + \frac{\omega d}{c - \omega R} \sin \omega t - \frac{\omega^2 d^2}{2(c - \omega R)^2} \cos \omega t + O(d^3) \tag{9.31}$$

Now define $k(d) := \sin \omega t_+^A$, so that $k(0) = \sin \omega t$ and

$$k^{(1)}(d) = \omega \cos \omega t_+^A \frac{\partial t_+^A}{\partial d} \,, \quad k^{(2)}(d) = -\omega^2 \sin \omega t_+^A \left(\frac{\partial t_+^A}{\partial d} \right)^2 + \omega \cos \omega t_+^A \frac{\partial^2 t_+^A}{\partial d^2} \,,$$

whence

$$k^{(1)}(0) = -\frac{\omega}{c - \omega R} \cos \omega t \,, \quad k^{(2)}(0) = -\frac{\omega^2}{(c - \omega R)^2} \sin \omega t \,,$$

using (9.18). Since

$$k(d) = k(0) + k^{(1)}(0)d + \frac{1}{2}k^{(2)}(0)d^2 + O(d^3) \,,$$

we thus find

$$\boxed{\sin \omega t_+^A = \sin \omega t - \frac{\omega d}{c - \omega R}\cos \omega t - \frac{\omega^2 d^2}{2(c - \omega R)^2}\sin \omega t + O(d^3)} \qquad (9.32)$$

Now define $l(d) := \cos(\omega t_+^B + \phi)$, so that

$$l^{(1)}(d) = - \left[\omega \frac{\partial t_+^B}{\partial d} + \phi^{(1)}\right]\sin(\omega t_+^B + \phi)$$

and

$$l^{(2)}(d) = - \left[\omega \frac{\partial^2 t_+^B}{\partial d^2} + \phi^{(2)}\right]\sin(\omega t_+^B + \phi) - \left[\omega \frac{\partial t_+^B}{\partial d} + \phi^{(1)}\right]^2 \cos(\omega t_+^B + \phi) \,.$$

By (9.25) and the fact that $\phi^{(2)}(0) = 0$, we have

$$l(0) = \cos \omega t \,, \quad l^{(1)}(0) = - \frac{c}{R(c + \omega R)}\sin \omega t \,, \quad l^{(2)}(0) = - \frac{c^2}{R^2(c + \omega R)^2}\cos \omega t \,,$$

and finally,

$$\boxed{\cos(\omega t_+^B + \phi) = \cos \omega t - \frac{cd}{R(c + \omega R)}\sin \omega t - \frac{c^2 d^2}{2R^2(c + \omega R)^2}\cos \omega t + O(d^3)}$$

$$(9.33)$$

Now define $m(d) := \sin(\omega t_+^B + \phi)$, so that

$$m^{(1)}(d) = \left[\omega \frac{\partial t_+^B}{\partial d} + \phi^{(1)}\right]\cos(\omega t_+^B + \phi)$$

and

$$m^{(2)}(d) = \left[\omega \frac{\partial^2 t_+^B}{\partial d^2} + \phi^{(2)}\right]\cos(\omega t_+^B + \phi) - \left[\omega \frac{\partial t_+^B}{\partial d} + \phi^{(1)}\right]^2 \sin(\omega t_+^B + \phi) \,.$$

Once again, by (9.25) and the fact that $\phi^{(2)}(0) = 0$, we have

$$m(0) = \sin \omega t \,, \quad m^{(1)}(0) = \frac{c}{R(c + \omega R)}\cos \omega t \,, \quad m^{(2)}(0) = - \frac{c^2}{R^2(c + \omega R)^2}\sin \omega t \,,$$

and finally,

$$\sin(\omega t_+^B + \phi) = \sin \omega t + \frac{cd}{R(c+\omega R)} \cos \omega t - \frac{c^2 d^2}{2R^2(c+\omega R)^2} \sin \omega t + O(d^3)$$

(9.34)

Now inserting (9.29) for $\cos(\omega t + \phi)$, (9.30) for $\sin(\omega t + \phi)$, (9.31) for $\cos \omega t_+^A$, and (9.32) for $\sin \omega t_+^A$ into the expression in (9.28) for the retarded displacement vector from A to B, we soon obtain

$$\mathbf{r}_+^{AB} = \frac{cd}{c - \omega R} \begin{pmatrix} -\sin \omega t \\ \cos \omega t \\ 0 \end{pmatrix} + \frac{(2\omega R - c)c}{2(c - \omega R)^2 R} d^2 \begin{pmatrix} \cos \omega t \\ \sin \omega t \\ 0 \end{pmatrix} + O(d^3)$$

(9.35)

Likewise, inserting (9.33) and (9.34) into the expression in (9.28) for the retarded displacement vector from B to A, we obtain

$$\mathbf{r}_+^{BA} = \frac{cd}{c + \omega R} \begin{pmatrix} \sin \omega t \\ -\cos \omega t \\ 0 \end{pmatrix} + \frac{c^2 d^2}{2(c + \omega R)^2 R} \begin{pmatrix} \cos \omega t \\ \sin \omega t \\ 0 \end{pmatrix} + O(d^3)$$

(9.36)

We can cross-check for mistakes at this point, as in Sect. 8.4. We estimate the lengths of these two vectors to $O(d^2)$ and check that we do indeed obtain $|\mathbf{r}_+^{AB}| = r_+^{AB}$ as given by (9.20), viz.,

$$r_+^{AB} = c(t - t_+^A) = \frac{c}{c - \omega R} d + O(d^3) ,$$

(9.37)

and $|\mathbf{r}_+^{BA}| = r_+^{BA}$ as given by (9.27), viz.,

$$r_+^{BA} = c(t - t_+^B) = \frac{c}{c + \omega R} d + O(d^3) .$$

(9.38)

Both $\mathbf{r}_+^{AB}$ and $\mathbf{r}_+^{BA}$ have the form

$$\mathbf{r} = \mathbf{p}d + \mathbf{q}d^2 + O(d^3) ,$$

and it was shown in Sect. 8.4 that this implies

$$|\mathbf{r}| = |\mathbf{p}|d \left(1 + \frac{\mathbf{p} \cdot \mathbf{q}}{\mathbf{p}^2} d\right) + O(d^3) .$$

In the case of $\mathbf{r}_+^{AB}$, we have

$$\mathbf{p} = \frac{c}{c - \omega R} \begin{pmatrix} -\sin \omega t \\ \cos \omega t \\ 0 \end{pmatrix} , \qquad \mathbf{q} = \frac{(2\omega R - c)c}{2(c - \omega R)^2 R} \begin{pmatrix} \cos \omega t \\ \sin \omega t \\ 0 \end{pmatrix} , \qquad (9.39)$$

so clearly,

$$|\mathbf{p}| = \frac{c}{c - \omega R} , \quad \mathbf{p} \cdot \mathbf{q} = 0 ,$$

whence we obtain confirmation that

$$|\mathbf{r}_+^{AB}| = \frac{c}{c - \omega R} d + O(d^3) ,$$

according to the expression (9.35), which agrees with (9.37). In the case of $\mathbf{r}_+^{BA}$, we have

$$\mathbf{p} = \frac{c}{c + \omega R} \begin{pmatrix} \sin \omega t \\ -\cos \omega t \\ 0 \end{pmatrix} , \qquad \mathbf{q} = \frac{c^2}{2(c + \omega R)^2 R} \begin{pmatrix} \cos \omega t \\ \sin \omega t \\ 0 \end{pmatrix} , \qquad (9.40)$$

so clearly,

$$|\mathbf{p}| = \frac{c}{c + \omega R} , \quad \mathbf{p} \cdot \mathbf{q} = 0 ,$$

whence we obtain confirmation that

$$|\mathbf{r}_+^{BA}| = \frac{c}{c + \omega R} d + O(d^3) ,$$

according to the expression (9.36), which agrees with (9.38).

When we come to examine the magnetic fields, we shall require the unit retarded displacement vectors $\mathbf{r}_+^{AB}/r_+^{AB}$ and $\mathbf{r}_+^{BA}/r_+^{BA}$. As we saw in (8.35) on p. 145, when a vector $\mathbf{r}$ has the form $\mathbf{r} = \mathbf{p} + \mathbf{q}d + O(d^2)$, we have

$$\frac{\mathbf{r}}{|\mathbf{r}|} = \frac{\mathbf{p}}{|\mathbf{p}|} + \frac{1}{|\mathbf{p}|} \left(\mathbf{q} - \frac{\mathbf{p} \cdot \mathbf{q}}{\mathbf{p}^2} \mathbf{p} \right) d + O(d^2) .$$

Now in both (9.39) and (9.40), we have $\mathbf{p} \cdot \mathbf{q} = 0$, so in both cases, the result reduces to

$$\frac{\mathbf{r}}{|\mathbf{r}|} = \frac{\mathbf{p} + \mathbf{q}d}{|\mathbf{p}|} + O(d^2) .$$

Applying this, we soon obtain

$$\frac{\mathbf{r}^{AB}_{+}}{r^{AB}_{+}} = \begin{pmatrix} -\sin\omega t \\ \cos\omega t \\ 0 \end{pmatrix} + \frac{2\omega R - c}{2(c - \omega R)R}d \begin{pmatrix} \cos\omega t \\ \sin\omega t \\ 0 \end{pmatrix} + O(d^2) \qquad (9.41)$$

and

$$\frac{\mathbf{r}^{BA}_{+}}{r^{BA}_{+}} = \begin{pmatrix} \sin\omega t \\ -\cos\omega t \\ 0 \end{pmatrix} + \frac{cd}{2(c + \omega R)R} \begin{pmatrix} \cos\omega t \\ \sin\omega t \\ 0 \end{pmatrix} + O(d^2) \qquad (9.42)$$

9.5 Power Series Expansion of Electric Self-Force

We come now to the calculation of the electrical fields generated by these point charges A and B in motion, obtaining only the leading order terms as usual. These are the terms $O(d^{-1})$ which diverge if we let $d \to 0$.

9.5.1 Electric Force of A on B

We shall find the electric field $\mathbf{E}^A(\text{at B})$ at B due to A as given by

$$\mathbf{E}^A(\text{at B}) \qquad (9.43)$$

$$= \frac{q_e}{8\pi\varepsilon_0} \cdot \frac{\left(\mathbf{r}^{AB}_{+} - \frac{r^{AB}_{+}\mathbf{v}^A_{+}}{c}\right)\left[1 - \frac{(\mathbf{v}^A_{+})^2}{c^2}\right] + \frac{r^{AB}_{+}}{c^2} \times \left[\left(\mathbf{r}^{AB}_{+} - \frac{r^{AB}_{+}\mathbf{v}^A_{+}}{c}\right) \times \frac{d\mathbf{v}^A}{dt}\Big|_{t=t^A_{+}}\right]}{\left(r^{AB}_{+} - \mathbf{r}^{AB}_{+}\cdot\mathbf{v}^A_{+}/c\right)^3} ,$$

then apply the usual rule to get the electric force

$$\mathbf{F}^A_{\text{elec}}(\text{on B}) = \frac{q_e}{2}\mathbf{E}^A(\text{at B}) . \qquad (9.44)$$

Let us begin with the expansion of $\mathbf{v}^A_{+} = \mathbf{v}^A(t^A_{+})$, keeping only terms up to $O(d)$. Recalling from (9.3) on p. 158 that

$$\mathbf{v}^A(t^A_{+}) = \omega R \begin{pmatrix} -\sin\omega t^A_{+} \\ \cos\omega t^A_{+} \\ 0 \end{pmatrix} ,$$

and substituting in the expansions (9.31) and (9.32) for $\cos\omega t^A_{+}$ and $\sin\omega t^A_{+}$, we obtain

$$
\mathbf{v}_+^A = \omega R \begin{pmatrix} -\sin\omega t \\ \cos\omega t \\ 0 \end{pmatrix} + \frac{\omega^2 Rd}{c-\omega R} \begin{pmatrix} \cos\omega t \\ \sin\omega t \\ 0 \end{pmatrix} d + O(d^2) \qquad (9.45)
$$

A short calculation now finds

$$
(\mathbf{v}_+^A)^2 = \omega^2 R^2 + O(d^2) ,
$$

whence

$$
1 - \frac{(\mathbf{v}_+^A)^2}{c^2} = \gamma^{-2} + O(d^2) \qquad (9.46)
$$

Further efforts using (9.35) for $\mathbf{r}_+^{AB}$, (9.37) for r_+^{AB}, and (9.45) for $\mathbf{v}_+^A$ lead to

$$
r_+^{AB} - \frac{\mathbf{r}_+^{AB} \cdot \mathbf{v}_+^A}{c} = d + O(d^3) \qquad (9.47)
$$

and

$$
\mathbf{r}_+^{AB} - \frac{r_+^{AB} \mathbf{v}_+^A}{c} = d \begin{pmatrix} -\sin\omega t \\ \cos\omega t \\ 0 \end{pmatrix} - \frac{R^2\omega^2 + (c-\omega R)^2}{2R(c-\omega R)^2} d^2 \begin{pmatrix} \cos\omega t \\ \sin\omega t \\ 0 \end{pmatrix} + O(d^3)
$$

$$(9.48)$$

We now have

$$
\left(\mathbf{r}_+^{AB} - \frac{r_+^{AB}\mathbf{v}_+^A}{c} \right)\left[1 - \frac{(\mathbf{v}_+^A)^2}{c^2} \right] = \frac{d}{\gamma^2} \begin{pmatrix} -\sin\omega t \\ \cos\omega t \\ 0 \end{pmatrix}
$$
$$
- \left[\frac{\omega^2 R\gamma^2}{2c^2}\left(1+\frac{\omega R}{c}\right)^2 + \frac{1}{2R\gamma^2} \right] d^2 \begin{pmatrix} \cos\omega t \\ \sin\omega t \\ 0 \end{pmatrix} + O(d^3)
$$

$$(9.49)$$

We still need to examine the term in $\mathbf{a}^A(t_+^A)$ in the numerator of (9.43). Now since $\mathbf{r}_+^{AB}$ and $\mathbf{r}_+^{AB} - r_+^{AB}\mathbf{v}_+^A/c$ are each $O(d)$, we only need the $O(d^0)$ term in $\mathbf{a}^A(t_+^A)$, which is just

$$
\mathbf{a}^A(t_+^A) = -\omega^2 R \begin{pmatrix} \cos\omega t \\ \sin\omega t \\ 0 \end{pmatrix} + O(d) .
$$

In this scenario, we obtain something of $O(d)$ for

$$\left(\mathbf{r}_+^{AB} - \frac{r_+^{AB}\mathbf{v}_+^A}{c}\right) \times \frac{d\mathbf{v}^A}{dt}\bigg|_{t=t_+^A} = \omega^2 R d \begin{pmatrix} 0 \\ 0 \\ 1 \end{pmatrix} + O(d^2),$$

whence finally,

$$\boxed{\frac{\mathbf{r}_+^{AB}}{c^2} \times \left[\left(\mathbf{r}_+^{AB} - \frac{r_+^{AB}\mathbf{v}_+^A}{c}\right) \times \frac{d\mathbf{v}^A}{dt}\bigg|_{t=t_+^A}\right] = \frac{\omega^2 R d^2}{c(c-\omega R)} \begin{pmatrix} \cos \omega t \\ \sin \omega t \\ 0 \end{pmatrix} + O(d^3)} \quad (9.50)$$

Adding (9.49) and (9.50) and dividing by d^3, we thus find

$$\boxed{\mathbf{E}^A(\text{at } B) = \frac{q_e}{8\pi\varepsilon_0}\left[\frac{1}{\gamma^2 d^2}\begin{pmatrix} -\sin \omega t \\ \cos \omega t \\ 0 \end{pmatrix} + \left(\frac{\omega^2 R}{c^2} - \frac{1}{2R}\right)\begin{pmatrix} \cos \omega t \\ \sin \omega t \\ 0 \end{pmatrix}\frac{1}{d} + O(d^0)\right]} \tag{9.51}$$

Then by (9.44), the electric force of A on B is

$$\boxed{\mathbf{F}_{\text{elec}}^A(\text{on } B) = \frac{e^2}{4}\left[\frac{1}{\gamma^2 d^2}\begin{pmatrix} -\sin \omega t \\ \cos \omega t \\ 0 \end{pmatrix} + \left(\frac{\omega^2 R}{c^2} - \frac{1}{2R}\right)\begin{pmatrix} \cos \omega t \\ \sin \omega t \\ 0 \end{pmatrix}\frac{1}{d} + O(d^0)\right]} \tag{9.52}$$

having inserted $e^2 := q_e^2/4\pi\varepsilon_0$.

9.5.2 Electric Force of B on A

We shall find the electric field $\mathbf{E}^B(\text{at } A)$ at A due to B as given by

$$\mathbf{E}^B(\text{at } A) \tag{9.53}$$

$$= \frac{q_e}{8\pi\varepsilon_0} \frac{\left(\mathbf{r}_+^{BA} - \frac{r_+^{BA}\mathbf{v}_+^B}{c}\right)\left[1 - \frac{(\mathbf{v}_+^B)^2}{c^2}\right] + \frac{r_+^{BA}}{c^2}\times\left[\left(\mathbf{r}_+^{BA} - \frac{r_+^{BA}\mathbf{v}_+^B}{c}\right)\times\frac{d\mathbf{v}^B}{dt}\bigg|_{t=t_+^B}\right]}{\left(r_+^{BA} - \mathbf{r}_+^{BA}\cdot\mathbf{v}_+^B/c\right)^3},$$

then apply the usual rule to get the electric force

$$\mathbf{F}^{B}_{elec}(\text{on A}) = \frac{q_e}{2}\mathbf{E}^{B}(\text{at A}) . \tag{9.54}$$

Let us begin with the expansion of $\mathbf{v}^{B}_{+} = \mathbf{v}^{B}(t^{B}_{+})$, keeping only terms up to $O(d)$. Recalling from (9.3) on p. 158 that

$$\mathbf{v}^{B}(t^{B}_{+}) = \omega R \begin{pmatrix} -\sin(\omega t^{B}_{+} + \phi) \\ \cos(\omega t^{B}_{+} + \phi) \\ 0 \end{pmatrix} ,$$

and substituting in the expansions (9.33) and (9.34) for $\cos \omega t^{B}_{+}$ and $\sin \omega t^{B}_{+}$, we obtain

$$\mathbf{v}^{B}_{+} = \omega R \begin{pmatrix} -\sin \omega t \\ \cos \omega t \\ 0 \end{pmatrix} - \frac{\omega c d}{c + \omega R} \begin{pmatrix} \cos \omega t \\ \sin \omega t \\ 0 \end{pmatrix} d + O(d^2) \tag{9.55}$$

A short calculation now finds

$$\left(\mathbf{v}^{B}_{+}\right)^2 = \omega^2 R^2 + O(d^2) ,$$

whence

$$1 - \frac{(\mathbf{v}^{B}_{+})^2}{c^2} = \gamma^{-2} + O(d^2) \tag{9.56}$$

Further efforts using (9.36) for $\mathbf{r}^{BA}_{+}$, (9.38) for r^{BA}_{+}, and (9.55) for $\mathbf{v}^{B}_{+}$ lead to

$$r^{BA}_{+} - \frac{\mathbf{r}^{BA}_{+} \cdot \mathbf{v}^{B}_{+}}{c} = d + O(d^3) \tag{9.57}$$

and

$$\mathbf{r}^{BA}_{+} - \frac{\mathbf{r}^{BA}_{+}\mathbf{v}^{B}_{+}}{c} = d \begin{pmatrix} \sin \omega t \\ -\cos \omega t \\ 0 \end{pmatrix} - \frac{c^2(c + 2\omega R)}{2Rc(c + \omega R)^2}d^2 \begin{pmatrix} \cos \omega t \\ \sin \omega t \\ 0 \end{pmatrix} + O(d^3) \tag{9.58}$$

We now have

$$\left(\mathbf{r}_+^{BA} - \frac{r_+^{BA} \mathbf{v}_+^{B}}{c} \right) \left[1 - \frac{(\mathbf{v}_+^{B})^2}{c^2} \right] = \frac{d}{\gamma^2} \begin{pmatrix} \sin \omega t \\ -\cos \omega t \\ 0 \end{pmatrix}$$

$$+ \frac{(c - \omega R)(c + 2\omega R)}{2Rc(c + \omega R)} d^2 \begin{pmatrix} \cos \omega t \\ \sin \omega t \\ 0 \end{pmatrix} + O(d^3)$$

(9.59)

We still need to examine the term in $\mathbf{a}^{B}(t_+^{B})$ in the numerator of (9.53). Now since $\mathbf{r}_+^{BA}$ and $\mathbf{r}_+^{BA} - r_+^{BA} \mathbf{v}_+^{B}/c$ are each $O(d)$, we only need the $O(d^0)$ term in $\mathbf{a}^{B}(t_+^{B})$, which is just

$$\mathbf{a}^{B}(t_+^{B}) = -\omega^2 R \begin{pmatrix} \cos \omega t \\ \sin \omega t \\ 0 \end{pmatrix} + O(d) .$$

Once again, we obtain something of $O(d)$ for

$$\left(\mathbf{r}_+^{BA} - \frac{r_+^{BA} \mathbf{v}_+^{B}}{c} \right) \times \left. \frac{d\mathbf{v}^{B}}{dt} \right|_{t=t_+^{B}} = -\omega^2 R d \begin{pmatrix} 0 \\ 0 \\ 1 \end{pmatrix} + O(d^2) ,$$

whence finally,

$$\frac{r_+^{BA}}{c^2} \times \left[\left(\mathbf{r}_+^{BA} - \frac{r_+^{BA} \mathbf{v}_+^{B}}{c} \right) \times \left. \frac{d\mathbf{v}^{B}}{dt} \right|_{t=t_+^{B}} \right] = \frac{\omega^2 R d^2}{c(c + \omega R)} \begin{pmatrix} \cos \omega t \\ \sin \omega t \\ 0 \end{pmatrix} + O(d^3) \quad (9.60)$$

Adding (9.59) and (9.60) and dividing by d^3, we thus find

$$\mathbf{E}^{B}(\text{at } A) = \frac{q_e}{8\pi\varepsilon_0} \left[\frac{1}{\gamma^2 d^2} \begin{pmatrix} \sin \omega t \\ -\cos \omega t \\ 0 \end{pmatrix} + \frac{1}{2R} \begin{pmatrix} \cos \omega t \\ \sin \omega t \\ 0 \end{pmatrix} \frac{1}{d} + O(d^0) \right] \quad (9.61)$$

Then by (9.54), the electric force of B on A is

$$\mathbf{F}_{\text{elec}}^{B}(\text{on } A) = \frac{e^2}{4} \left[\frac{1}{\gamma^2 d^2} \begin{pmatrix} \sin \omega t \\ -\cos \omega t \\ 0 \end{pmatrix} + \frac{1}{2R} \begin{pmatrix} \cos \omega t \\ \sin \omega t \\ 0 \end{pmatrix} \frac{1}{d} + O(d^0) \right] \quad (9.62)$$

having inserted $e^2 := q_e^2/4\pi\varepsilon_0$.

9.5.3 Electric Self-Force

We now simply add together the results in (9.52) and (9.62) to obtain

$$
\mathbf{F}_{\text{elec}}^{\text{self}} = \frac{e^2}{4c^2 d} \omega^2 R \begin{pmatrix} \cos \omega t \\ \sin \omega t \\ 0 \end{pmatrix} + O(d^0)
\tag{9.63}
$$

The first thing to note is that, as usual, the $O(d^{-2})$ terms, which are basically Coulomb terms, cancel one another. Secondly, it is worth comparing this with the $O(d^{-1})$ term of (7.53) on p. 132, for the scenario in which the system is accelerating along a straight line parallel to its axis. Note that $\omega^2 R$ is the magnitude of the acceleration of either A or B in the present case, and $\mathbf{F}_{\text{elec}}^{\text{self}}$ is directed radially outward, i.e., in the direction opposite to the acceleration. However, in this case, $\mathbf{F}_{\text{elec}}^{\text{self}}$ does not contain any factors involving γ. Regarding the comments in Sect. 7.5, we note that the same γ factors are associated with A and B, so the disappearance of γ from $\mathbf{F}_{\text{elec}}^{\text{self}}$ looks rather disappointing on the face of things. The problem is resolved below.

Another difference with (7.53) is that it contains a factor of $e/2c^2 d$, whereas (9.63) contains the factor $e/4c^2 d$. However, there are magnetic fields in the present case and they are going to make a contribution.

9.6 Power Series Expansion of Magnetic Self-Force

According to the formula (2.62) given on p. 21 of Chap. 2, the magnetic field at B due to A is given by

$$
\mathbf{B}^{\text{A}}(\text{at B}) = \frac{\mathbf{r}_+^{\text{AB}} \times \mathbf{E}^{\text{A}}(\text{at B})}{c r_+^{\text{AB}}} ,
\tag{9.64}
$$

where $\mathbf{E}^{\text{A}}(\text{at B})$ is given by (9.51), and the magnetic force of A on B is then

$$
\mathbf{F}_{\text{mag}}^{\text{A}}(\text{on B}) = \frac{q_e}{2} \mathbf{v}^{\text{B}}(t) \times \mathbf{B}^{\text{A}}(\text{at B}) .
\tag{9.65}
$$

We have already calculated the unit retarded displacement vector $\mathbf{r}_+^{\text{AB}}/r_+^{\text{AB}}$ in (9.41) on p. 169, and a short calculation gives

$$
\mathbf{B}^{\text{A}}(\text{at B}) = \frac{q_e \omega}{16 \pi \varepsilon_0 c^2 d} \begin{pmatrix} 0 \\ 0 \\ 1 \end{pmatrix} + O(d^0)
\tag{9.66}
$$

Note that there is no $O(d^{-2})$ term and also that this vector is perpendicular to the page in the view of Fig. 9.1.

Now according to (9.3),

$$\mathbf{v}^B(t) = \omega R \begin{pmatrix} -\sin(\omega t + \phi) \\ \cos(\omega t + \phi) \\ 0 \end{pmatrix} = \omega R \begin{pmatrix} -\sin \omega t \\ \cos \omega t \\ 0 \end{pmatrix} + O(d) ,$$

and this is accurate enough because there is no $O(d^{-2})$ term in the magnetic field (9.66). Finally, we obtain

$$\mathbf{F}^A_{\text{mag}}(\text{on B}) = \frac{e^2 \omega^2 R}{8c^2 d} \begin{pmatrix} \cos \omega t \\ \sin \omega t \\ 0 \end{pmatrix} + O(d^0) \tag{9.67}$$

having made the usual definition $e^2 := q_e/4\pi\varepsilon_0$.

The magnetic field at A due to B is given by

$$\mathbf{B}^B(\text{at A}) = \frac{\mathbf{r}^{BA}_+ \times \mathbf{E}^B(\text{at A})}{c r^{BA}_+} , \tag{9.68}$$

where $\mathbf{E}^B(\text{at A})$ is given by (9.61), and the magnetic force of B on A is then

$$\mathbf{F}^B_{\text{mag}}(\text{on A}) = \frac{q_e}{2} \mathbf{v}^A(t) \times \mathbf{B}^B(\text{at A}) . \tag{9.69}$$

We have already calculated the unit retarded displacement vector $\mathbf{r}^{BA}_+/r^{BA}_+$ in (9.42) on p. 169, and a short calculation gives

$$\mathbf{B}^B(\text{at A}) = \frac{q_e \omega}{16\pi\varepsilon_0 c^2 d} \begin{pmatrix} 0 \\ 0 \\ 1 \end{pmatrix} + O(d^0) \tag{9.70}$$

which turns out to be identical with (9.66) to this order. Once again, as viewed in Fig. 9.1, this is perpendicular to the page. Now according to (9.3),

$$\mathbf{v}^A(t) = \omega R \begin{pmatrix} -\sin \omega t \\ \cos \omega t \\ 0 \end{pmatrix} ,$$

and we soon find

$$\mathbf{F}^{B}_{mag}(\text{on A}) = \frac{e^2 \omega^2 R}{8c^2 d} \begin{pmatrix} \cos \omega t \\ \sin \omega t \\ 0 \end{pmatrix} + O(d^0) \tag{9.71}$$

We can now write down the magnetic self-force, simply taking the sum of (9.67) and (9.71) to obtain

$$\mathbf{F}^{self}_{mag} = \frac{e^2}{4c^2 d} \omega^2 R \begin{pmatrix} \cos \omega t \\ \sin \omega t \\ 0 \end{pmatrix} + O(d^0) \tag{9.72}$$

This is radially outward.

9.7 Power Series Expansion of Total Self-Force and Interpretation

We now obtain the power series expansion of the total self-force by simply adding (9.63) and (9.72) to obtain

$$\mathbf{F}^{self} = \frac{e^2}{2c^2 d} \omega^2 R \begin{pmatrix} \cos \omega t \\ \sin \omega t \\ 0 \end{pmatrix} + O(d^0) \tag{9.73}$$

If one is interested in renormalising the inertial mass of the system by absorbing this leading order term into whatever else constitutes its inertia, there are several features about this that look helpful. The first is that it is radially outward, directly opposing the acceleration of either A or B. Secondly, it is proportional to the acceleration $\omega^2 R$ of either particle. And finally, we recover the same factor of $e^2/2c^2 d$ that seems to tag along with this system when its velocity is parallel to its axis. In other words, we have the same contrast between (8.68) on p. 153 and (9.73) here as we did between (7.54) and (7.55) on p. 133.

This confirms the interesting feature of the electromagnetic mass, as derived from the self-force, that it depends which way the object is moving relative to its own geometry. This is not something one normally expects of the inertial mass, and yet it is a fact that, in bound systems that do not have spherical symmetry, the inertia of the system will depend which way it moves relative to its own geometry. This is what we have just confirmed for an electromagnetic binding force in a different situation to the one described in Chaps. 6 and 7.

The only possibly disappointing thing about (9.73) is that there is no γ factor to help us renormalise in the context of the relativistic Lorentz force law. However, we can get one factor of γ back in by observing that the length d of the system will be contracted to approximately d/γ, where γ is the factor appropriate to either A or B, as in the above calculation. Note that this is indeed an approximation, because neither A nor B is ever moving exactly parallel to the system axis (the line joining A and B). However, correction would only introduce a term of $O(d)$, so to $O(d^{-1})$, we do have

$$\mathbf{F}^{\text{self}} = \frac{e^2}{2c^2d}\omega^2 R\gamma \begin{pmatrix} \cos\omega t \\ \sin\omega t \\ 0 \end{pmatrix} + O(d^0) \tag{9.74}$$

Interestingly, apart from the factor of 2, this is now identical with (8.68) for the third scenario, and the γ factor is perfect for renormalisation purposes.

Should this factor of γ be included? The answer is affirmative. We are trying to consider what happens to a system that has length d when at rest in an inertial frame. In Chap. 7, we assumed that the system had to continually adjust its length as it accelerated along its axis, so logically, we do need to allow for a constant contraction here, relative to whatever inertial frame we refer the rotational motion.

9.8 General Conclusion for the Four Scenarios

Each of the four scenarios we have described is characterised by the configuration of three vectors, namely, the velocity and acceleration, and the vector specifying the system axis. Even in the relativistic limit, when factors of γ are relevant, it seems that the magnitude of the self-force-derived electromagnetic mass (SFDM) is determined by the relative orientation of the velocity and axis vectors ($e^2/2c^2d$ when they are parallel and $e^2/4c^2d$ when they are perpendicular), while the direction of the $O(d^{-1})$ term in the self-force, which is the only divergent term if one lets the system size tend to zero, always opposes the acceleration vector.

It is a very easy matter to see that, if we replace the like charges $q_e/2$ at A and $q_e/2$ at B by unlike charges $q_e/2$ at A and $-q_e/2$ at B, the electromagnetic self-forces switch sign in each case, to actually assist the acceleration. This is no surprise. In physics, it is well known that the negative binding energy of a bound particle contributes negatively to the inertial mass. The mass renormalisations are all negative then.

Chapter 10
Summary of Results

Let us take stock of the all the results obtained in the preceding chapters, regarding both the spherical charge shell and the charge dumbbell. Table 10.1 shows the calculated values of the electromagnetic mass for these models in the various situations considered. It is interesting that one has to make a distinction between longitudinal and transverse motion for the charge dumbbell. Longitudinal motion refers to the fact that the velocity is along the system axis, while transverse motion implies that the velocity is perpendicular to the system axis.

But it is also interesting that the direction of acceleration relative to the system axis is irrelevant, so that one can group the result of Chap. 6 for transverse linear acceleration with the result of Chap. 8 for transverse rotational motion. The acceleration is perpendicular to the system axis in the first, and along it in the second, but the velocity is perpendicular to the system axis in both. Likewise, we can group the result of Chap. 7 for axial linear acceleration with the result of Chap. 9 for longitudinal rotational motion. The acceleration is along the system axis in the first, and perpendicular to it in the second, but the velocity is along the system axis in both.

We can conclude that the contribution to the EM mass from a spatial charge structure may depend on the direction of the velocity relative to the spatial orientation of that structure. One must wonder what would happen to this result when quantum theoretical models are made.

Another point is that, in all four cases where we have calculated the self-force on the charge dumbbell, the highest order term, going as $1/d$, is always opposite in direction to the imposed acceleration, and hence contributes to the inertia of the system. Given the complexity of the calculations with Maxwell's theory in order to arrive at such a clear and consistent result, this must be telling us something. It must be possible to get a completely general result of this kind for abitrary motion.

But in a sense, we do have such a result, although the method used to obtain it is very different, and in fact less explicit. For this is precisely the result which allows classical mass renormalisation when we take the point particle limit. This is what Dirac showed in [3], a modern version of which can be found in [8]. Classical mass renormalisation is possible precisely because the divergent term arising for a point particle has the *right form* to be absorbed into another term in the equation of motion

Lyle, S.N.: *Summary of Results*. Lect. Notes Phys. **796**, 179–181 (2010)
DOI 10.1007/978-3-642-04785-5_10   © Springer-Verlag Berlin Heidelberg 2010

Table 10.1 Results for the electromagnetic mass in various scenarios. $m_{\mathrm{EM}}^{\mathrm{SFDM}}$ is the self-force-derived EM mass, $m_{\mathrm{EM}}^{\mathrm{MDM}}$ the momentum-derived mass, and $m_{\mathrm{EM}}^{\mathrm{EDM}}$ the energy-derived mass. These are rest mass values, i.e., the relativistic γ factor has not been included

Model	$m_{\mathrm{EM}}^{\mathrm{SFDM}}$	$m_{\mathrm{EM}}^{\mathrm{MDM}}$	$m_{\mathrm{EM}}^{\mathrm{EDM}}$
Spherical shell (radius a)	$\dfrac{2}{3}\dfrac{e^2}{ac^2}$	$\dfrac{2}{3}\dfrac{e^2}{ac^2}$	$\dfrac{1}{2}\dfrac{e^2}{ac^2}$
Charge dumbbell (length d) in longitudinal motion	$\dfrac{e^2}{2c^2d}$	$\dfrac{e^2}{2c^2d}$	$\dfrac{e^2}{4c^2d}$
Charge dumbbell (length d) in transverse motion	$\dfrac{e^2}{4c^2d}$	$\dfrac{e^2}{4c^2d}$	$\dfrac{e^2}{4c^2d}$

and then forgotten. But it is because we get the right form for the self-force in the above calculations that we can consider them to contribute to inertia. These are two facets of the same result.

Moreover, as we have noted in the results of Chaps. 6–9, the leading term in our self-force always has exactly the right form relativistically speaking, since it contains the appropriate power of γ in every case. When the velocity and acceleration of the system are parallel, we require a factor of γ^3, and when they are orthogonal, we require a factor of γ.

The connection between the fact that EM self-forces contribute to inertia and the possibility of mass renormalisation for EM effects looks interesting when one considers self-forces due to other fundamental forces, i.e., the weak and strong force, or any other forces operating on smaller length scales not yet investigated. We now know that all gauge theories are renormalisable [34]. Perhaps this means that (the leading contributions to) self-forces due to the weak and strong forces, presently modelled by gauge theories, must inevitably have the right form to contribute to inertia.

The conjecture is therefore that any particle made up of sub-particles that are themselves sources for some gauge field will exert an inertial self-force on itself when accelerated (where 'inertial' just means 'aligned with the acceleration'). If such a result is true, it should be possible to obtain some general theorem rather simply. Since the weak and strong forces have no classical model, one would have to make a quantum theoretical version of the self-force idea, and use the general arguments that show that gauge field theories are always renormalisable.

Another crucial result here is the explanation of EM radiation effects through the second term in the self-force, something we only demonstrated for transverse linear acceleration of our system (see Chap. 6). By the second term in the self-force, we mean the one that is actually independent of the system dimensions, i.e., order d^0 for the charge dumbbell. This helpful piece of understanding for the classical theory is lost in point particle models. When we move to quantum theory, however, we do not need it any more. Quantum theory deals with radiation in a quite different way. In fact, a much more abstract way, with Hilbert space vectors and operators acting on them. The whole particle ontology is up in the air in quantum theory (but not in the extremely interesting and convincing theory of Bohmian mechanics [35]).

Another key point about the results of all the calculations with the charge dumb-bell is the change of sign that occurs when the two point charges (or small spherical charge shells) at the ends of the dumbbell have opposite charges. In this case, the highest order contribution to the self-force from the d level structure actually helps any imposed acceleration, i.e., reduces the inertia of the system. Likewise, the momentum in the EM fields is reduced by this structure level. This concords perfectly with the adage from special relativity that binding energy should be added in to the inertial mass of any bound system. The difference here is that we have an explanation for what usually comes down to us as a principle today, an idea exposed more carefully in Sect. 13.1.

There is one other important and striking result in Table 10.1. For two of the scenarios, viz., linear acceleration of the spherical charge shell and longitudinal motion of the charge dumbbell, there is a discrepancy between the self-force-derived and momentum-derived EM masses on the one hand, and the energy-derived EM mass on the other. This is what we need to investigate in the next chapter. The answer, as we shall see, lies in the binding forces. The third line in the above table, which shows that there is no discrepancy for transverse motion of the charge dumbbell, will confirm our explanation.

Chapter 11
Reconciling Energy- and Momentum-Derived EM Masses

The discrepancy between the energy- and momentum-derived EM masses first discussed for the spherical charge shell in Sect. 3.3, and then rediscovered for the charge dumbbell in the preceding chapters, has led to a considerable debate in the literature. We shall concentrate on four papers:

- First Rohrlich's [21], in which he redefines the four-momentum of the EM fields around a spherical charge shell in order to make it into a four-vector, transforming by a Lorentz transformation under change of inertial frame.
- We then discuss Boyer's challenge to this [23], in which he criticises what is effectively an ad hoc redefinition of the energy and momentum density in the EM fields, and in doing so, brings in the role of the binding forces in the charge shell, which ensure Lorentz covariance of the theory without the need for ad hoc adjustments to definitions.
- Rohrlich's attempted reply [22] to Boyer's paper is examined in order to demonstrate just how arbitrary Rohrlich's redefinition actually is.
- Along the way, we also consider a more recent paper by Moylan [24], which supports the Rohrlich redefinition, and at the same time illustrates the risks of neat mathematical accounts replacing real physical understanding.

11.1 Energy and Momentum in the Electron EM Fields

This section is inspired by [21], but follows its own path to expose the origin of the discrepancy. Rohrlich deals with a spherical charge shell of radius a, but his solution to the problem, as made explicit in [24], is to patch things up with an ad hoc change in the definition of the field energy and momentum, rather than to take the finite electron as a real possibility in which binding forces must be taken into account. Rohrlich was approaching the problem from the angle of quantum electrodynamics where such a redefinition is part and parcel of the usual process of renormalisation.

So let us also consider the spherical charge shell of radius a, and review the way we have found the energy and momentum in the EM fields from a more global stand-

Lyle, S.N.: *Reconciling Energy- and Momentum-Derived EM Masses.* Lect. Notes Phys. **796**, 183–262 (2010)
DOI 10.1007/978-3-642-04785-5_11   © Springer-Verlag Berlin Heidelberg 2010

point. On the basis of the discussion in Sect. 2.4, we have been using the following formulas for the energy W and momentum p^k of the fields surrounding an electron moving at constant velocity:

$$W = \int u d^3 x \,, \qquad p^k = \frac{1}{c^2} \int S^k d^3 x \,, \tag{11.1}$$

where the Latin index runs over $\{1,2,3\}$. These can be written in the form

$$W = -\int T^{00} d^3 x \,, \qquad p^k = -\frac{1}{c} \int T^{0k} d^3 x \,. \tag{11.2}$$

As we know, when the electron has zero velocity, so that the field is just the Coulomb field, this leads to

$$W = \frac{e^2}{2a} \,, \qquad \mathbf{p} = 0 \,, \tag{11.3}$$

for an electron which is a spherical shell of charge (with zero field inside). The electromagnetic mass has to be defined as

$$m_{\text{EM}}^{\text{EDM}} := \frac{e^2}{2ac^2} \,. \tag{11.4}$$

If the whole setup here were relativistically covariant, we know what we should find when the electron is in uniform motion at speed v along the x axis. We merely Lorentz boost the result in (11.3). Then we should have

$$W' = \gamma m_{\text{EM}}^{\text{EDM}} c^2 \,, \qquad \mathbf{p}' = \gamma m_{\text{EM}}^{\text{EDM}} \mathbf{v} \,. \tag{11.5}$$

However, when we actually do our calculations using the second relation of (11.1), we obtain a factor of 4/3 in the momentum [see, for example, (3.27) on p. 40], whence there must have been something that was not relativistically covariant in our formula (11.1).

So clearly the formulas in (11.2) are not relativistically covariant. Rohrlich asserts [21] that the unique covariant version of (11.2) is

$$p^\mu = \left(\frac{1}{c} W, \mathbf{p} \right) = -\frac{1}{c} \int T^{\mu\nu} d\sigma_\nu \,, \tag{11.6}$$

where $d\sigma_\nu$ is a covector-valued measure on the spacelike hypersurface in which we consider the electron charge to be distributed from our frame of reference, relative to which the electron is moving. We shall need to think carefully about the measure. We are told that we can always write it in the form

$$d\sigma_\nu = n_\nu d\sigma \,, \tag{11.7}$$

where n_ν is a covector field and $d\sigma$ is an invariant measure. However, there seems to be a problem with the 'invariant' measure, discussed further below, and the picture described here would appear to miss a key point when it claims that (11.6) is manifestly covariant. These issues will be the subject of the present chapter.

For the moment, note that in the stationary frame,

$$n_\mu = (1,0,0,0) , \tag{11.8}$$

in order to obtain (11.2). Under the Lorentz boost, this becomes

$$n_\mu = v_\mu/c = \gamma(1, -\mathbf{v}/c) , \quad \text{where} \quad \mathbf{v} = \frac{d\mathbf{r}}{dt} . \tag{11.9}$$

In other words, apart from the factor of c, n^μ has to be the 4-velocity, which is constant over the spacetime.

As an aside, note here that we are integrating a vector-valued 3-form over a three-dimensional hypersurface to obtain a vector field. This is not so easy to achieve in general relativity. If it is possible here, it is because our transformations (the Lorentz transformations) are limited to transformations that are linear, i.e., constant over the whole spacetime. This will also be discussed later (see Sect. 11.2.2).

Let us now examine Rohrlich's proposed formulas for the field energy and momentum when the electron is moving with constant 4-velocity v^μ. According to (11.6),

$$p^\mu = -\frac{1}{c^2} \int T^{\mu\nu} v_\nu d\sigma , \tag{11.10}$$

whence

$$W = cp^0 = -\frac{1}{c} \int T^{0\nu} v_\nu d\sigma$$
$$= \gamma \int u d\sigma - \frac{\gamma}{c^2} \int \mathbf{S} \cdot \mathbf{v} d\sigma , \tag{11.11}$$

and

$$p^k = -\frac{1}{c^2} \int T^{k\nu} v_\nu d\sigma = -\frac{1}{c^2} \int T^{k0} v_0 d\sigma - \frac{1}{c^2} \int T^{kj} v_j d\sigma , \tag{11.12}$$

whence

$$\mathbf{p} = \frac{\gamma}{c^2} \int \mathbf{S} d\sigma + \frac{\gamma}{c^2} \int \mathbf{T} \cdot \mathbf{v} d\sigma , \tag{11.13}$$

where T is defined by (2.34) on p. 12. This differs from Rohrlich's expressions by some factors of c, so let us check the dimensions.

Note from (2.33) that our energy–momentum tensor is dimensionally homogeneous, i.e., all components have dimensions of $[\varepsilon_0][\mathbf{E}^2]$, since $[\mathbf{E}] = [c\mathbf{B}]$. This has to be the case for the definition (11.10) to be physically possible, because v_μ is dimensionally homogeneous. We also know that $\varepsilon_0 E^2$ has dimensions of an energy density, viz., $ML^{-1}T^{-2}$, hence $[u] = ML^{-1}T^{-2}$. The vector $\mathbf{S}/c$ has the same dimensions as u, so $[\mathbf{S}] = MT^{-3}$, whence it is $\mathbf{S}/c^2$ that has the dimensions of a momentum density. The first term in (11.11) has dimensions of energy and the second term has dimensions

$$\frac{T^2}{L^2} MT^{-3} \frac{L}{T} L^3 = ML^2 T^{-2} ,$$

the dimensions of energy again. Likewise, the components of the 3×3 matrix T have dimensions of energy density and we can soon check that the two terms of (11.13) have the same dimensions.

There do remain two important questions: what is the 'invariant' measure and what spacelike hypersurface do we integrate over? It may seem odd to ask the second question here. The spacelike hypersurface must be the hyperplane of simultaneity (HOS) of the observer who sees the charge shell as moving. The reader should be warned, however, that the two papers [21,22] are not clear at all about this issue. For example, in [21], Rohrlich suggests that the problems are caused by the surface of the shell appearing to be ellipsoidal for a relatively moving observer, then states the related but irrelevant fact that the worldline of its center is not relativistically orthogonal to the HOS of that observer, claiming that Lorentz covariance requires this orthogonality. Elsewhere [22], he appears to claim that, with his redefinition, one is in fact always integrating over a HOS in the rest frame of the shell.

Worse, regarding the measure, the two papers [21] and [22] seem to contradict one another. The first would have it that $\gamma d\sigma = d^3 x$, which would appear to be untenable, while the second seems to revert to the more likely $d\sigma = \gamma d^3 x$. The level of vagueness is such, however, that the author's real intentions remain open to discussion.

Let us suppose for a moment that we can put $\gamma d\sigma = d^3 x$, as claimed by Rohrlich in [21], without justification. Then according to this theory, the correct formulas for the energy and momentum in the fields are

$$W = cp^0 = \int u d^3 x - \frac{1}{c^2} \int \mathbf{S} \cdot \mathbf{v} d^3 x , \qquad (11.14)$$

and

$$\mathbf{p} = \frac{1}{c^2} \int \mathbf{S} d^3 x + \frac{1}{c^2} \int \mathsf{T} \cdot \mathbf{v} d^3 x , \qquad (11.15)$$

where the integration is taken over a spacelike hypersurface $t = $ constant for the observer who sees the electron as moving. Naturally, this observer integrates over the region of that hypersurface outside the ellipsoid occupied by the charge. Each formula includes an extra term compared with (11.1).

In the non-relativistic limit $v \ll c$ ($\gamma \approx 1$), it turns out that the second term in (11.14) can be neglected so that this calculation would have worked out correctly if we had done it. However, the second term in (11.15) makes a significant contribution to the momentum calculations, even in the non-relativistic limit. In any case, it is immediately obvious that

$$W = cp^0 \neq \int u \mathrm{d}^3 x , \tag{11.16}$$

and also

$$\mathbf{p} \neq \frac{1}{c^2} \int \mathbf{S} \mathrm{d}^3 x , \tag{11.17}$$

where $\mathrm{d}^3 x$ is the usual volume measure on a spacelike hypersurface $t = $ constant for the observer who sees the electron as moving. In this view the formulas we proposed in (11.1) are not correct precisely because the field energy and momentum have to be calculated by *integrating* a second rank tensor. This was a mistake that arose purely through not observing the rules of relativistic covariance.

In his paper [21], Rohrlich goes on to show that, in the non-relativistic approximation $v \ll c$,

$$\frac{1}{c^2} \int \varepsilon_0 \mathsf{T} \cdot \mathbf{v} \mathrm{d}^3 x = -\frac{e^2 \mathbf{v}}{6ac^2} = -\frac{1}{3} m_{\mathrm{EM}}^{\mathrm{EDM}} \mathbf{v} , \tag{11.18}$$

and since we know that

$$\frac{1}{c^2} \int \mathbf{S} \mathrm{d}^3 x = \frac{4}{3} m_{\mathrm{EM}}^{\mathrm{EDM}} \mathbf{v} ,$$

the final result is the desired

$$\mathbf{p} = m_{\mathrm{EM}}^{\mathrm{EDM}} \mathbf{v} .$$

Naturally, the final four-momentum p^μ is the right Lorentz transformation of the one in (11.3), if Rohrlich's construction via (11.6) really is Lorentz covariant.

Rohrlich's Correction for the Spherical Electron

For the record, let us include this step. For nonrelativistic motion $v \ll c$, we have

$$\mathbf{E} \approx e \frac{\mathbf{r}}{r^3} , \qquad \mathbf{B} \approx \frac{\mathbf{v} \times \mathbf{E}}{c^2} . \tag{11.19}$$

This is the usual approximation to the full Lienard–Wiechert expression for the fields due to a moving spherically symmetric charge.

Let us first check that the correction to p^0 is negligible. We have

$$\mathbf{S} = \varepsilon_0 c^2 \mathbf{E} \times \mathbf{B}$$
$$= \varepsilon_0 c^2 e \frac{\mathbf{r}}{r^3} \times \frac{e}{c^2} \left(\mathbf{v} \times \frac{\mathbf{r}}{r^3} \right)$$
$$= \frac{\varepsilon_0 e^2}{r^6} \mathbf{r} \times (\mathbf{v} \times \mathbf{r})$$
$$= \frac{\varepsilon_0 e^2}{r^6} \left[r^2 \mathbf{v} - (\mathbf{r} \cdot \mathbf{v}) \mathbf{r} \right] \ ,$$

which is clearly $O(v)$. The correction term to p^0 is

$$\frac{1}{c^2} \int \mathbf{S} \cdot \mathbf{v} \mathrm{d}^3 x \sim \frac{v^2}{c^2} \ , \tag{11.20}$$

which we can neglect.

Now the correction to $\mathbf{p}$ is

$$\frac{1}{c^2} \int \mathsf{T} \cdot \mathbf{v} \mathrm{d}^3 x = \frac{\varepsilon_0}{c^2} \int \left[\mathbf{E}(\mathbf{E} \cdot \mathbf{v}) + c^2 \mathbf{B}(\mathbf{B} \cdot \mathbf{v}) - \frac{1}{2} \mathbf{v}(E^2 + c^2 B^2) \right] \mathrm{d}^3 x$$
$$\approx \frac{\varepsilon_0}{c^2} \int \left[\mathbf{E}(\mathbf{E} \cdot \mathbf{v}) - \frac{1}{2} \mathbf{v} E^2 \right] \mathrm{d}^3 x$$
$$= \frac{\varepsilon_0 e^2}{c^2} \int_{r=a}^{\infty} \left[\frac{\mathbf{r}(\mathbf{r} \cdot \mathbf{v})}{r^6} - \frac{\mathbf{v}}{2r^4} \right] \mathrm{d}^3 x \ ,$$

where we dropped the term $\mathbf{B}(\mathbf{B} \cdot \mathbf{v})$ because it is zero, and we have also dropped the term in B^2 because it is $O(v^2)$. We choose $\mathbf{v} = (v, 0, 0)$. We thus have

$$\frac{1}{c^2} \int \mathsf{T} \cdot \mathbf{v} \mathrm{d}^3 x = \frac{\varepsilon_0 e^2 v}{c^2} \int_{r=a}^{\infty} \left[\frac{x\mathbf{r}}{r^6} - \frac{1}{2r^4} \begin{pmatrix} 1 \\ 0 \\ 0 \end{pmatrix} \right] \mathrm{d}^3 x$$
$$= \frac{\varepsilon_0 e^2 v}{c^2} \int_{r=a}^{\infty} \int_{\theta=0}^{\pi} \int_{\phi=0}^{2\pi} \left[\frac{x\mathbf{r}}{r^4} - \frac{1}{2r^2} \begin{pmatrix} 1 \\ 0 \\ 0 \end{pmatrix} \right] \sin\theta \, \mathrm{d}\theta \mathrm{d}\phi \mathrm{d}r \ .$$

The second term is

$$-\frac{\varepsilon_0 e^2 v}{c^2} \int_{r=a}^{\infty} \int_{\theta=0}^{\pi} \int_{\phi=0}^{2\pi} \frac{\sin\theta}{2r^2} \begin{pmatrix} 1 \\ 0 \\ 0 \end{pmatrix} \mathrm{d}\theta \mathrm{d}\phi \mathrm{d}r = -\frac{\varepsilon_0 e^2 \mathbf{v}}{2c^2} \int_{r=a}^{\infty} \frac{2\pi}{r^2} \left[-\cos\theta \right]_0^{\pi} \mathrm{d}r$$
$$= -\frac{2\pi \varepsilon_0 e^2 \mathbf{v}}{ac^2} \ .$$

Concerning the other term, note that there is only an x component. The other two components are zero by symmetry. The correction to $\mathbf{p}$ is thus in the direction of $\mathbf{v}$. We have to evaluate

$$\frac{\varepsilon_0 e^2 \mathbf{v}}{c^2} \int_{r=a}^{\infty} \frac{x^2}{r^6} \mathrm{d}^3 x = \frac{\varepsilon_0 e^2 \mathbf{v}}{c^2} \int_{r=a}^{\infty} \int_{\theta=0}^{\pi} \int_{\phi=0}^{2\pi} \frac{x^2}{r^4} \sin\theta \mathrm{d}\theta \mathrm{d}\phi \mathrm{d}r$$

$$= \frac{\varepsilon_0 e^2 \mathbf{v}}{c^2} \int_{r=a}^{\infty} \int_{\theta=0}^{\pi} \int_{\phi=0}^{2\pi} \frac{\cos^2\theta \sin\theta}{r^2} \mathrm{d}\theta \mathrm{d}\phi \mathrm{d}r$$

$$= \frac{\varepsilon_0 e^2 \mathbf{v}}{c^2 a} \left[-\frac{1}{3} \cos^3\theta \right]_0^{\pi} 2\pi$$

$$= \frac{4\pi\varepsilon_0 e^2 \mathbf{v}}{3c^2 a} .$$

Finally,

$$\frac{1}{c^2} \int \mathsf{T} \cdot \mathbf{v} \mathrm{d}^3 x = -\frac{2\pi\varepsilon_0 e^2 \mathbf{v}}{3c^2 a} . \tag{11.21}$$

To obtain Rohrlich's result, we put $\varepsilon_0 \to 1/4\pi$, which leads to

$$\frac{1}{c^2} \int \mathsf{T} \cdot \mathbf{v} \mathrm{d}^3 x = -\frac{e^2 \mathbf{v}}{6c^2 a} \qquad \text{(Rohrlich)} . \tag{11.22}$$

Rohrlich's formula does therefore deliver the right correction, as he claims, at least for the case of a spherical charge distribution and in the non-relativistic approximation.

One may also check that the right correction is obtained in the non-relativistic approximation for the charge dumbbell, when it moves along its axis, for which there was a discrepancy (see Sect. 5.2), and when it moves normally to its axis, for which there was no discrepancy (see Sect. 5.3), and indeed Rohrlich's prescription leads to no adjustment.

This may look like a convincing argument in favour of this approach, but it should be remembered that this kind of redefinition is made explicitly to render the result covariant under Lorentz transformation. Indeed, we shall see a general proof later, inspired by [22], that a prescription of this kind does deliver a Lorentz covariant four-momentum for the EM fields around the charge shell moving at uniform velocity, in the fully relativistic case where we do not assume $v \ll c$. But once again, the reader should be warned: in the later paper [22], the 'invariant' measure would appear to have become $\mathrm{d}\sigma = \gamma \mathrm{d}^3 x$ rather than $\gamma \mathrm{d}\sigma = \mathrm{d}^3 x$, while the author claims there to be integrating over a HOS in the rest frame of the shell, rather than a HOS in a frame relative to which the shell is moving.

11.2 Measure and Integration Space

So we shall have more to say later about what is going on in the Rohrlich construction, but for the moment, let us just consider where a relation like $\gamma \mathrm{d}\sigma = \mathrm{d}^3 x$ might come from. Why should this be so? We said earlier that $\mathrm{d}\sigma$ had to be an invariant measure. However, when we change inertial frame, the new $d\sigma$ is a measure on a different spacelike hypersurface in the Minkowski spacetime! To understand the relation $\gamma d\sigma = d^3x$, we would have to carry out a Lorentz transformation of our quantity on the right-hand side of (11.10) to see why this works.

However, we shall adopt a slightly different road, first of all to expose exactly what leads to the discrepancy between energy-derived and momentum-derived EM masses. To set the scene, let us reconsider what happens when we define charge and 4-momentum from a current density 4-vector and an energy–momentum tensor that are conserved everywhere. This will remind us of what is usually done in a simpler situation.

11.2.1 Charge for a Conserved Current Density

We only consider flat spacetime here. The point of considering a current density is only to illustrate the kind of problem that is involved in defining something from a conserved quantity.

Let J^μ be a conserved current density so that $J^\mu_{\ ,\mu} = 0$. The usual definition of the total charge is

$$Q := \int_{x^0=0} J^0 d^3x \,, \tag{11.23}$$

integrating over the whole of the hyperplane of simultaneity (HOS) $x^0 = 0$ in our chosen frame. Strictly speaking, there are several things to show here:

- Q is independent of time in the sense that it does not depend on the choice $x^0 = 0$. In other words, if we define

$$Q(t) := \int_{x^0=ct} J^0 d^3x \,, \tag{11.24}$$

  for any value of t, it turns out that Q is actually independent of t.
- Q is a Lorentz scalar in the sense that, if we define

$$Q' := \int_{x'^0=0} J'^0 d^3x' \,, \tag{11.25}$$

  in some other inertial frame, then it turns out that $Q' = Q$.

In fact both of these results are proven using the fact that the charge is conserved.

Conservation of Charge

No sophisticated techniques are required here. We note that

$$J^0 d^3 x = J^\mu d\sigma_\mu , \quad \text{where} \quad d\sigma_\mu = (d^3 x, 0, 0, 0) . \tag{11.26}$$

In fact, we have written $d\sigma_\mu$ in the form $n_\mu d\sigma$, where $d\sigma = d^3 x$ and n_μ is a normal form with respect to the hypersurface of integration. We shall achieve this in a more sophisticated way shortly. Now we have, for $t_1 < t_2$,

$$
\begin{aligned}
Q(t_2) - Q(t_1) &= \int_{x^0 = ct_2} J^0 d^3 x - \int_{x^0 = ct_1} J^0 d^3 x \\
&= \int_{\text{surface}} J^\mu d\sigma_\mu \\
&= \int_{\text{volume}} J^\mu{}_{,\mu} d\tau \\
&= 0 ,
\end{aligned}
$$

where the surface is the surface of the whole slab of spacetime between $x^0 = ct_1$ and $x^0 = ct_2$ and the volume is the volume of that same slab, with the usual measure on Minkowski spacetime, viz., $d\tau = d^4 x$. We have just applied Gauss' theorem. This does require us to make sure that we are in the conditions of Gauss' theorem. We also assume that the surface integrals at large spatial distances tend to zero.

We can show the same result in the form of a differentiation:

$$
\begin{aligned}
\frac{dQ}{dt} &= \lim_{\delta t \to 0} \frac{1}{\delta t} \left[\int_{x^0 = c(t + \delta t)} J^\mu d\sigma_\mu - \int_{x^0 = ct} J^\mu d\sigma_\mu \right] \\
&= \lim_{\delta t \to 0} \frac{1}{\delta t} \left[\int_{\text{slab surface}} J^\mu d\sigma_\mu \right] \\
&= \lim_{\delta t \to 0} \frac{1}{\delta t} \left[\int_{\text{slab volume}} J^\mu{}_{,\mu} d\tau \right] \\
&= 0 .
\end{aligned}
$$

Naturally, this is in essence the same as the first argument.

One sometimes sees

$$\frac{dQ}{dt} = \int J^0{}_{,0} d^3 x . \tag{11.27}$$

This hides the fact that Q depends on time through the choice of spacelike hypersurface, which has not clearly become part of the differentiation process here. In actual fact, the above statement means something like this:

$$\partial_t Q = \lim_{\delta t \to 0} \frac{1}{\delta t} \left[\int_{t+\delta t} J^0(t+\delta t, x) \mathrm{d}^3 x - \int_t J^0(t, x) \mathrm{d}^3 x \right]$$

$$= \lim_{\delta t \to 0} \frac{1}{\delta t} \int_t \left[J^0(t+\delta t, x) - J^0(t, x) \right] \mathrm{d}^3 x \, .$$

Presumably this is a valid step. The rest of the argument is more convincing. We have $J^0{}_{,0} = -J^i{}_{,i}$, where i is summed over $\{1,2,3\}$. This is a divergence in the spacelike hypersurface. We now use Gauss' theorem in 3D to convert the other way, not from a surface integral to a volume integral, but from a volume integral (in 3D) to a surface integral at infinity, which we assume to be zero.

Lorentz Invariance of Charge

The approach here is essentially the same, but we can and will be a little more sophisticated, in preparation for the problem at hand. We now consider the difference

$$Q' - Q = \int_{x'^0=0} J'^0 \mathrm{d}^3 x' - \int_{x^0=0} J^0 \mathrm{d}^3 x \, . \tag{11.28}$$

What we would like is to write this in the form

$$Q' - Q = \int_{\text{primed HOS}} J'^\mu \mathrm{d}\sigma'_\mu - \int_{\text{unprimed HOS}} J^\mu \mathrm{d}\sigma_\mu \, , \tag{11.29}$$

then say that $J'^\mu \mathrm{d}\sigma'_\mu = J^\mu \mathrm{d}\sigma_\mu$ because this is 'manifestly' a scalar. Thereafter, we include integrations over surfaces at infinity to close our surface integrations around two wedges of spacetime between the primed HOS and the unprimed HOS. These convert to two 4-volume integrals over the interiors of the wedges using Gauss' theorem, and the integrand is of course $J^\mu{}_{,\mu}$, which we assume at the outset to be zero.

The crucial thing here is to see why we should have $J'^\mu \mathrm{d}\sigma'_\mu = J^\mu \mathrm{d}\sigma_\mu$ and understand how Gauss' theorem works in Minkowski spacetime. The following discussion is based on the kind of integration theory to be found in standard textbooks on manifold theory, such as [13, pp. 26–31] or [25, pp. 47–50].

What is the correct definition of our integrals? As explained in [25], we can integrate a vector field J^a over a 3D submanifold $\mathscr{S}$ of spacetime by contracting it with the canonical 4-form

$$\eta_{abcd} := |g|^{1/2} \varepsilon_{abcd} = \varepsilon_{abcd} \, , \tag{11.30}$$

where we have used the fact that the metric is $g_{ij} = \eta_{ij}$ with $|g| = 1$. Here ε_{abcd} is the totally antisymmetric symbol with $\varepsilon_{0123} = +1$. We still have to define the integral

$$Q := \int_{\mathscr{S}} J^a \eta_{abcd} \, . \tag{11.31}$$

As explained in [13], this involves parametrising $\mathscr{S}$ with some map

$$S : \mathbb{R}^3 \hookrightarrow \mathbb{R}^4 \quad \text{with} \quad S(\mathbb{R}^3) = \mathscr{S} . \tag{11.32}$$

In the present case, this is very easy to do for the unprimed HOS $x^0 = 0$. An appropriate map is

$$S(y^1, y^2, y^3) = (0, y^1, y^2, y^3) . \tag{11.33}$$

For a different hyperplane of simultaneity $x^0 = ct_1$, we merely put

$$S(y^1, y^2, y^3) = (ct_1, y^1, y^2, y^3) , \tag{11.34}$$

whereupon the following argument is unaffected.

The next task is to pull back the 3-form on spacetime to a 3-form on the parametrisation space ($\mathbb{R}^3$ in this case). Now our 3-form on spacetime is

$$\frac{1}{3!} J^a \varepsilon_{abcd} \mathrm{d}x^b \wedge \mathrm{d}x^c \wedge \mathrm{d}x^d ,$$

with pullback

$$S^* \left(\frac{1}{3!} J^a \varepsilon_{abcd} \mathrm{d}x^b \wedge \mathrm{d}x^c \wedge \mathrm{d}x^d \right) = \frac{1}{3!} J^a \varepsilon_{abcd} S^*(\mathrm{d}x^b) \wedge S^*(\mathrm{d}x^c) \wedge S^*(\mathrm{d}x^d) , \tag{11.35}$$

using a well-known result concerning the pullback of an exterior product [13, Chap. 2]. It is not difficult to show that

$$S^*(\mathrm{d}x^0) = 0 , \qquad S^*(\mathrm{d}x^i) = \mathrm{d}y^i , \quad i = 1, 2, 3 . \tag{11.36}$$

This is done by turning to the definitions of the pullback and tangent map on manifolds.

We are thus integrating

$$\frac{1}{3!} J^a \varepsilon_{abcd} S^*(\mathrm{d}x^b) \wedge S^*(\mathrm{d}x^c) \wedge S^*(\mathrm{d}x^d) = \frac{1}{3!} J^0 \varepsilon_{0ijk} \mathrm{d}y^i \wedge \mathrm{d}y^j \wedge \mathrm{d}y^k$$
$$= J^0 \mathrm{d}y^1 \wedge \mathrm{d}y^2 \wedge \mathrm{d}y^3 ,$$

whence

$$Q := \int_{\mathscr{S}} J^a \eta_{abcd} = \int_{\mathbb{R}^3} J^0 \mathrm{d}^3 x . \tag{11.37}$$

All this merely shows that our original definition of Q is indeed the same as the sophisticated integral definition! Note that we do indeed get something with the form $J^\mu n_\mu \mathrm{d}\sigma$, where n_μ is a normal form with respect to the surface in question, as claimed. We can now fully justify the above proof that $Q(t)$ is in fact independent

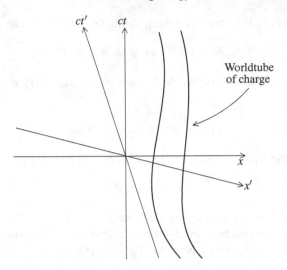

Fig. 11.1 Two coordinate systems for specifying how much charge there is

of t, because we now have the precise context in which Gauss' theorem is valid, as explained in [25].

At last we come to a more challenging case. We consider a Lorentz transformation to primed coordinates. In fact we shall consider a primed observer moving to the left along the x axis at speed v. In the unprimed picture, the primed time axis is sloping up to the left and the primed HOS $x'^0 = 0$ is sloping down to the right, below the unprimed HOS (see Fig. 11.1). This observer sees the unprimed system moving off to the right. The problem is to get a more precise understanding of the expression

$$Q' = \int_{x'^0=0} J'^0 \mathrm{d}^3 x' . \tag{11.38}$$

For reference, we have

$$\begin{cases} x^0 = \gamma\left(x'^0 - \dfrac{v}{c}x'^1\right) , \\ x^1 = \gamma\left(x'^1 - \dfrac{v}{c}x'^0\right) , \end{cases} \qquad \begin{cases} x'^0 = \gamma\left(x^0 + \dfrac{v}{c}x^1\right) , \\ x'^1 = \gamma\left(x^1 + \dfrac{v}{c}x^0\right) . \end{cases} \tag{11.39}$$

Let us view everything in the unprimed system. Then the primed HOS $x'^0 = 0$ is given by the equation $x^0 = -vx^1/c$. Let us denote it by $\mathcal{R}$. We hope to integrate over this hypersurface. This means seeking an injection

$$R : \mathbb{R}^3 \hookrightarrow \mathbb{R}^4 \quad \text{with} \quad R(\mathbb{R}^3) = \mathcal{R} . \tag{11.40}$$

We define

$$R : (y^1, y^2, y^3) \longmapsto \left(-\frac{\gamma v}{c} y^1, \gamma y^1, y^2, y^3\right) . \tag{11.41}$$

Now what should we integrate? Let us try evaluating

$$Q' := \int_{\mathscr{R}} J'^a \eta'_{abcd} . \tag{11.42}$$

By the perfect parallel with what was shown just now, we expect to find

$$\int_{\mathscr{R}} J'^a \eta'_{abcd} = \int_{\mathscr{R}} J'^0 d^3 x' . \tag{11.43}$$

Let us see how this works.
  We have

$$\eta'_{abcd} = |\eta'|^{1/2} \varepsilon_{abcd} = \varepsilon_{abcd} . \tag{11.44}$$

The volume element is unchanged under (proper, homogeneous) Lorentz transformations. Note that ε_{abcd} is itself invariant under these transformations. In fact, the 3-form we intend to integrate over $\mathscr{R}$ is just

$$\frac{1}{3!} J'^a \varepsilon'_{abcd} dx'^b \wedge dx'^c \wedge dx'^d = \frac{1}{3!} J^a \varepsilon_{abcd} dx^b \wedge dx^c \wedge dx^d . \tag{11.45}$$

This is very important, in fact. We are integrating the same 3-form as we integrated over the unprimed HOS earlier. This has to be so if we are to have any chance of applying Gauss' theorem. It is also in this sense that $J'^\mu d\sigma'_\mu = J^\mu d\sigma_\mu$.
  We now pull back the 3-form to $\mathbb{R}^3$ with the result

$$R^* \left(\frac{1}{3!} J^a \varepsilon_{abcd} dx^b \wedge dx^c \wedge dx^d\right) = \frac{1}{3!} J^a \varepsilon_{abcd} R^*(dx^b) \wedge R^*(dx^c) \wedge R^*(dx^d) . \tag{11.46}$$

It is not difficult to show that

$$R^*(dx^0) = -\frac{\gamma v}{c} dy^1 , \qquad R^*(dx^1) = \gamma dy^1 , \tag{11.47}$$

$$R^*(dx^2) = dy^2 , \qquad R^*(dx^3) = dy^3 . \tag{11.48}$$

We can now rewrite the pullback of the 3-form we hope to integrate. On the right-hand side of (11.46), dx^0 and dx^1 both pull back to multiples of dy^1, and the wedge products of their pullbacks would therefore be zero. We drop terms in the sum with $\{0, 1\} \subset \{b, c, d\}$. The set $\{b, c, d\}$ must therefore be either $\{0, 2, 3\}$ or $\{1, 2, 3\}$. We then sum over all permutations of its three elements. The first sum goes with $J^1 \varepsilon_{1bcd}$, whilst the second goes with $J^0 \varepsilon_{0bcd}$. The result is

$$\frac{1}{3!}J^a \varepsilon_{abcd}R^*(\mathrm{d}x^b) \wedge R^*(\mathrm{d}x^c) \wedge R^*(\mathrm{d}x^d)$$

$$= \gamma J^0 \mathrm{d}y^1 \wedge \mathrm{d}y^2 \wedge \mathrm{d}y^3 + \frac{\gamma v}{c}J^1 \mathrm{d}y^1 \wedge \mathrm{d}y^2 \wedge \mathrm{d}y^3$$

$$= \gamma\left(J^0 + \frac{v}{c}J^1\right)\mathrm{d}y^1 \wedge \mathrm{d}y^2 \wedge \mathrm{d}y^3 . \qquad (11.49)$$

We now have the integral in the form

$$Q' := \int_{\mathscr{R}} J'^a \eta'_{abcd} = \int_{\mathbb{R}^3} \gamma\left(J^0 + \frac{v}{c}J^1\right)\mathrm{d}^3 y . \qquad (11.50)$$

In the integrand, the quantities J^0 and J^1 are

$$J^0 = J^0\left(-\frac{\gamma v}{c}y^1, \gamma y^1, y^2, y^3\right) , \qquad J^1 = J^1\left(-\frac{\gamma v}{c}y^1, \gamma y^1, y^2, y^3\right) . \qquad (11.51)$$

Now by comparison with (11.39), we note that

$$\gamma\left(J^0 + \frac{v}{c}J^1\right) = J'^0 . \qquad (11.52)$$

It is in this sense that

$$Q' := \int_{\mathscr{R}} J'^0 \mathrm{d}^3 x' , \qquad (11.53)$$

since we can equate $\mathrm{d}^3 y$ with $\mathrm{d}^3 x'$ and we evaluate J'^0 at points (x^0, x^1, x^2, x^3) with $x^0 = -\gamma v y^1/c$ and $x^1 = \gamma y^1$, which ensures that $x'^0 = 0$.

What is the normal 1-form to the hypersurface $x'^0 = 0$ (which we called $\mathscr{R}$ above)? In the unprimed system it is

$$n_\mu = \gamma(1, v/c, 0, 0) . \qquad (11.54)$$

We have to check that $n(e) = 0$ for all vectors e lying in the hypersurface. This is clear when e is in the space spanned by $\partial/\partial x^2$ and $\partial/\partial x^3$. If e is in the space spanned by $\partial/\partial x^0$ and $\partial/\partial x^1$, it must be tangent to a curve

$$\tau(s) = \left(-\frac{v}{c}s, s, 0, 0\right) .$$

It thus has the form

$$e = \left.\frac{\partial}{\partial s}\right|_\tau = -\frac{v}{c}\frac{\partial}{\partial x^0} + \frac{\partial}{\partial x^1} .$$

But then

$$n(e) = \gamma\left(-\frac{v}{c} + \frac{v}{c}\right) = 0 .$$

This is the proof that n is the normal form to the hypersurface. Note that it is normalised and $n_\mu = u_\mu/c$, where $u^\mu = \gamma(c, -v, 0, 0)$ is the time axis of the primed frame as viewed from the unprimed frame, i.e., the 4-velocity of the primed frame as viewed from the unprimed frame. It is quite clear that this must be normal to the hypersurface $x'^0 = 0$.

We now observe that our integrand is

$$\gamma\left(J^0 + \frac{v}{c}J^1\right) = J^\mu n_\mu . \tag{11.55}$$

This too is obvious because the right-hand side is a scalar, hence equal to $J'^\mu n'_\mu$, and $n'_\mu = (1, 0, 0, 0)$. Fortunately, everything fits together. We can write

$$Q' = \int_\mathscr{R} J^\mu n_\mu \mathrm{d}\sigma , \tag{11.56}$$

where $\mathrm{d}\sigma$ refers to the measure d^3x' on $\mathscr{R}$. It is important to note that the relevant 4-velocity is not the 4-velocity v^μ of the unprimed frame, but the 4-velocity u^μ of the primed frame. After all, it is this that is normal to the hypersurface $x'^0 = 0$ over which the integration is carried out.

We shall soon extend this analysis to a 4-momentum vector defined from an energy–momentum tensor and we shall find the same results. One should be a little suspicious of the Rohrlich method, which seems to operate too naively when it comes to changing the hypersurface of integration. It is not so clear why Rohrlich's recipe works. However, it is crucial to know whether it is just a recipe or a natural definition, because a recipe would be inadequate here. If we want a Lorentz covariant 4-momentum for the fields outside a spherical charge distribution, we can just Lorentz boost the simple 4-momentum obtained for the stationary distribution. But this would prove nothing. However, Rohrlich's method does deliver precisely the same result as that tactic. Later we shall see exactly why it achieves its aim so perfectly. On the other hand, using the above type of integration theory, we shall demonstrate very clearly why the integrations do not, and should not, give a Lorentz covariant result for the spherical electron.

Returning briefly to the training ground provided by the example in this section, we can now prove that the charge as defined by (11.25) is Lorentz invariant. The proof is:

$$\begin{aligned}
Q' - Q &= \int_{x'^0=0} J'^0 \mathrm{d}^3x' - \int_{x^0=0} J^0 \mathrm{d}^3x \\
&= \int_{x'^0=0} J^a \eta_{abcd} - \int_{x^0=0} J^a \eta_{abcd} \\
&= \int_{\text{wedge surface}} J^a \eta_{abcd} \\
&= \int_{\text{wedge volume}} J^a_{,a} \mathrm{d}\tau \\
&= 0 .
\end{aligned} \tag{11.57}$$

It is the same proof as before, but it now clarifies the application of Gauss' theorem as described in [25].

It is worth making a last note here, concerning the case of a charge distribution that is perfectly stationary in the unprimed frame. Hence,

$$J^\mu = (c\rho, 0, 0, 0) \ . \tag{11.58}$$

In a moving frame where the charge appears to be moving,

$$J'^\mu = \gamma(c\rho, \mathbf{v}\rho) = \rho v'^\mu \ , \tag{11.59}$$

where v'^μ is the 4-velocity of the charge in this frame. The new prescription for the charge is

$$Q' = \frac{1}{c} \int_{t'=0} J'^0 \, d^3x' = \frac{1}{c} \int_{t'=0} \gamma c\rho \, d^3x' \ . \tag{11.60}$$

Now let us visualise ρ as a variable function of (x^1, x^2, x^3) for fixed x^0 in the sense that it is constant over some finite region and zero outside that region. Let us also imagine that it is constant in the rest frame time at any point (x^1, x^2, x^3). Recalling the parametrisation (11.41) of the $x'^0 = 0$ hypersurface, we can in this case write

$$\frac{1}{c} \int_{t'=0} \gamma c\rho \, d^3x' = \int_{t=0} \rho \, d^3x \ . \tag{11.61}$$

There are special circumstances here for changing the integral over the primed hypersurface to one over the unprimed hypersurface without appealing to Gauss' theorem. We cannot normally do this. It was essential that ρ was constant in the way described. The best we can do in our general case is

$$\int_{x'^0=0} J'^a \eta'_{abcd} = \int_{\mathbb{R}^3} \gamma \left[J^0 \left(-\frac{\gamma v}{c} y^1, \gamma y^1, y^2, y^3 \right) + \frac{v}{c} J^1 \left(-\frac{\gamma v}{c} y^1, \gamma y^1, y^2, y^3 \right) \right] d^3 y$$

$$= \int_{\mathbb{R}^3} \left[J^0 \left(-\frac{v}{c} y^1, y^1, y^2, y^3 \right) + \frac{v}{c} J^1 \left(-\frac{v}{c} y^1, y^1, y^2, y^3 \right) \right] d^3 y \ ,$$

where we have made a change of variable in the last step. It is quite clear why we cannot proceed further: J^0 and J^1 can vary with the value of their first argument. Only Gauss' theorem can save us. However, when $J^\mu = (c\rho, 0, 0, 0)$ with ρ constant in time, the last displayed formula does indeed reduce to

$$\int_{x'^0=0} J'^a \eta'_{abcd} = \int_{\mathbb{R}^3} c\rho(0, y^1, y^2, y^3) d^3 y \ . \tag{11.62}$$

The change of variable that removes the γ factor is now clear.

It is in fact the above situation that justifies Rohrlich's definition in the end, in the sense that his definition gives a Lorentz covariant quantity. Indeed, the energy–momentum tensor in the rest frame of the electron is constant in the rest frame time. However, we shall have more to say later about what Rohrlich's definition really achieves.

11.2.2 Four-Momentum for a Conserved Energy–Momentum Tensor

We shall now apply the above integration theory to the definition

$$P^\alpha = \int_{x^0=0} T^{\alpha 0} d^3 x \,, \tag{11.63}$$

where $T^{\alpha\beta}$ is conserved everywhere, i.e.,

$$T^{\alpha\beta}{}_{,\beta} = 0 \,. \tag{11.64}$$

We would like to show something like

$$\int_{x^0=0} T^{\alpha 0} d^3 x = \int_{x^0=0} T^{ab} \eta_{bcde} \,. \tag{11.65}$$

The object $T^{ab} \eta_{bcde}$ is no longer a 3-form but can be considered as a vector-valued 3-form in the context of special relativity, where we make only linear coordinate changes.

We have to be sure that Gauss' theorem applies in the obvious way. In general relativity or when using curvilinear coordinates in a flat spacetime, the connection coefficients are not generally zero, and all partial derivatives with respect to coordinates have to be replaced by covariant derivatives, so the situation becomes much more complex. For example, it is the covariant derivative that comes into Gauss' theorem and this would mess things up because $T^{\alpha\beta}{}_{;\beta}$ is not like $J^\beta{}_{;\beta}$. Indeed, the former contains two terms involving the connection coefficients and the latter only one. However, it is only $T^{\alpha\beta}{}_{,\beta}$ that turns up here and we can envisage the following ploy to establish the necessary results. (We only mention it here and then drop it, as an unnecessary sophistication.)

We can contract with a one-form field of our choice W to apply the integration results. For example, we can take W consecutively to be $W^{(0)} = dx^0$, then each of $W^{(i)} = dx^i$, for $i = 1, 2, 3$. This gives back precisely the case studied above in four cases:

$$J^{(0)b} = W_a^{(0)} T^{ab} \,, \qquad J^{(i)b} = W_a^{(i)} T^{ab} \,.$$

With a choice like these $W^{(\mu)}$, we have

$$W_{i,j}^{(\mu)} = 0 \,, \quad \forall i, j, \mu \,.$$

Note in passing the extra complexity in a curved spacetime or when using curvilinear coordinates in a flat spacetime, where the simple coordinate derivative gets replaced by a covariant derivative, and even the constancy of the components of a one-form would not guarantee the constancy of the components of its covariant derivative. In the present flat spacetime context, with Minkowski coordinates, we can define the corresponding component of the 4-momentum for each vector field $J^{(\mu)}$, viz.,

$$P^\mu = \int_{x^0=0} J^{(\mu)0} d^3x \,, \tag{11.66}$$

and

$$J^{(\mu)i}{}_{,i} = \left[W_a^{(\mu)} T^{ai}\right]_{,i} = W_a^{(\mu)} T^{ai}{}_{,i} = 0 \,. \tag{11.67}$$

Each 'current vector' is conserved.

From here on we shall not be concerned with this detail. The extra index on the energy–momentum tensor will just tag along. However, it will be useful to reproduce the whole argument of Sect. 11.2.1, in essentially the same form as for the 4-current vector, just so that we can apply it to the case at hand. We thus start with

$$P^\alpha = \int_{x^0=0} T^{\alpha 0} d^3x \,, \tag{11.68}$$

and show that it is the same as

$$\int_{x^0=0} T^{ab} \eta_{bcde} \,, \tag{11.69}$$

where $x^0 = 0$ is our 3D submanifold and $T^{ab} \eta_{bcde}$ is a vector-valued 3-form.

As before, $\eta_{bcde} = \varepsilon_{bcde}$ and we use the injection

$$S : \mathbb{R}^3 \hookrightarrow \mathbb{R}^4 \quad \text{with} \quad S(\mathbb{R}^3) = \mathscr{S} \,, \tag{11.70}$$

where

$$S(y^1, y^2, y^3) = (0, y^1, y^2, y^3) \,, \tag{11.71}$$

to parametrise the spacelike hypersurface $\mathscr{S} = \{x^0 = 0\}$. We then pull back the vector-valued 3-form

$$\frac{1}{3!} T^{ab} \varepsilon_{bcde} dx^c \wedge dx^d \wedge dx^e \,,$$

ignoring the extra vector index a. The result is

$$S^* \left(\frac{1}{3!} T^{ab} \varepsilon_{bcde} dx^c \wedge dx^d \wedge dx^e \right) = \frac{1}{3!} T^{ab} \varepsilon_{bcde} S^*(dx^c) \wedge S^*(dx^d) \wedge S^*(dx^e) \,, \tag{11.72}$$

using a well-known result concerning the pullback of an exterior product [13, Chap. 2]. As before, we show that

$$S^*(dx^0) = 0 \,, \qquad S^*(dx^i) = dy^i \,, \quad i = 1, 2, 3 \,. \tag{11.73}$$

We are thus integrating

$$\frac{1}{3!}T^{ab}\varepsilon_{bcde}S^*(\mathrm{d}x^c)\wedge S^*(\mathrm{d}x^d)\wedge S^*(\mathrm{d}x^e) = \frac{1}{3!}T^{a0}\varepsilon_{0ijk}\mathrm{d}y^i\wedge\mathrm{d}y^j\wedge\mathrm{d}y^k$$
$$= T^{a0}\mathrm{d}y^1\wedge\mathrm{d}y^2\wedge\mathrm{d}y^3\,,$$

This is the proof we need that

$$P^a = \int_{x^0=\text{constant}} T^{ab}\eta_{bcde} = \int_{x^0=\text{constant}} T^{a0}\mathrm{d}^3x\,. \qquad (11.74)$$

What happens when we choose different values for x^0, say ct_1 and ct_2, where $t_1 < t_2$? We have

$$P^a(t_2) - P^a(t_1) = \int_{x^0=ct_2} T^{ab}\eta_{bcde} - \int_{x^0=ct_1} T^{ab}\eta_{bcde}$$
$$= \int_{\text{slab surface}} T^{ab}\eta_{bcde}$$
$$= \int_{\text{slab volume}} T^{ab}_{,b}$$
$$= 0\,, \qquad (11.75)$$

where the slab is the piece of spacetime between the two spatial hypersurfaces and we have applied Gauss' theorem to get line three.

The Case of the Spherical Electron

Let us view the spherical electron in its rest frame. We define

$$P^\alpha(t_1) = \int_{\substack{x^0=ct_1 \\ \text{outside electron}}} T^{a0}\mathrm{d}^3x\,. \qquad (11.76)$$

This is independent of t_1 because we are dealing with a static field! But we also know that T^{ab} is conserved outside the electron, so we should be able to apply Gauss' theorem in the spacetime region bounded by the two spacelike hypersurfaces $x^0 = ct_1$ and $x^0 = ct_2$ and the electron worldtube $\mathcal{W}$ where it lies between them. The latter is $\mathcal{W} = [ct_1, ct_2] \times \{r = a\}$, where a is the radius of the electron. We have to stop the integration for $r < a$ because the energy–momentum tensor is discontinuous on the electron surface where the charge is accumulated. Indeed, the fields inside the electron, and the energy–momentum tensor with them, are all zero.

The interesting thing here is to evaluate the integral over the electron worldtube. By Gauss' theorem, it has to be zero, since we know that $P(t_1) = P(t_2)$ in this case, so we hope to show that

$$\int_{\mathcal{W}} T^{ab}\eta_{bcde} = 0\,. \qquad (11.77)$$

Recall that the energy–momentum tensor is

$$T^{ab} = \varepsilon_0 \begin{pmatrix} -\dfrac{1}{2}E^2 & 0 \\ 0 & \mathbf{EE} - \dfrac{1}{2}E^2 \end{pmatrix} , \tag{11.78}$$

where

$$\mathbf{E} = \frac{e\mathbf{r}}{r^3} , \tag{11.79}$$

up to the usual problem of constants. (With our conventions, there should be a factor of $1/\sqrt{4\pi\varepsilon_0}$ in the last relation, but as we hope to get zero, we shall deal in proportionalities here.) Hence,

$$E^2 = \frac{e^2}{r^4} , \qquad \mathbf{EE} = \frac{e^2}{r^6} \begin{pmatrix} x^2 & xy & xz \\ xy & y^2 & yz \\ xz & yz & z^2 \end{pmatrix} . \tag{11.80}$$

The next problem is to parametrise the electron worldtube

$$[t_1, t_2] \times \{r = a\} = \left\{ (ct, x, y, z) : t \in [t_1, t_2], x^2 + y^2 + z^2 = a^2 \right\} .$$

Heuristically, the normal form n_a for this region should have zero component n_0 equal to zero and spatial part radial with respect to the origin. It is interesting to see how this works out. We take the parametrisation

$$S : (y^0, y^1, y^2) \longmapsto (y^0, a\sin y^1 \cos y^2, a\sin y^1 \sin y^2, a\cos y^1) , \tag{11.81}$$

where

$$ct_1 < y^0 < ct_2 , \quad 0 < y^1 < \pi , \quad 0 < y^2 < 2\pi \tag{11.82}$$

is the cuboidal domain of integration in the y-space. We have to calculate the pullbacks $S^* dx^0$, $S^* dx^1$, $S^* dx^2$, and $S^* dx^3$. It is intuitively obvious that we should get

$$S^* dx^0 = dy^0 , \tag{11.83}$$

$$S^* dx^1 = a\cos y^1 \cos y^2 \, dy^1 - a\sin y^1 \sin y^2 \, dy^2 , \tag{11.84}$$

$$S^* dx^2 = a\cos y^1 \sin y^2 \, dy^1 + a\sin y^1 \cos y^2 \, dy^2 , \tag{11.85}$$

$$S^* dx^3 = -a\sin y^1 \, dy^1 , \tag{11.86}$$

and it is not difficult to prove. We can now analyse

$$\frac{1}{3!}T^{ab}\varepsilon_{bcde}S^*(dx^c)\wedge S^*(dx^d)\wedge S^*(dx^e) \tag{11.87}$$

$$= T^{a0}\left(S^*dx^1\wedge S^*dx^2\wedge S^*dx^3\right)+T^{a1}\left(S^*dx^2\wedge S^*dx^0\wedge S^*dx^3\right)$$
$$+T^{a2}\left(S^*dx^0\wedge S^*dx^1\wedge S^*dx^3\right)+T^{a3}\left(S^*dx^1\wedge S^*dx^0\wedge S^*dx^2\right),$$

where it is quite clear that

$$S^*dx^1\wedge S^*dx^2\wedge S^*dx^3=0, \tag{11.88}$$

because we have a wedge product of three forms in the 2D subspace spanned by dy^1 and dy^2. The other pullbacks are

$$S^*dx^2\wedge S^*dx^0\wedge S^*dx^3=-a^2\sin^2 y^1\cos y^2\,dy^0\wedge dy^1\wedge dy^2, \tag{11.89}$$

$$S^*dx^0\wedge S^*dx^1\wedge S^*dx^3=-a^2\sin^2 y^1\sin y^2\,dy^0\wedge dy^1\wedge dy^2, \tag{11.90}$$

and

$$S^*dx^1\wedge S^*dx^0\wedge S^*dx^2=-a^2\cos y^1\sin y^1\,dy^0\wedge dy^1\wedge dy^2. \tag{11.91}$$

Summing these terms, we have

$$\frac{1}{3!}T^{ab}\varepsilon_{bcde}S^*(dx^c)\wedge S^*(dx^d)\wedge S^*(dx^e) \tag{11.92}$$

$$= -a^2\sin y^1\mathbf{T}^a\cdot\begin{pmatrix}\sin y^1\cos y^2\\\sin y^1\sin y^2\\\cos y^1\end{pmatrix}dy^0\wedge dy^1\wedge dy^2,$$

where, for each $a=0,1,2,3$, the vector $\mathbf{T}^a$ is the vector

$$\mathbf{T}^a=(T^{a1},T^{a2},T^{a3}),$$

taking three components from each row of the energy–momentum tensor.

We can now express the integral in (11.77) in the form

$$\int_{\mathcal{W}}T^{ab}\eta_{bcde}=\int_{\text{cuboid in }\mathbb{R}^3}T^{ab}n_b d\sigma, \tag{11.93}$$

where T^{ab} is evaluated at $(y^0,a\sin y^1\cos y^2,a\sin y^1\sin y^2,a\cos y^1)$, the form n_b is given by

$$n_b=(0,-\sin y^1\cos y^2,-\sin y^1\sin y^2,-\cos y^1), \tag{11.94}$$

and

$$d\sigma=a^2\sin y^1\,dy^0dy^1dy^2. \tag{11.95}$$

The spatial part of the form n is indeed radial from the origin and the time component of n is zero, as predicted on heuristic grounds. There is just a question about the sign of the spatial part: should it point towards or away from the origin? Intuitively, it should point towards the origin in this case, because we integrate over the region outside the electron. This has to be checked by referring to the orientation of the submanifold. The measure $d\sigma$ is precisely the measure we normally use on a spherical surface of radius a.

We have to check that each component of (11.93) is zero. The zero component is clearly zero because $\mathbf{T}^0 = 0$ for the energy–momentum tensor in (11.78). What about the other three components? We have

$$\mathbf{T}^1 \propto a^2 \begin{pmatrix} \sin^2 y^1 \cos^2 y^2 - 1/2 \\ \sin^2 y^1 \sin y^2 \cos y^2 \\ \sin y^1 \cos y^1 \cos y^2 \end{pmatrix} = x\mathbf{r} - \frac{1}{2}r^2 \begin{pmatrix} 1 \\ 0 \\ 0 \end{pmatrix}. \qquad (11.96)$$

The 1-integral is an integral of something proportional to the scalar product of the last thing with $\mathbf{r}$, viz.,

$$x\mathbf{r} \cdot \mathbf{r} - \frac{1}{2}r^2 \begin{pmatrix} 1 \\ 0 \\ 0 \end{pmatrix} \cdot \mathbf{r} = \frac{1}{2}xr^2.$$

Note that there are other factors of r^2 in the integrand, but $r^2 = a^2$ all over the cuboid of integration specified in (11.82). Now the integral of $x = a \sin y^1 \cos y^2$ over our cuboid of integration is obviously zero. The other two integrals in (11.93) are shown to be zero in the same way.

The point of this was more an exercise in integration theory than anything else! The fields here are static so it is clear that the 4-momentum of the fields has to be constant with this definition. We are merely checking the application of Gauss' theorem and seeing how to do an integration over the electron worldtube.

Lorentz Covariance of the Four-Momentum

Returning to the case of an energy–momentum tensor that is conserved everywhere, we can now understand why the definition (11.63) on p. 199 gives a Lorentz covariant quantity. The analysis is very similar to the one made for a conserved 4-current in Sect. 11.2.1 [see (11.57) on p. 197].

We begin with

$$P'^b = \int_{x'^0=0} T'^{b0} \, \mathrm{d}^3 x' = \int_{x'^0=0} T'^{bc} \eta'_{cdef}, \qquad (11.97)$$

meaning that we integrate the vector-valued 3-form

$$\frac{1}{3!} T'^{bc} \eta'_{cdef} dx'^d \wedge dx'^e \wedge dx'^f \tag{11.98}$$

over the primed HOS. Hence, $(L^{-1})^a_b P'^b$ is the integral over $x'^0 = 0$ of the vector-valued 3-form

$$\frac{1}{3!} T^{ac} \eta_{cdef} dx^d \wedge dx^e \wedge dx^f . \tag{11.99}$$

We can conclude that

$$
\begin{aligned}
(L^{-1})^a_b P'^b - P^a &= \int_{\substack{\text{wedge} \\ \text{surface}}} T^{ac} \eta_{cdef} \\
&= \int_{\substack{\text{wedge} \\ \text{volume}}} T^{ac}{}_{,c} d\tau \\
&= 0 .
\end{aligned}
$$

Note that we could also write the first line here in the form

$$(L^{-1})^a_b P'^b - P^a = \int_{x'^0=0} T^{ab} d\sigma_b - \int_{x^0=0} T^{ab} d\sigma_b . \tag{11.100}$$

However, this involves showing that we can give this interpretation of our integrals in terms of normal forms and it seems less direct.

The Case of the Spherical Electron

Let us see what happens to this analysis when we cannot integrate inside the spacetime region $\mathbb{R} \times \{r < a\}$ because the energy–momentum tensor is discontinuous across the electron surface where the charge is supposed to be accumulated. We still have that the energy–momentum tensor is conserved outside the electron. We can therefore usefully apply Gauss' theorem there, and the volume integral will be zero, but we pick up an integral over the electron worldtube where it lies between the two spacelike hypersurfaces $x^0 = 0$ and $x'^0 = 0$. It is precisely the value of this integral that breaks the Lorentz covariance of our definitions in this case.

So the argument starts as before. We make the following picture in the unprimed frame which is the electron rest frame (see Fig. 11.2). Relative to the x^0 and x^1 axes, the x'^0 axis slopes up to the left of the x^0 axis and the x'^1 axis slopes down to the right below the x^1 axis. Two wedges are formed between the x^1 and x'^1 axis (unprimed and primed HOS). Our definitions of the 4-momentum in the primed and unprimed frames require us to integrate something over the bits of these hypersurfaces that lie outside the electron worldtube $\mathbb{R} \times \{r < a\}$. If we consider $(L^{-1})^a_b P'^b - P^a$, then we are integrating the same thing, viz., $T^{ab} \eta_{bcde}$, over the relevant surfaces. We can close the wedges at infinity with impunity because the surface integrals there will tend to zero. If we add in the integrals over the relevant parts of the electron worldtube, we obtain the integral of $T^{ab} \eta_{bcde}$ over a closed surface and Gauss' theorem applies.

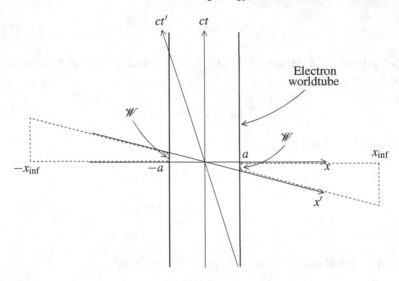

Fig. 11.2 Two coordinate systems for specifying the four-momentum. The *dotted lines* enclose the region of spacetime over which Gauss' theorem is applied, and $\mathscr{W}$ is the part of that surface that coincides with the surface of the electron worldtube, i.e., the part of the surface that leads to the infamous discrepancy between energy-derived and momentum-derived EM masses for this system

The only problem then is to subtract off the surface integrals we added in. We shall now evaluate this surface integral over part of the electron worldtube, i.e., we shall evaluate

$$\int_{\mathscr{W}} T^{ab} \eta_{bcde} \,,$$ (11.101)

where $\mathscr{W}$ is the relevant portion of the surface of the electron worldtube. The integrand is the vector-valued 3-form

$$\frac{1}{3!} T^{ab} \varepsilon_{bcde} \mathrm{d}x^c \wedge \mathrm{d}x^d \wedge \mathrm{d}x^e$$ (11.102)

and we must parametrise the surface $\mathscr{W}$ by some injection of a region $\mathscr{U} \subset \mathbb{R}^3$ onto $\mathscr{W}$, viz.,

$$W : \mathscr{U} \hookrightarrow \mathbb{R}^4 \,, \qquad W(\mathscr{U}) = \mathscr{W} \,.$$ (11.103)

We then use W to pull back $\mathrm{d}x^0$, $\mathrm{d}x^1$, $\mathrm{d}x^2$, and $\mathrm{d}x^3$ and carry out a standard integration over $\mathscr{U}$.

First of all, what points of spacetime are on the surface $\mathscr{W}$? In fact,

$$\mathscr{W} = \left\{ (x^0, x^1, x^2, x^3) : (x^1)^2 + (x^2)^2 + (x^3)^2 = a^2 , \ x^0 \in \left[0, -\frac{vx^1}{c^2}\right], \ x^1 < 0 \right\}$$

$$\cup \left\{ (x^0, x^1, x^2, x^3) : (x^1)^2 + (x^2)^2 + (x^3)^2 = a^2 , \ x^0 \in \left[-\frac{vx^1}{c^2}, 0\right], \ x^1 > 0 \right\} .$$

We shall therefore define the map W by

$$W : (y^0, y^1, y^2) \longmapsto (y^0, a \sin y^1 \cos y^2, a \sin y^1 \sin y^2, a \cos y^1) . \tag{11.104}$$

The only remaining problem is to establish the domain of integration $\mathscr{U}$.

This is no longer a cuboid in y-space because of the constraint

$$-\frac{va}{c^2} \sin y^1 \cos y^2 < y^0 < 0, \quad \text{for } \cos y^2 > 0 ,$$

$$0 < y^0 < -\frac{va}{c^2} \sin y^1 \cos y^2, \quad \text{for } \cos y^2 < 0 ,$$

whilst $0 \le y^1 < \pi$ and $0 \le y^2 < 2\pi$. We picture the following region $\mathscr{U}$ in y-space (see Fig. 11.3). It intersects the plane $y^2 = 0$ in the region between the y^1 axis and the first hump of a negative sine curve going down to $-va/c^2$. As we move along the y^2 axis, this sine hump decreases in depth to zero at $y^2 = \pi/2$. As we move further along the y^2 axis, the sine hump grows positively to reach a peak of va/c^2 at $y^2 = \pi$ and then drops back down to zero at $y^2 = 3\pi/2$ before falling down to its original trace on the plane $y^2 = 2\pi$, with a maximum depth of $-va/c^2$. In other words, this movement of the sine peak follows a cosinusoidal variation as we move along the y^2 axis.

There are three regions here: one with positive y^0 and two with negative y^0. The latter are actually joined in spacetime. y^1 and y^2 correspond to the usual polar angles θ and ϕ, respectively. Moving along the y^2 axis corresponds to moving round the x^3 axis from the x^1 axis through the x^2 axis when $y^2 = \pi/2$ to the negative x^1 axis when $y^2 = \pi$, the negative x^2 axis when $y^2 = 3\pi/2$ and finally back to the x^1 axis when $y^2 = 2\pi$. Now y^0 is negative in the zones $0 < y^2 < \pi/2$ and $3\pi/2 < y^2 < 2\pi$ and it is positive in the zone $\pi/2 < y^2 < 3\pi/2$, i.e., it is negative between the negative x^2 axis and the positive x^2 axis and positive between the positive x^2 axis and the negative x^2 axis.

Let us carry out this integral for the spherical electron. We have the pullbacks in (11.83) on p. 202, viz.,

$$W^* dx^0 = dy^0 , \tag{11.105}$$

$$W^* dx^1 = a \cos y^1 \cos y^2 dy^1 - a \sin y^1 \sin y^2 dy^2 , \tag{11.106}$$

$$W^* dx^2 = a \cos y^1 \sin y^2 dy^1 + a \sin y^1 \cos y^2 dy^2 , \tag{11.107}$$

$$W^* dx^3 = -a \sin y^1 dy^1 , \tag{11.108}$$

and the energy–momentum tensor on p. 202, viz.,

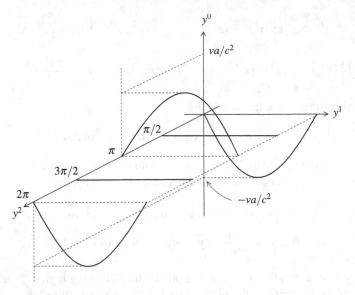

Fig. 11.3 Domain of integration in y space

$$T^{ab} = \varepsilon_0 \begin{pmatrix} -\frac{1}{2}E^2 & 0 \\ 0 & \mathbf{EE} - \frac{1}{2}E^2 \end{pmatrix},$$ (11.109)

where

$$\mathbf{E} = \frac{e\mathbf{r}}{r^3},$$ (11.110)

up to a constant (in fact, we have have dropped a factor of $1/\sqrt{4\pi\varepsilon_0}$), with

$$E^2 = \frac{e^2}{r^4}, \qquad \mathbf{EE} = \frac{e^2}{r^6} \begin{pmatrix} x^2 & xy & xz \\ xy & y^2 & yz \\ xz & yz & z^2 \end{pmatrix}.$$ (11.111)

It should not be forgotten that $r = a$ all over the domain of integration. We can say that

$$T^{ab} = -\frac{\varepsilon_0 e^2}{2a^4} \begin{pmatrix} 1 & 0 & 0 & 0 \\ 0 & 1 & 0 & 0 \\ 0 & 0 & 1 & 0 \\ 0 & 0 & 0 & 1 \end{pmatrix} + \frac{\varepsilon_0 e^2}{a^6} \begin{pmatrix} 0 & 0 \\ 0 & \mathbf{rr} \end{pmatrix},$$ (11.112)

where

$$\mathbf{rr} = \begin{pmatrix} x^2 & xy & xz \\ xy & y^2 & yz \\ xz & yz & z^2 \end{pmatrix} . \tag{11.113}$$

We have already worked out the pullback in (11.92) on p. 203:

$$\frac{1}{3!} T^{ab} \varepsilon_{bcde} W^*(\mathrm{d}x^c) \wedge W^*(\mathrm{d}x^d) \wedge W^*(\mathrm{d}x^e) \tag{11.114}$$

$$= -a^2 \sin y^1 \mathbf{T}^a \cdot \begin{pmatrix} \sin y^1 \cos y^2 \\ \sin y^1 \sin y^2 \\ \cos y^1 \end{pmatrix} \mathrm{d}y^0 \wedge \mathrm{d}y^1 \wedge \mathrm{d}y^2 ,$$

where, for each $a = 0, 1, 2, 3$, the vector $\mathbf{T}^a$ is the vector

$$\mathbf{T}^a = (T^{a1}, T^{a2}, T^{a3}) ,$$

taking three components from each row of the energy–momentum tensor. As before, we have $\mathbf{T}^0 = 0$. This says that

$$\int_{\mathscr{W}} T^{0b} \eta_{bcde} = 0 . \tag{11.115}$$

For $i = 1, 2, 3$,

$$\int_{\mathscr{W}} T^{ib} \eta_{bcde} = \frac{\varepsilon_0 e^2}{a^6} \int_{\mathscr{U}} -a \sin y^1 (\mathbf{rr}) \cdot \mathbf{r} \, \mathrm{d}y^0 \mathrm{d}y^1 \mathrm{d}y^2 - \frac{\varepsilon_0 e^2}{2a^4} \int_{\mathscr{U}} -a \sin y^1 \mathbf{r} \, \mathrm{d}y^0 \mathrm{d}y^1 \mathrm{d}y^2$$

$$= -\frac{\varepsilon_0 e^2}{2a^2} \int_{\mathscr{U}} \begin{pmatrix} \sin^2 y^1 \cos y^2 \\ \sin^2 y^1 \sin y^2 \\ \sin y^1 \cos y^1 \end{pmatrix} \mathrm{d}y^0 \mathrm{d}y^1 \mathrm{d}y^2 . \tag{11.116}$$

There is a question of sign here, which depends on the orientation of the surface in spacetime and the fact that we are using Gauss' theorem outside the electron worldtube rather than inside it.

We now write the integral as a sum for different ranges of the y^2 variable. It is important to note that, because the two spacelike hypersurfaces $x^0 = 0$ and $x'^0 = 0$ cross at the origin, there is a relative sign between the integral obtained for $y^2 \in [\pi/2, 3\pi/2]$ and the integral obtained for $y^2 \in [0, \pi/2] \cup [3\pi/2, 2\pi]$. As regards the overall sign, we shall just hope that it is right for the moment. We have

$$\int_{\mathscr{W}} T^{ib} \eta_{bcde} = \frac{\varepsilon_0 e^2}{2a^2} \int_0^\pi \mathrm{d}y^1 \left(\int_0^{\pi/2} \mathrm{d}y^2 + \int_{3\pi/2}^{2\pi} \mathrm{d}y^2 \right) \int_{-(va/c^2)\sin y^1 \cos y^2}^0 \mathrm{d}y^0 \sin y^1 \mathbf{r}$$

$$+ \frac{\varepsilon_0 e^2}{2a^2} \int_0^\pi \mathrm{d}y^1 \int_{\pi/2}^{3\pi/2} \mathrm{d}y^2 \int_0^{(va/c^2)\sin y^1 \cos y^2} \mathrm{d}y^0 \sin y^1 \mathbf{r} . \tag{11.117}$$

The y^0 integrals are easy to do because the integrand is independent of y^0. We thus obtain

$$\int_{\mathscr{W}} T^{1b} \eta_{bcde} = \frac{\varepsilon_0 e^2}{2a^2} \int_0^\pi dy^1 \left(\int_0^{\pi/2} dy^2 + \int_{3\pi/2}^{2\pi} dy^2 \right) \frac{va}{c^2} \sin y^1 \cos y^2 \sin y^1 \mathbf{r}$$

$$+ \frac{\varepsilon_0 e^2}{2a^2} \int_0^\pi dy^1 \int_{\pi/2}^{3\pi/2} dy^2 \frac{va}{c^2} \sin y^1 \cos y^2 \sin y^1 \mathbf{r}$$

$$= \frac{\varepsilon_0 e^2}{2a^2} \int_0^\pi dy^1 \int_0^{2\pi} dy^2 \frac{va}{c^2} \sin y^1 \cos y^2 \sin y^1 \mathbf{r} . \tag{11.118}$$

Let us evaluate

$$\int_0^{2\pi} dy^2 \sin y^1 \cos y^2 \sin y^1 \mathbf{r} = \begin{pmatrix} \sin^3 y^1 \int_0^{2\pi} dy^2 \cos^2 y^2 \\ \sin^3 y^1 \int_0^{2\pi} dy^2 \cos y^2 \sin y^2 \\ \sin^2 y^1 \cos y^1 \int_0^{2\pi} dy^2 \cos y^2 \end{pmatrix}$$

$$= \pi \sin^3 y^1 \begin{pmatrix} 1 \\ 0 \\ 0 \end{pmatrix} . \tag{11.119}$$

Since it is straightforward to show that

$$\int_0^\pi \sin^3 y^1 \, dy^1 = \frac{4}{3} , \tag{11.120}$$

we can put everything together to show that

$$\int_{\mathscr{W}} T^{1b} \eta_{bcde} = \frac{2\varepsilon_0 e^2 \pi}{3ac^2} \mathbf{v} . \tag{11.121}$$

Reinstating the factor of $1/\sqrt{4\pi\varepsilon_0}$ that should come with every appearance of e, the final result is

$$\int_{\mathscr{W}} T^{1b} \eta_{bcde} = \pm \frac{e^2}{6ac^2} \mathbf{v} . \tag{11.122}$$

The choice of sign arises partly because we have dropped a sign in the definition of P^a itself and partly because we have not checked the question of orientation of the surface.

What is significant is that this result is precisely the discrepancy discovered by Rohrlich in his equation (11.18), up to a sign. Of course, it had to be. We know this is the discrepancy destroying Lorentz covariance in our definition of P^a. The point about the above, somewhat lengthy exercise is to show this explicitly with absolutely standard integration theory. The problem occurs precisely because the purely electromagnetic energy–momentum tensor we are using is only conserved outside the charge shell, and not actually on the shell, and this in turn happens because we have not included the binding forces and the energy–momentum tensor that describes them.

Note that Rohrlich's result (11.22) on p. 189, obtained in [21], is an approximation, whereas the calculation leading to (11.122) was exact in every way. The above calculation used the rest frame energy–momentum tensor. Let us make the connection between the two results absolutely clear. In the above discussion we began with the 4-momentum P'^β as we would have worked it out in Chap. 5, in the frame relative to which the electron has 4-velocity $v'^\mu = \gamma(c, \mathbf{v})$. We then showed that

$$(L^{-1})^\alpha{}_\beta P'^\beta = P^\alpha + \frac{e^2}{6ac^2}(0, \mathbf{v}) , \qquad (11.123)$$

where L is the Lorentz transformation from the rest frame of the shell to a frame in which it appears to be moving, and we take the positive sign for the correction term here to get the agreement below. (We do not need to check that this is the right sign, because we know from the general theory what adjustment we need to get Lorentz covariance.) This result is exact. Rohrlich estimates in the primed frame (the calculation described on pp. 187 ff.) and gets the same correction to the result in the sense that he claims to evaluate the real $L^a{}_b P^b$ and finds

$$L^a{}_b P^b = P'^a - \frac{e^2}{6ac^2}(0, \mathbf{v}) .$$

Hence, he claims that

$$(L^{-1})^a{}_b P'^b = P^a + \frac{e^2}{6ac^2}(L^{-1})^a{}_b(0, \mathbf{v}) .$$

This is not exactly the same as (11.123), but it is the same within the approximation $v \ll c$. We have, according to (11.39) on p. 194,

$$L^{-1} = \begin{pmatrix} \gamma & -\gamma v/c & 0 & 0 \\ -\gamma v/c & \gamma & 0 & 0 \\ 0 & 0 & 1 & 0 \\ 0 & 0 & 0 & 1 \end{pmatrix} , \qquad (11.124)$$

whence

$$L^{-1}\begin{pmatrix} 0 \\ \mathbf{v} \end{pmatrix} = \begin{pmatrix} -\gamma v^2/c \\ \gamma \mathbf{v} \end{pmatrix} \approx \begin{pmatrix} 0 \\ \mathbf{v} \end{pmatrix} , \qquad (11.125)$$

as required.

So it still remains to see exactly what is going on with Rohrlich's succinct redefinition of the four-momentum in the fields, and we shall do that in Sect. 11.4. But first, let us get a better understanding of why the binding forces must be taken into account.

11.3 Role of the Binding Forces

The present discussion is based on the article [23] by Boyer, entitled *Classical Model of the Electron and the Definition of Electromagnetic Field Momentum*. This paper opposes the covariant extension advocated by Rohrlich, considered in Sect. 11.1. In Boyer's version, the usual calculation of the momentum in the fields is correct and, although the Rohrlich formula leads to a Lorentz covariant definition, it has no physical interpretation (see Sect. 11.4).

Boyer approaches the problem of the 4/3 factor by examining the assembly of a charged spherical shell as seen in two different inertial frames. In fact, he imagines the assembly of the classical model of the electron in terms of a thin spherical shell of total mechanical mass m and charge e sent rushing inwards from spatial infinity with the initial kinetic energy $mc^2(\gamma - 1)$ at spatial infinity equal to the final electrostatic potential energy $U_{em} = e^2/2a$. Since the shell is perfectly spherically symmetric, there is no radiation loss and all the initial kinetic energy at spatial infinity is converted into electrostatic potential energy when the shell comes momentarily to rest at radius a. Just at this instant when the spherical shell comes to rest, the stabilising forces are applied. These forces prevent the reexpansion of the shell. The external forces are applied simultaneously at zero velocity and hence transfer neither energy nor momentum to the spherical shell. We thus assemble our classical electron as a thin shell of charge of energy

$$U_{\text{tot}} = mc^2 + \frac{e^2}{2a}$$

and zero momentum $\mathbf{P}_{\text{tot}} = 0$. Note that in Boyer's discussion the electron has a 'mechanical mass'. In other words, there is no attempt to make a self-contained theory of inertial mass.

Concerning the comment about radiation losses, one must consider the question of the fields outside the collapsing shell. On the face of it, this seems to be a rather complex problem, given the complexity of the source. However, we have Gauss' theorem based on the Maxwell equation which relates $\nabla \cdot \mathbf{E}$ to the charge density. It turns out that the field outside the sphere at any instant of time is just the static Coulomb field due to a point charge at the sphere centre. Likewise the fields are all identically zero inside the sphere. This is clearly an idealisation, just as the spherical shell itself is an idealisation. However, it will be important later and we shall discuss this in some detail.

The problem then will be to view the above assembly process from a primed frame S' moving with 3-velocity $-\mathbf{v}$ relative to the initial frame S, so that the charge shell appears to be moving to the right with 3-velocity $\mathbf{v}$ (see Fig. 11.4). When the shell has infinite radius, all points lie within it, so the fields are zero everywhere. All the particle energy and momentum comes from the 'mechanical mass'. The energy and momentum of this 'mechanical mass' transform as a Lorentz four-vector. Hence, initially, the system momentum, which is all mechanical momentum, is given by

$$\mathbf{P}'_{\text{tot}} = \mathbf{P}'_{\text{mech}}(t' \to -\infty) = \mathbf{v}\gamma U_{\text{tot}}/c^2 \,.$$

As time goes by, the electromagnetic fields increase from their initial zero values and part of the mechanical momentum is converted into electromagnetic momentum. The total momentum is conserved as long as there are no external forces on the system.

In frame S, the stabilising forces are applied simultaneously. Consequently, the net external force on the system is zero and there is no change in the momentum of the system. Contrastingly, in the S' frame, the external stabilising forces are applied, beginning at some time t'_s and continuing through some period which actually depends on v. Indeed, they are applied at different times to different parts of the spherical shell. Thus, from the moment (in the S' frame) when the first force is applied and until the moment (in the S' frame) when all the external forces have been applied, there is a net external force on the shell, and hence a net momentum is transferred to the shell. So we deduce that

$$\mathbf{v}\gamma U_{\text{tot}}/c^2 = \mathbf{P}'_{\text{tot}} = \mathbf{P}'_{\text{mech}} + \mathbf{P}'_{\text{em}} \qquad (t' < t'_s) \,, \tag{11.126}$$

but after this time, the total momentum of the shell and fields is changed from the value $\mathbf{P}'_{\text{tot}} = \mathbf{v}\gamma U_{\text{tot}}/c^2$ which prevailed before the external forces were applied.

The change in momentum $\Delta\mathbf{P}'$ of the charge shell as seen in the S' frame is equal to the net impulse $\mathbf{I}'$ delivered by the external stabilising forces as seen in the S' frame. The problem now is to compute $\mathbf{I}'$.

11.3.1 Dynamics of a Collapsing Charged Spherical Shell

We shall say that the charge shell is at infinity at time $t = -\infty$ in the unprimed frame (in which the final assembly is motionless), and reaches its final radius a at time $t = 0$. The binding forces are thus switched on at time $t = 0$. When we view this from the primed frame, in which the final assembly has speed v along the positive x axis, the binding forces will affect different parts of the shell at different primed times.

We make the following picture (see Fig. 11.4). Draw the ct' and x' axes at right angles with the ct axis slanting off to the right from the origin and the x axis rotated up slightly from the x' axis. Draw the electron world tube with rear wall going through the point $(x, ct) = (-a, 0)$ and front wall going through the point $(x, ct) = (a, 0)$ and both walls parallel to the ct axis, slanting off to the right of the ct' axis. We can imagine that the stabilising forces are applied at time $t = 0$ in the unprimed frame. Then they appear to begin at the event whose primed coordinates correspond to $(-a, 0)$ and end at the event whose primed coordinates correspond to $(a, 0)$. Using the conversion

$$x'^0 = \gamma\left(x^0 + \frac{v}{c}x^1\right) \,, \qquad x'^1 = \gamma\left(x^1 + \frac{v}{c}x^0\right) \,,$$

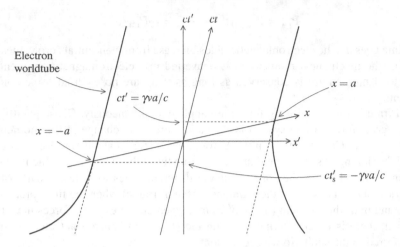

Fig. 11.4 Worldtube of the charge shell when viewed from a primed frame moving to the left along the x axis at speed v

we find the correspondence

$$(-a, 0) \longleftrightarrow \left(-\gamma a, -\frac{\gamma v}{c} a\right) , \qquad (a, 0) \longleftrightarrow \left(\gamma a, \frac{\gamma v}{c} a\right) . \tag{11.127}$$

We introduce parameters θ and ϕ. We take the axis $\theta = 0$ to be the x or x' axis with ϕ measured axially around it. These angles can be taken in the S frame, because this is where we evaluate the force, before transforming to the S' frame. We divide the shell into rings of charge that are symmetrical around this axis, of area $2\pi a^2 d\theta \sin \theta$. Since the surface charge density of the sphere is $e/4\pi a^2$, the charge in the ring at θ is

$$\frac{1}{2} e \sin \theta \, d\theta . \tag{11.128}$$

Note that, if we discussed rings of charge as viewed in the primed frame, we would divide the shell into rings of charge that were symmetrical around the x' axis, of area $2\pi (a^2/\gamma) d\theta \sin \theta$. Since the surface charge density of the ellipsoid is $e\gamma/4\pi a^2$, the charge in the ring at θ would be the same as in (11.128).

Viewed from the S frame, at each x value between $-a$ and a, we can say that a force $F(\theta, \phi)$ is applied towards the spatial origin. In the S' frame, this appears to act from the time

$$t' = \frac{\gamma v}{c^2} x \quad \text{until the time} \quad \frac{\gamma v}{c^2} a .$$

We find the force on the ring in the S frame and Lorentz transform. This force, summed over all the ϕ values, has resultant parallel to the x axis in space. Let us call it

$$F^i = \begin{pmatrix} 0 \\ F(\theta) \\ 0 \\ 0 \end{pmatrix}.$$

(11.129)

The first component is zero because there is no motion in the unprimed frame. (Recall that 4-velocity and 4-force are orthogonal.)

How can we work out the force on the ring. One plausible argument is that the rest of the shell acts like a point charge e at the spatial origin. If this were valid, the x component of the force on the ring would be

$$F(\theta) = -\frac{1}{2a^2} e \sin \theta \, d\theta \times e \times \cos \theta = -\frac{e^2}{4a^2} \sin 2\theta \, d\theta .$$

(11.130)

The minus sign says that we are acting against the repulsion from the spatial origin due to the Coulomb force. The Lorentz transformation to the primed frame is

$$F'^i = \begin{pmatrix} \gamma & \gamma v/c \\ \gamma v/c & \gamma \end{pmatrix} \begin{pmatrix} 0 \\ F(\theta) \end{pmatrix} = \begin{pmatrix} \frac{\gamma v}{c} F(\theta) \\ \gamma F(\theta) \end{pmatrix}.$$

(11.131)

However, we are interested in F'^1/γ. This is because we need

$$\frac{d\mathbf{P}'}{dt'} = \frac{d}{dt'}(\gamma m_0 \mathbf{u}') = \mathbf{f}' ,$$

where

$$F' = (\gamma f_4', \gamma \mathbf{f}') ,$$

using the obvious notation.

We can now integrate over all the rings making up the shell. The result is

$$\Delta \mathbf{P}' = -\frac{e^2}{4a^2} \int_{\theta=0}^{\pi} \sin 2\theta \, d\theta \frac{\gamma v}{c^2} (a - x) .$$

(11.132)

We are summing the forces multiplied by the times they act as measured in the primed frame in order to get the impulse in the primed frame. (Note that the forces act for ever more, but we only consider the time during which they appear to be unbalanced in the primed frame, which ends at $t' = \gamma v a/c^2$.) Since $x = a \cos \theta$, the result is therefore

$$\Delta \mathbf{P'}\,(\text{incorrect}) = -\frac{e^2}{4a^2} \int_{\theta=0}^{\pi} \sin 2\theta \, d\theta \, \frac{\gamma v a}{c^2} (1 - \cos \theta)$$

$$= \frac{e^2 \gamma v}{4ac^2} \int_{\theta=0}^{\pi} 2 \sin \theta \cos^2 \theta \, d\theta$$

$$= \frac{e^2 \gamma v}{2ac^2} \left[-\frac{1}{3} \cos^3 \theta \right]_0^{\pi}$$

$$= \frac{e^2 \gamma v}{3ac^2} . \tag{11.133}$$

However, this is not correct. What we should obtain is

$$\Delta \mathbf{P'}\,(\text{correct}) = \frac{v}{3c^2} \gamma U_{\text{em}} = \frac{e^2 \gamma v}{6ac^2} , \tag{11.134}$$

so we have a disagreement by a factor of 2. This can be traced to the fact that the electromagnetic force density on the shell is actually

$$f_{\text{em}}^{\mu} = \left(0, \frac{e}{4\pi a^2} \delta(r-a) \frac{e\hat{\mathbf{r}}}{2a^2} \right) . \tag{11.135}$$

In (11.135), $e/4\pi a^2$ is the surface charge density, but a factor of $1/2$ has crept into the 4-force. We need to see how this follows from the formula

$$\mathbf{E} = \theta(r-a) \frac{e\mathbf{r}}{r^3} \tag{11.136}$$

for the electrostatic field in the unprimed frame, where θ is the usual step function, and we shall do this in Sect. 11.3.3.

Of course, the delicate part of the above argument is the idea that, when we consider our ring of charge, the rest of the shell should act like a point charge at the origin. We ought to prove this in some rigorous manner. We have the remarkable fact that the fields are always zero inside the collapsing shell and that they instantaneously go to the appropriate Coulomb value as the shell passes. But if we start thinking about retarded times in this dynamic situation, it is not nearly so obvious that the field acting on the ring at any given time should be the result of taking the whole distribution as a point charge at the spatial origin.

11.3.2 EM Fields of a Collapsing Charged Spherical Shell

We know that the force density on the shell is given by (see Sect. 2.4)

$$f_{\text{em}}^{\mu} = \partial_{\nu} \Theta_{\text{em}}^{\mu\nu} , \tag{11.137}$$

where $\Theta_{\text{em}}^{\mu\nu}$ comes from the electric field

$$E = \theta(r-a)\frac{e\mathbf{r}}{r^3} ,\tag{11.138}$$

up to the usual factor of $1/\sqrt{4\pi\varepsilon_0}$. To model this situation, we introduce the monotonically decreasing function $a(t)$ which gives the radius of the sphere at time t, with $a(0) = a$ and $\dot{a}(0) = 0$. Then we can propose that the electric field at time t is

$$E = \theta\left[r-a(t)\right]\frac{e\mathbf{r}}{r^3} .\tag{11.139}$$

The magnetic field will be zero everywhere.

Let us check that Maxwell's equations are satisfied! We have to check that

$$\nabla \times E = 0 , \qquad \frac{\partial E}{\partial t} = -4\pi \mathbf{J} , \qquad \nabla \cdot E = 4\pi\rho .\tag{11.140}$$

Note that these differ in the obvious way from the version (2.1) of Maxwell's equations, due to the different units for the charge. We need expressions for ρ and $\mathbf{J}$. The charge density at time t is

$$\rho = \frac{e}{4\pi a(t)^2}\delta\left[r-a(t)\right] .\tag{11.141}$$

The current density is

$$\begin{aligned}\mathbf{J} &= \rho\mathbf{v}_{\text{collapse}}\\ &= \frac{e}{4\pi a(t)^2}\delta\left[r-a(t)\right]\dot{a}(t)\hat{\mathbf{r}} ,\end{aligned}\tag{11.142}$$

since $\mathbf{v}_{\text{collapse}} = \dot{a}(t)\hat{\mathbf{r}}$. Note that $\dot{a}(t) < 0$.

It may appear that there is something missing from these expressions for the 4-current density, namely, a factor of $\gamma(\dot{a})$. In the relativistic context, the zero component of the 4-current density is normally the local proper charge density multiplied by the γ factor for whatever speed the charged fluid may have there. Likewise, the spatial part of the 4-current density is proportional to the coordinate 3-velocity with the associated γ factor. The above expressions thus imply that the local proper charge density is in this case

$$\rho_{\text{proper}} = \frac{e}{4\pi a(t)^2}\delta\left[r-a(t)\right]\gamma^{-1}\left[\dot{a}(t)\right] .\tag{11.143}$$

This has a slightly unnatural appearance. However, it is right. It is the presence of the delta function in the radial direction that leads to this odd-looking formula.

In fact, we can see directly that (11.141) and (11.142) are correct. To begin with, they have the right ratio, which is just the coordinate 3-velocity. Secondly, (11.141) gives the right amount of charge when integrated over spatial regions in some hyperplane of simultaneity of this coordinate system. Let us explain this. Consider a small region δS on the shell, defined by angles subtended at the origin. Enclose in an open region of thickness δr, defined by these same angles. Then (11.141) says that the charge in this region, as observed in the present frame of reference which is not moving with the charge, is just

$$\rho \delta S \delta r = \frac{e}{4\pi a(t)^2} \delta S \, .$$

We have used the fact that the presence of $\delta[r - a(t)] \delta r$ just evaluates the rest of the expression at $r = a(t)$. This is indeed, the amount of charge we expect to find in the given region. The total charge is then e.

Now (11.143) is the charge density in a frame moving with the charge locally, which is actually a scalar field on the manifold. But the coordinate r appearing in the expression is not the coordinate of an observer that would be moving with the charge! When we make the appropriate transformation of the variable appearing in the delta function, this will introduce a γ factor, precisely because it is the radial coordinate that appears in the delta function, and the difference in motion between the first observer (using coordinate r) and the second observer moving with the charge is actually radial.

All this is very fortunate, because it is obvious that (11.139) is the right electric field here, and it actually implies the charge density (11.141) via the Maxwell equations. Let us now check that Maxwell's equations are satisfied. Note that we have

$$\nabla \times \frac{\mathbf{r}}{r^3} = 0 \, . \tag{11.144}$$

Now

$$\nabla \times \mathbf{E} = e\nabla \times \frac{\theta[r - a(t)]\mathbf{r}}{r^3} = e \begin{vmatrix} \mathbf{i} & \mathbf{j} & \mathbf{k} \\ \dfrac{\partial}{\partial x} & \dfrac{\partial}{\partial y} & \dfrac{\partial}{\partial z} \\ \dfrac{\theta x}{r^3} & \dfrac{\theta y}{r^3} & \dfrac{\theta z}{r^3} \end{vmatrix}$$

$$= e \begin{pmatrix} \dfrac{\partial \theta}{\partial y}\dfrac{z}{r^3} - \dfrac{\partial \theta}{\partial z}\dfrac{y}{r^3} \\ \dfrac{\partial \theta}{\partial z}\dfrac{x}{r^3} - \dfrac{\partial \theta}{\partial x}\dfrac{z}{r^3} \\ \dfrac{\partial \theta}{\partial x}\dfrac{y}{r^3} - \dfrac{\partial \theta}{\partial y}\dfrac{x}{r^3} \end{pmatrix} = e\delta[r - a(t)] \begin{pmatrix} \dfrac{\partial r}{\partial y}\dfrac{z}{r^3} - \dfrac{\partial r}{\partial z}\dfrac{y}{r^3} \\ \dfrac{\partial r}{\partial z}\dfrac{x}{r^3} - \dfrac{\partial r}{\partial x}\dfrac{z}{r^3} \\ \dfrac{\partial r}{\partial x}\dfrac{y}{r^3} - \dfrac{\partial r}{\partial y}\dfrac{x}{r^3} \end{pmatrix}$$

$$= 0 \, .$$

Furthermore,

$$\frac{\partial \mathbf{E}}{\partial t} = -\dot{a}(t)\delta[r - a(t)]\frac{e\mathbf{r}}{r^3} = -4\pi \mathbf{J} \, ,$$

by comparison with (11.142). Concerning the last equation above, we have

$$\nabla \cdot \mathbf{E} = e\theta\left(r - a(t)\right)\nabla \cdot \frac{\mathbf{r}}{r^3} + \frac{e\mathbf{r}}{r^3} \cdot \nabla \theta\left(r - a(t)\right)$$

$$= \frac{e\mathbf{r}}{r^3} \cdot \begin{pmatrix} \dfrac{\partial}{\partial x}\theta\left(r - a(t)\right) \\[2mm] \dfrac{\partial}{\partial y}\theta\left(r - a(t)\right) \\[2mm] \dfrac{\partial}{\partial z}\theta\left(r - a(t)\right) \end{pmatrix}$$

$$= \frac{e\mathbf{r}}{r^3} \cdot \begin{pmatrix} \partial r/\partial x \\ \partial r/\partial y \\ \partial r/\partial z \end{pmatrix} \delta\left(r - a(t)\right)$$

$$= \frac{e\mathbf{r}}{r^3} \cdot \frac{\mathbf{r}}{r}\delta\left(r - a(t)\right)$$

$$= \frac{e\delta\left(r - a(t)\right)}{r^2}$$

$$= 4\pi\rho .$$

This shows that, whatever the function $a(t)$ describing the collapse, the fields

$$\mathbf{E} = \theta\left[r - a(t)\right]\frac{e\mathbf{r}}{r^3} , \qquad \mathbf{B} = 0 \qquad (11.145)$$

satisfy Maxwell's equations for the sources

$$\rho = \frac{e}{4\pi a(t)^2}\delta\left(r - a(t)\right) , \qquad \mathbf{J} = \frac{e}{4\pi a(t)^2}\delta\left(r - a(t)\right)\dot{a}(t)\hat{\mathbf{r}} . \qquad (11.146)$$

This means that, whatever the function $a(t)$, these are the fields generated by the collapsing charged spherical shell.

The function $a(t)$ remains undetermined by Maxwell's equations. We can get the shell to collapse at different rates, and Maxwell's equations merely tell us what fields will be generated. On the other hand, the Lorentz force law tells us how these fields will act on the shell. In order to impose some particular $a(t)$, we will need to overcome the electromagnetic forces by applying, in general, other forces. Note, however, that this is not the reason why we do not expect the associated electro-magnetic energy–momentum tensor to be conserved. This tensor is not conserved where there is matter, because we must then add a mass-energy tensor and it is the total that is conserved. Indeed, it is the conservation of this total that implies the Lorentz force law (see Sect. 2.4 and also [8, p. 115]).

For the sake of completeness, let us obtain the equation for $a(t)$ by applying the Lorentz force law. This might be compared with Sect. 5.2 of Parrott's excellent book [8], where he discusses the general spherically symmetric situation. [But note a subtle difference with his equation (17) on p. 179 which is due to the fact that the charged dust is infinitely compressed in the radial direction! Indeed, it highlights the fact that we are still using a purely mathematical notion here to model our electron, namely the notion of an infinitely thin surface of charge.]

We have to borrow the results of Sect. 11.3.3 below here to get the 4-force density on the shell. Starting with (11.157), viz.,

$$
\Theta_{em} = \varepsilon_0 \theta \left[r - a(t) \right]
\begin{pmatrix}
-\dfrac{e^2}{2r^4} & 0 \\[2ex]
0 & \dfrac{e^2}{2r^6}
\begin{pmatrix}
x^2 & xy & xz \\
yx & y^2 & yz \\
zx & zy & z^2
\end{pmatrix}
- \dfrac{e^2}{2r^4}\mathbb{I}
\end{pmatrix} ,
\tag{11.147}
$$

we calculate

$$
f_{em}^{\mu} = \Theta_{em,\nu}^{\mu\nu} = \frac{e^2}{8\pi r^4} \delta\left[r - a(t)\right] (\dot{a}, \hat{\mathbf{r}}) .
\tag{11.148}
$$

This is obtained by the same arguments as are used below to get (11.162), except that we have not used $\dot{a}(0) = 0$ here. In other words, we are obtaining the 4-force density at a general time t.

Now consider the force on a small element of the shell with area δS, multiplied by a radial thickness δr. The latter will evaluate the integrand at $r = a(t)$ via the delta function when we calculate the force on the element. Hence, the Lorentz force law gives

$$
m\frac{du}{d\tau} = \frac{e^2}{8\pi a^4} (\dot{a}, \hat{\mathbf{r}}) \delta S .
\tag{11.149}
$$

On the left-hand side, m is the mass of the element. Hence,

$$
m = \frac{m_e}{4\pi a^2} \delta S ,
\tag{11.150}
$$

where m_e is the mass of the electron. Furthermore, the 4-velocity of the element is

$$
u = \gamma(\dot{a}) (1, \dot{a}\hat{\mathbf{r}}) .
\tag{11.151}
$$

The Lorentz force law now reads

$$
m_e \frac{d}{d\tau} \left[\gamma(\dot{a}) (1, \dot{a}\hat{\mathbf{r}}) \right] = \frac{e^2}{2a^2} (\dot{a}, \hat{\mathbf{r}}) .
\tag{11.152}
$$

We recall that

$$
\frac{d}{d\tau} = \gamma \frac{d}{dt} .
$$

Hence, after a little calculation,

$$
m_e \gamma^4 \ddot{a} (\dot{a}, \hat{\mathbf{r}}) = \frac{e^2}{2a^2} (\dot{a}, \hat{\mathbf{r}}) .
\tag{11.153}
$$

This reduces to one equation due to the proportionality of the vectorial aspect. Hence, finally,

$$\gamma^4 \ddot{a} a^2 = \frac{e^2}{2m_e} . \tag{11.154}$$

Note that γ is a function of $\dot{a}$, making this a very nonlinear second order differential equation!

The idea now would be to solve this equation starting with some initial conditions for a shell collapsing from infinity, to find out how big it would be at the instant when charge repulsion causes the collapse to stop, just prior to rebound. We shall not attempt this. What is important for the moment is to see that it is indeed the Lorentz force law that determines $a(t)$, even though Maxwell's equations alone were satisfied for any function $a(t)$. The latter fact just means that we can tamper with the collapse (provided that it remains spherically symmetric) via other forces and we still know what the electromagnetic fields will be. Effectively, the Lorentz force law tells us how we must tamper with the collapse, i.e., what radial forces must be supplied, in order to obtain some arbitrary function $a(t)$.

Let us compare (11.154) briefly with Parrott's equation (17) on p. 179 of his book [8], viz.,

$$\frac{d^2 r}{dt^2} = k^{-1} \left[1 - \left(\frac{dr}{dt} \right)^2 \right]^{3/2} \frac{Q}{r^2} .$$

The function $r(t)$ corresponds to $a(t)$ and $k := m_e/e$. In the context of Parrott's discussion, $Q(r)$ is a radially symmetric function on $\mathbb{M}$ which basically gives the amount of charge contained in a sphere of radius r. The main difference with (11.154) is that the latter contains a factor of γ^4 rather than γ^3. This arises because the charged dust is infinitely thin in the radial direction! See the discussion just after (11.141) and (11.142). Another difference, undoubtedly of the same origin, is a factor of 2.

Parrott points out that the equation he gives has no easy solutions. What about (11.154)? It is not quite the same in structure as Parrott's equation, due to the extra γ factor, so it might just be soluble.

11.3.3 Energy–Momentum Tensor for Collapsing Charged Spherical Shell

Let us now write down the energy–momentum tensor according to our definition (2.32) on p. 11. By this definition,

$$\Theta_{em} = \varepsilon_0 \begin{pmatrix} -\frac{1}{2}E^2 & 0 \\ 0 & \mathbf{EE} - \frac{1}{2}E^2 \mathbb{I} \end{pmatrix} . \tag{11.155}$$

Let us just show that the covariant divergence of this is zero away from the shell. When $r < a(t)$, the fields are zero and this is obvious. When $r > a(t)$, we have

$$E^2 = \frac{e^2}{r^4}, \qquad \mathbf{EE} = \frac{e^2}{r^6} \begin{pmatrix} x^2 & xy & xz \\ yx & y^2 & yz \\ zx & zy & z^2 \end{pmatrix}. \tag{11.156}$$

Hence,

$$\Theta_{\text{em},\nu}^{\mu\nu} = \varepsilon_0 e^2 \begin{pmatrix} -\frac{1}{2}\partial_0 \frac{1}{r^4} \\ \partial_x \frac{x^2}{r^6} + \partial_y \frac{xy}{r^6} + \partial_z \frac{xz}{r^6} - \frac{1}{2}\partial_x \frac{1}{r^4} \\ \partial_x \frac{xy}{r^6} + \partial_y \frac{y^2}{r^6} + \partial_z \frac{yz}{r^6} - \frac{1}{2}\partial_y \frac{1}{r^4} \\ \partial_x \frac{zx}{r^6} + \partial_y \frac{zy}{r^6} + \partial_z \frac{z^2}{r^6} - \frac{1}{2}\partial_z \frac{1}{r^4} \end{pmatrix}.$$

A straightforward calculation shows that all four components are zero, something we knew anyway. This just says that the energy–momentum tensor is conserved outside the shell.

The interesting part of the calculation of $\Theta_{\text{em},\nu}^{\mu\nu}$ is the part which deals with the discontinuity at the charge shell. Since the step function θ squares to itself,

$$\Theta_{\text{em}} = \varepsilon_0 \theta \left[r - a(t) \right] \begin{pmatrix} -\dfrac{e^2}{2r^4} & 0 \\ 0 & \dfrac{e^2}{2r^6} \begin{pmatrix} x^2 & xy & xz \\ yx & y^2 & yz \\ zx & zy & z^2 \end{pmatrix} - \dfrac{e^2}{2r^4}\mathbb{I} \end{pmatrix}, \tag{11.157}$$

with the obvious notation. Apart from the step function in front, this is the same as we had just above. This gives the pattern

$$\Theta_{\text{em},\nu}^{\mu\nu} = \varepsilon_0 e^2 \begin{pmatrix} -\dfrac{1}{2r^4}\partial_0 \theta \left[r - a(t) \right] \\ (\partial_x \theta)\left[\dfrac{x^2}{r^6} - \dfrac{1}{2r^4}\right] + (\partial_y \theta)\dfrac{xy}{r^6} + (\partial_z \theta)\dfrac{xz}{r^6} \\ (\partial_x \theta)\dfrac{xy}{r^6} + (\partial_y \theta)\left[\dfrac{y^2}{r^6} - \dfrac{1}{2r^4}\right] + (\partial_z \theta)\dfrac{yz}{r^6} \\ (\partial_x \theta)\dfrac{zx}{r^6} + (\partial_y \theta)\dfrac{zy}{r^6} + (\partial_z \theta)\left[\dfrac{z^2}{r^6} - \dfrac{1}{2r^4}\right] \end{pmatrix}. \tag{11.158}$$

Now

$$\partial_0 \theta \left[r - a(t) \right] = -\dot{a}(t)\delta \left[r - a(t) \right], \tag{11.159}$$

and this is zero when $t = 0$ because $\dot{a}(0) = 0$. This alone shows that the dynamic aspect of the situation, i.e., the collapse, will not influence our arguments, provided we consider the situation at $t = 0$, when the stabilising forces are applied.

We also have

$$\partial_x \theta = \delta[r - a(t)]\frac{\partial r}{\partial x} = \frac{x}{r}\delta[r - a(t)], \qquad (11.160)$$

with similar results for $\partial_y \theta$ and $\partial_z \theta$. Evaluating at $t = 0$, we have

$$\Theta_{em}^{\mu\nu}{}_{,\nu}\big|_{t=0} = \varepsilon_0 e^2 \delta(r-a) \begin{pmatrix} 0 \\ \dfrac{x}{r}\left[\dfrac{x^2}{r^6} - \dfrac{1}{2r^4}\right] + \dfrac{xy^2}{r^7} + \dfrac{xz^2}{r^7} \\ \dfrac{x^2 y}{r^7} + \dfrac{y}{r}\left[\dfrac{y^2}{r^6} - \dfrac{1}{2r^4}\right] + \dfrac{yz^2}{r^7} \\ \dfrac{zx^2}{r^7} + \dfrac{zy^2}{r^7} + \dfrac{z}{r}\left[\dfrac{z^2}{r^6} - \dfrac{1}{2r^4}\right] \end{pmatrix}$$

$$= \frac{\varepsilon_0 e^2}{2r^5}\delta(r-a)\begin{pmatrix} 0 \\ \mathbf{r} \end{pmatrix}. \qquad (11.161)$$

This is to be compared with the result (11.135) we claimed on p. 216, viz.,

$$f_{em}^{\mu} = \left(0, \frac{e}{4\pi a^2}\delta(r-a)\frac{e\hat{\mathbf{r}}}{2a^2}\right). \qquad (11.162)$$

We have the same result if we make the usual replacement $\varepsilon_0 \to 1/4\pi$. We do now have the factor of 1/2.

It is interesting to see exactly how the factor of 1/2 arises here. Let us rederive the above relation for the EM force density by considering the force on a charge q with 4-velocity v^i due to electromagnetic fields F^{ij} in the form

$$\phi^i = qF^i{}_j v^j. \qquad (11.163)$$

This is the Lorentz force law. Note that qv^j is the 4-current. If we have a charge distribution with local rest frame density ρ and 4-velocity field v^i, the 4-current density of the distribution is $J^i = \rho v^i$. The 4-force density on this distribution is

$$f^i = F^i{}_k J^k. \qquad (11.164)$$

In the present case,

$$\rho = \frac{e}{4\pi a^2}\delta(r-a) \quad \text{and} \quad v^i = (c,0,0,0), \qquad (11.165)$$

at $t = 0$. At earlier times, we can replace a by $a(t)$ and we can write an expression for the 4-velocity field in terms of $a(t)$, viz.,

$$v^\mu = \gamma(\dot{a})\big(c, \dot{a}(t)\hat{\mathbf{r}}\big) . \tag{11.166}$$

At $t = 0$, other forces come into play in Boyer's scenario, and Maxwell's equations no longer determine the whole game. The function $a(t)$ is not allowed to follow its natural course under the sole influence of the electromagnetic effects. Instead, it remains constant at the value a. But note that the energy–momentum tensor will never be conserved at $r = a(t)$, according to (11.161). Mechanical momentum is converting into electromagnetic momentum all the time in this scenario.

We can evaluate the electromagnetic 4-force density at $t = 0$ using (11.164), which gives

$$F^i{}_k J^k = \begin{pmatrix} 0 & E_1/c & E_2/c & E_3/c \\ E_1/c & 0 & 0 & 0 \\ E_2/c & 0 & 0 & 0 \\ E_3/c & 0 & 0 & 0 \end{pmatrix} \begin{pmatrix} J^0 \\ 0 \\ 0 \\ 0 \end{pmatrix} = \begin{pmatrix} 0 \\ J^0 E_1/c \\ J^0 E_2/c \\ J^0 E_3/c \end{pmatrix}$$

$$= \frac{e}{4\pi a^2}\delta(r-a)(0, \mathbf{E}) . \tag{11.167}$$

Now we know that

$$\mathbf{E} = \theta(r-a)\frac{e\hat{\mathbf{r}}}{r^2} , \tag{11.168}$$

implying that

$$F^i{}_k J^k = \frac{e}{4\pi a^2}\delta(r-a)\theta(r-a)\frac{e\hat{\mathbf{r}}}{a^2} . \tag{11.169}$$

It is the product of the delta function and the step function which gives rise to the factor of 1/2, i.e., it is a standard result from distribution theory that

$$\theta(x)\delta(x) = \frac{1}{2}\delta(x) . \tag{11.170}$$

Returning to the calculation (11.169), we now have

$$F^i{}_k J^k = \frac{e}{4\pi a^2}\delta(r-a)\frac{e\hat{\mathbf{r}}}{2a^2} . \tag{11.171}$$

11.3.4 The Main Argument

We now return to the main argument, continuing from where we left off on p. 216. Given the result (11.162), we insert a factor of 1/2 into our (11.130). We now find that

$$\Delta \mathbf{P}' = \frac{\mathbf{v}}{3c^2}\gamma U_{\text{em}} = \frac{e^2 \gamma \mathbf{v}}{6ac^2} \,. \tag{11.172}$$

The total system momentum after all the external stabilising forces have been applied has been changed in the S' frame from the value

$$\mathbf{P}'_{\text{tot}} = \frac{\mathbf{v}}{c^2}\gamma U_{\text{tot}}$$

to

$$\mathbf{P}_{\text{tot}}^{\prime(\text{after})} = \frac{\mathbf{v}}{c^2}\gamma U_{\text{tot}} + \frac{\mathbf{v}}{3c^2}\gamma U_{\text{em}}$$

$$= \mathbf{v}\gamma m + \frac{4}{3}\frac{\mathbf{v}}{c^2}\gamma U_{\text{em}} \,. \tag{11.173}$$

This corresponds to the momentum of the mechanical mass m and exactly the electromagnetic momentum found by integrating over the traditional field momentum density [noting the different units compared to (2.39) on p. 12]

$$\mathbf{g}' = \frac{1}{4\pi c}\mathbf{E}' \times \mathbf{B}' \,. \tag{11.174}$$

Boyer calculates this as follows (contrast with our method in Sect. 3.2). We have to evaluate

$$\mathbf{P}'_{\text{em}} := \frac{1}{4\pi c}\int \mathrm{d}^3 x' \mathbf{E}' \times \mathbf{B}' \,. \tag{11.175}$$

We express $\mathbf{E}'$ and $\mathbf{B}'$ in terms of $\mathbf{E}$ and $\mathbf{B}$ using the standard formulas [2, Chap. 26]

$$\mathbf{E}' = \gamma(\mathbf{E} - c\boldsymbol{\beta} \times \mathbf{B}) - \frac{\gamma^2}{\gamma+1}\boldsymbol{\beta}(\boldsymbol{\beta} \cdot \mathbf{E}) \,, \tag{11.176}$$

$$\mathbf{B}' = \gamma\left(\mathbf{B} + \frac{1}{c}\boldsymbol{\beta} \times \mathbf{E}\right) - \frac{\gamma^2}{\gamma+1}\boldsymbol{\beta}(\boldsymbol{\beta} \cdot \mathbf{B}) \,, \tag{11.177}$$

where $\boldsymbol{\beta} = \mathbf{v}/c$. Fortunately, $\mathbf{B} = 0$, so

$$\mathbf{E}' = \gamma\mathbf{E} - \frac{\gamma^2}{(\gamma+1)c^2}\mathbf{v}(\mathbf{v} \cdot \mathbf{E}) \,, \tag{11.178}$$

$$\mathbf{B}' = \frac{\gamma}{c^2}\mathbf{v} \times \mathbf{E} \,. \tag{11.179}$$

Hence,

$$\mathbf{E}' \times \mathbf{B}' = \frac{\gamma^2}{c^2}\mathbf{E} \times (\mathbf{v} \times \mathbf{E}) - \frac{\gamma^3}{(\gamma+1)c^4}(\mathbf{v} \cdot \mathbf{E})\mathbf{v} \times (\mathbf{v} \times \mathbf{E})$$

$$= \frac{\gamma^2}{c^2}\left[E^2\mathbf{v} - (\mathbf{v} \cdot \mathbf{E})\mathbf{E}\right] - \frac{\gamma^3}{(\gamma+1)c^4}(\mathbf{v} \cdot \mathbf{E})\left[(\mathbf{v} \cdot \mathbf{E})\mathbf{v} - v^2\mathbf{E}\right] \,.$$

Because $\mathbf{E}$ and hence the whole integrand is constant in t (we are now considering the static charge shell), we can replace the integral

$$\int_{\text{HOS}'} d^3x' \longrightarrow \int_{\text{HOS}} \frac{d^3x}{\gamma} \; .$$

Now

$$E^2\mathbf{v} - (\mathbf{v}\cdot\mathbf{E})\mathbf{E} = (E_x^2 + E_y^2 + E_z^2)\begin{pmatrix} v \\ 0 \\ 0 \end{pmatrix} - \begin{pmatrix} vE_x^2 \\ vE_xE_y \\ vE_xE_z \end{pmatrix}$$

$$= \mathbf{v}(E_y^2 + E_z^2) - v\begin{pmatrix} 0 \\ E_xE_y \\ E_xE_z \end{pmatrix} \; .$$

Both nonzero components of the second term integrate to zero by symmetry. We also have to consider

$$(\mathbf{v}\cdot\mathbf{E})^2\mathbf{v} - v^2(\mathbf{v}\cdot\mathbf{E})\mathbf{E} = v^2E_x^2\begin{pmatrix} v \\ 0 \\ 0 \end{pmatrix} - v^3\begin{pmatrix} E_x^2 \\ E_xE_y \\ E_xE_z \end{pmatrix}$$

$$= -v^3\begin{pmatrix} 0 \\ E_xE_y \\ E_xE_z \end{pmatrix} \; ,$$

which integrates to zero by symmetry. We are left with

$$\frac{1}{4\pi c}\int d^3x'\,\mathbf{E}'\times\mathbf{B}' = \frac{\gamma}{4\pi c^3}\int_{\text{HOS}} d^3x\,\mathbf{v}(E_y^2 + E_z^2) \; . \tag{11.180}$$

By the spherical symmetry of the fields in the unprimed frame S, we can replace

$$E_y^2 + E_z^2 \longrightarrow \frac{2}{3}E^2$$

in the integrand, and hence

$$\frac{1}{4\pi c}\int d^3x'\,\mathbf{E}'\times\mathbf{B}' = \frac{4}{3}\frac{\mathbf{v}}{c^2}\gamma U_{\text{em}} \; , \tag{11.181}$$

as required above.

11.3.5 Conclusion Regarding the Collapsing Charge Shell

The factor of 4/3 in the electromagnetic momentum is by no means an anomaly. It is needed to maintain the validity of the force–momentum balance in the S' frame.

A physical particle or system will in general involve contributions to the total momentum from both the electromagnetic fields and other sources. In our example the mechanical momentum of the shell at spatial infinity is converted into electromagnetic momentum as the shell rushes inward. Only the total momentum can be expected to satisfy covariant behaviour when transformed between different inertial frames. So the factor of 4/3 may not actually be an embarrassment that should be removed.

One method for removing the factor involves redefining the electromagnetic momentum of the system so that it is not the integral of the density

$$\mathbf{g} := \frac{1}{4\pi c} \mathbf{E} \times \mathbf{B} \,,$$

but the integral of something rather different. This is what we shall investigate shortly (see Sect. 11.4). Such redefinitions are not what is advocated in this book, but they need to be given due consideration.

The view here is that Boyer's arguments in [23] do indeed hold water. He claims that the usual ideas of force, energy, and momentum hold together properly with the traditional definition and not with the use of the density function to be considered shortly [see (11.199) on p. 235]. The latter may eliminate the factor of 4/3, but what relation does it bear to momentum as we know it? Consider the example of the spherical charge shell: if the laws of physics are to hold for all inertial frames in such an open system in which non-electromagnetic external forces are applied, then the electromagnetic field momentum should not transform as a Lorentz four-vector and the factor of 4/3 is a consistent reflection of this fact.

Boyer considers that it is an error to take seriously as a model for a point charged particle the electromagnetic energy and momentum behaviour of the classical model of the electron despite the non-electromagnetic forces required for stability in the classical model [23]. The non-electromagnetic stabilising forces play a crucial role and the attempts to circumvent the role played by these forces by redefining the electromagnetic momentum density only destroy the conceptual simplicity of the traditional view of classical electrodynamics. Naturally, Rohrlich does not agree [22].

Note the essential role of the mechanical mass for the collapsing shell model. Indeed, it could not be set to zero for this construction of the electron from infinity (which is not intended to be realistic anyway). The point is that, at the beginning of the construction, all the energy and momentum of the system is in the mechanical mass and its motion. Viewed in the inertial frame in which the electron is finally at rest, part of this mechanical energy–momentum is gradually converted to electrostatic field energy. However, this observation should not lead the reader to think that we have here an argument against the idea promoted in this book that all inertial mass might be of bootstrap type. It is simply a quirk of the present demonstration, which merely aimed to show that the 4/3 factor should be present in our formulas, or put another way, that the energy and momentum of the EM fields outside a charge shell should not form a Lorentz covariant four-vector.

In a moment, we shall reconsider Rohrlich's formulas (11.11) and (11.13) on p. 185 and see exactly why they give a Lorentz covariant four-vector. But first, let us attempt to apply the Boyer collapse argument to the charge dumbbell. This will be an opportunity to understand it in a different context, but one that turns out to be physically more sophisticated than the one provided by the charge shell. At the same time, it will give us a clue as to why there is no discrepancy between the energy- and momentum-derived EM masses for transverse motion of the system (see Table 10.1).

11.3.6 Collapse Model for the Charge Dumbbell

Consider a model in which two charges $q_e/2$ move in from infinity along a straight line to meet one another (or almost) at the origin. They each start out with the same speed, which is exactly the right value to ensure that, when they are just a distance d apart (each a distance $d/2$ from the origin), they are both stationary in the inertial frame from which we view this (the center of mass frame). At this point the binding forces of the system are switched on. The two charges subsequently remain motionless in the given frame.

The binding force we have to switch on at the crucial instant (just one instant of time for this frame!) is the Coulomb force $e^2/4d^2$, where $e^2 := q_e^2/4\pi\varepsilon_0$. Now imitating the collapsing shell model, one would like to say that the EM fields thereafter have energy

$$U_{em} = \frac{e^2}{4d} ,$$

as calculated in Sect. 5.1 [see (5.4) on p. 74]. But, of course, something goes wrong here, because that would only be the energy if the two charges had always been sitting at their final resting places. The fact that they were just previously in motion spoils this picture. The beauty of the spherically symmetric charge shell model is that the fields outside it are always instantaneously the Coulomb fields, as shown in Sect. 11.3.2.

In this frame, the initial kinetic energy of the charges is

$$\text{initial kinetic energy} = \sum_{\text{charges}} m_{\text{charge}} c^2 (\gamma - 1) ,$$

where each charge is assumed to have a mechanical mass m_{charge}, and γ is the usual relativistic function of the required initial speed. That is fine, but unfortunately, there is always a great deal of EM energy in the fields, even initially. Once again, the beauty of the collapsing shell model is that the fields are identically zero within it, so if we assume that the whole of space is initially within it, then there is no EM energy to begin with.

Note also that, in the case of the spherical charge shell, the magnetic field was always zero outside the shell during collapse, so there could be no radiation loss. This meant that all the initial kinetic energy at infinity could be converted into electrostatic potential energy when the shell came momentarily to rest. In the present case, there are definitely magnetic fields around, and there will definitely be EM radiation. These considerations make a collapse scenario much less informative for the charge dumbbell.

However, there is an interesting point here, which will eventually explain why there is no discrepancy between the energy- and momentum-derived EM masses for transverse motion of the system, as is evident from Table 10.1. In Sect. 11.6.1, we shall consider a system (the charge shell again) that has always been stabilised by some cohesive forces, and we shall show how the binding forces contribute to the total four-momentum of the system in a frame in which the system appears to be in motion.

The reason is just this. When the system is moving, the cohesive forces act for different lengths of time on different parts of the system, in such a way that they do not always balance one another. This transpires directly from examination of the calculations in Sect. 11.6.1, and the point is discussed explicitly in Sect. 11.6.2. The idea is already present in the collapse model for the shell, and it perdures even in the case where, rather than switching on the cohesive forces at some specific set of spacetime events, these forces have always been in operation.

Now consider the charge dumbbell in transverse motion along the x axis. We see immediately that, if the binding force were switched on at some specific time in the rest frame, it would nevertheless affect both charges at the same time in the moving frame, just because their positions have the same x coordinate at any instant of time in the moving frame. We would never therefore expect to see an imbalance of the cohesive forces that could lead to a momentum contribution, and this perdures to the ever-stabilised case.

But when we consider the same dumbbell in longitudinal motion along the x axis, switching on a binding force at some specific rest frame time amounts in the moving frame to switching on the force on the left-hand particle first. The binding force would only come into effect on the right-hand particle a time $2\gamma va/c^2$ later, where $a = d/2$, so the force to the right on the left-hand particle would be unbalanced for a time $\gamma vd/c^2$. In the rest frame system, the force to be countered by the binding is just the Coulomb force $e^2/4d^2$. In the moving frame, this gives a four-force

$$F' = \begin{pmatrix} \gamma & \gamma v/c \\ \gamma v/c & \gamma \end{pmatrix} \begin{pmatrix} 0 \\ e^2/4d^2 \end{pmatrix} = \begin{pmatrix} \dfrac{\gamma v}{c}\dfrac{e^2}{4d^2} \\ \dfrac{\gamma e^2}{4d^2} \end{pmatrix}.$$

We are interested in $F'/\gamma = e^2/4d^2$, as explained on p. 215, so the impulse here is just the product, viz.,

$$\text{impulse due to binding forces} = \frac{\gamma vd}{c^2}\frac{e^2}{4d^2} = \frac{e^2}{4c^2d}\gamma v,$$

and this is in the direction of $\mathbf{v}$. This corresponds precisely to the discrepancy between the energy-derived EM mass $e^2/4dc^2$ and the momentum-derived EM mass

$e^2/2dc^2$ for this case (see Table 10.1). The latter should indeed be twice the former, because we have to add in the impulse from the binding forces, when viewing from this frame.

Strictly speaking, this analysis needs to be reformulated for the case of an ever-stabilised charge dumbbell, with the kind of analysis used in Sects. 11.6.1 and 11.6.2 for the charge shell. This can be left as the proverbial exercise for the reader (see in particular the end of Sect. 11.6.2).

11.4 Why the Redefined Four-Momentum Is Lorentz Covariant

We shall use the notation x_R for rest frame coordinates and x without a subscript for coordinates in the frame moving to the left in such a way that the electron world-line moves off to the right, with four-velocity components v^μ. The subscript R will generally indicate that a quantity is found relative to the electron rest frame. Note also that we do not take $\gamma d\sigma = d^3x$ in Rohrlich's formulas, but rather $d\sigma := \gamma d^3x$. One of the assertions made here is that we must in fact make this definition in order to obtain covariance. Finally, we use an observation similar to the one in (11.61) on p. 198, wherein the rest frame energy–momentum tensor is seen to be constant with respect to the rest frame time t_R.

We thus take Rohrlich's definition to be

$$P^\mu = -\frac{1}{c} \int_{\substack{\text{HOS} \\ \text{outside electron}}} T^{\mu\nu} v_\nu \gamma d^3x . \tag{11.182}$$

Let $L^a{}_b$ be the Lorentz transformation from the rest frame of the electron to the frame in which it appears to be moving to the right. Then the argument is this:

$$L^a{}_b P^b_R = -\frac{1}{c} \int_{\substack{\text{HOS(R)} \\ \text{outside electron}}} L^a{}_b T^{bc}_R v^R_c d^3x_R$$

$$= -\frac{1}{c} \int_{\substack{\text{HOS} \\ \text{outside electron}}} L^a{}_b T^{bc}_R v^R_c \gamma d^3x$$

$$= -\frac{1}{c} \int_{\substack{\text{HOS} \\ \text{outside electron}}} T^{ac} v_c \gamma d^3x .$$

In the second step we move from an integration over the rest frame HOS to an integration over the HOS in the moving frame, using the fact that everything in the integrand is constant with respect to the rest frame time (although not with respect to the time in the moving frame).

This is the exact parallel of what is expressed in (11.61). We are using something rather accidental, i.e., the fact that the electron is decreed a static structure in its rest frame, so that the energy–momentum tensor in that frame is constant with respect to time in that frame. Furthermore, this manoeuvre has nothing of the standard integration theory. This is illustrated by the fact that the covector-valued measure $v_\mu \gamma d^3x$ is not normal to the surface of integration in the definition (11.182). Indeed, it is normal to the rest frame HOS. In other words, whatever HOS we integrate over, we always reduce to an integration over the rest frame HOS, as witnessed by the explicit appearance of v_μ in the formula.

The above argument naturally works backwards:

$$
\begin{aligned}
(L^{-1})^a{}_b P^b &= -\frac{1}{c} \int_{\substack{\text{HOS} \\ \text{outside electron}}} (L^{-1})^a{}_b T^{bc} v_c \gamma d^3 x \\
&= -\frac{1}{c} \int_{\substack{\text{HOS} \\ \text{outside electron}}} T_R^{ac} v_c^R \gamma d^3 x \\
&= -\frac{1}{c} \int_{\substack{\text{HOS(R)} \\ \text{outside electron}}} T_R^{ac} v_c^R d^3 x_R .
\end{aligned}
$$

In the last step, we use the fact that the integrand is constant with respect to the rest frame time t_R. This shows quite sharply why Rohrlich's definition gives a Lorentz covariant quantity. Let us now see the connection with the well known Lorentz invariant $F_{mn}F^{mn}$.

Connection with $F_{mn}F^{mn}$

We have just seen that, although Rohrlich achieves Lorentz covariance, his definition (11.182) is somewhat contrived: we use a covector field v_m that is not normal to the relevant spacelike hypersurface in spacetime and we stick in a factor of γ. What is more, it only works because the electron has a static structure in its rest frame, so that the energy–momentum tensor of its fields is constant with respect to time in that frame. Let us therefore aim for a slightly more elegant formulation in terms of the well known Lorentz invariant $F_{mn}F^{mn}$.

With the definition (2.12) of F_{mn} on p. 8, it is easy to check that

$$
F_{mn}F^{mn} = -\frac{2}{c^2}\left(E^2 - c^2 B^2\right) . \tag{11.183}
$$

Let us first look at the situation in the electron rest frame. Here we have $\mathbf{B} = 0$ and we note that

$$
-\frac{\varepsilon_0 c^2}{4} F_{mn}^R F_R^{mn} = \frac{\varepsilon_0}{2} E_R^2 = u_R . \tag{11.184}
$$

Since we have the argument

$$
\begin{aligned}
P_R^0 &= \int_{\substack{\text{HOS(R)} \\ \text{outside electron}}} u_R d^3 x_R \\
&= \int_{\substack{\text{HOS} \\ \text{outside electron}}} -\frac{\varepsilon_0 c^2}{4} F_{mn}^R F_R^{mn} \gamma d^3 x \\
&= \int_{\substack{\text{HOS} \\ \text{outside electron}}} -\frac{\varepsilon_0 c^2}{4} F_{mn} F^{mn} \gamma d^3 x ,
\end{aligned}
$$

it might on the face of it look as though the other term in the energy–momentum tensor, i.e., $-\varepsilon_0 c^2 F^i{}_l F^{lk}$, does not contribute. In fact, this is not so because of the signs associated with the terms. We have

$$T^{ik} = -\varepsilon_0 c^2 \left(F^i{}_l F^{lk} + \frac{1}{4} F_{mn} F^{mn} \eta^{ik} \right) ,$$ (11.185)

and Rohrlich defines

$$P^i = -\frac{1}{c} \int_{\substack{\text{HOS} \\ \text{outside electron}}} T^{ij} v_j \gamma \mathrm{d}^3 x .$$ (11.186)

The point is that we have a minus sign in front of the integral, which means that the second term in the energy–momentum tensor actually gives

$$-\int_{\substack{\text{HOS} \\ \text{outside electron}}} -\frac{\varepsilon_0 c^2}{4} F_{mn} F^{mn} \gamma \mathrm{d}^3 x ,$$

in the rest frame where $v_j = (c,0,0,0)$. This means that the term $-\varepsilon_0 c^2 F^i{}_l F^{lk}$ does in fact contribute. However, we do have the key to our present reduction here!

Once again, it is easy to check from the definition (2.12) of F_{mn} on p. 8 that

$$\varepsilon_0 c^2 F^i_{Rl} F^{lk}_R v^R_k = \varepsilon_0 \begin{pmatrix} E^2_R & 0 \\ 0 & -\mathbf{EE} \end{pmatrix} \begin{pmatrix} c \\ 0 \end{pmatrix} = \varepsilon_0 \begin{pmatrix} cE^2_R \\ 0 \end{pmatrix} .$$ (11.187)

We have seen that

$$\frac{\varepsilon_0 c^2}{4} F^R_{mn} F^{mn}_R = -\frac{1}{2} \varepsilon_0 E^2_R ,$$ (11.188)

whereupon

$$T^{ik}_R v^R_k = -\varepsilon_0 \left[\begin{pmatrix} cE^2_R \\ 0 \end{pmatrix} - \frac{1}{2} \begin{pmatrix} cE^2_R \\ 0 \end{pmatrix} \right] = -c \frac{\varepsilon_0}{2} E^2_R \begin{pmatrix} 1 \\ 0 \end{pmatrix} .$$ (11.189)

Naturally, this leads to

$$-\frac{1}{c} \int_{\substack{\text{HOS(R)} \\ \text{outside electron}}} T^{ik}_R v^R_k \mathrm{d}^3 x_R = \int_{\substack{\text{HOS(R)} \\ \text{outside electron}}} \frac{\varepsilon_0}{2} E^2_R \mathrm{d}^3 x_R \begin{pmatrix} 1 \\ 0 \end{pmatrix} = \begin{pmatrix} P^0_R \\ 0 \end{pmatrix} .$$ (11.190)

The key is this: both terms in the energy–momentum tensor contribute and furthermore

$$c^2 F^i_{Rl} F^{lk}_R v^R_k = \begin{pmatrix} cE^2_R \\ 0 \end{pmatrix} = -2 \times -\frac{1}{2} E^2_R \begin{pmatrix} c \\ 0 \end{pmatrix}$$

$$= -2 \times \frac{c^2}{4} F^R_{mn} F^{mn}_R v^i_R$$

$$= -\frac{c^2}{2} F^R_{mn} F^{mn}_R v^i_R .$$

This tensor relation must hold in every frame! We have the quite general result

$$F^i_l F^{lk} v_k = -\frac{1}{2} F_{mn} F^{mn} v^i . \qquad (11.191)$$

This only requires the electron structure to be static in the electron rest frame, so that the magnetic fields in that frame are zero.

We now have a miraculous transformation of Rohrlich's definition. Since

$$T^{ik} v_k = -\varepsilon_0 c^2 \left(F^i_l F^{lk} v_k + \frac{1}{4} F_{mn} F^{mn} v^i \right) = \frac{\varepsilon_0 c^2}{4} F_{mn} F^{mn} v^i \qquad (11.192)$$

in any frame, it follows that Rohrlich's definition becomes

$$P^i = -\frac{1}{c} \int_{\substack{\text{HOS} \\ \text{outside electron}}} \frac{\varepsilon_0 c^2}{4} F_{mn} F^{mn} v^i \gamma d^3 x , \qquad (11.193)$$

or again,

$$P^i = -\frac{1}{c} v^i \int_{\substack{\text{HOS} \\ \text{outside electron}}} \frac{\varepsilon_0 c^2}{4} F_{mn} F^{mn} \gamma d^3 x . \qquad (11.194)$$

The integrand can be rewritten $F^R_{mn} F^{mn}_R$, since this is evidently a scalar, and this is constant with respect to time in the rest frame. We replace the integral over the HOS in the moving frame by an integral over HOS(R), replacing $\gamma d^3 x$ by $d^3 x_R$.

This makes it very clear how Lorentz covariance comes about. It is the factor of v^i outside the integral that now holds the key. The integral part is merely a scalar. Indeed, it is evidently the electromagnetic mass of the electron up to a factor! We expect

$$P^i = m_{em} c v^i . \qquad (11.195)$$

This does indeed work correctly, since

$$-\frac{1}{c^2} \int_{\substack{\text{HOS(R)} \\ \text{outside electron}}} \frac{\varepsilon_0 c^2}{4} F^R_{mn} F^{mn}_R d^3 x_R = \frac{1}{c^2} \int_{\substack{\text{HOS(R)} \\ \text{outside electron}}} u_R d^3 x_R = m_{em} . \qquad (11.196)$$

This is clearly a highly contrived definition of the 4-momentum of the electromagnetic fields with a view to achieving Lorentz covariance. We can now say that Lorentz covariance is achieved because of the fortuitous reduction (11.192) combined with the assumption that the fields are static in the electron rest frame.

Connection with Earlier Calculations

Let us just refer back to (11.11) and (11.13) on p. 185, viz.,

$$W = \gamma \int u \, d\sigma - \frac{\gamma}{c^2} \int \mathbf{S} \cdot \mathbf{v} \, d\sigma , \qquad (11.197)$$

and

$$\mathbf{p} = \frac{\gamma}{c^2} \int \mathbf{S} \, d\sigma + \frac{\gamma}{c^2} \int \mathsf{T} \cdot \mathbf{v} \, d\sigma . \qquad (11.198)$$

What do these tell us about calculation results like (3.9) on p. 35? In fact, if as Rohrlich claims one can replace $\gamma d\sigma$ by d^3x, this tells us that the first terms in (11.197) and (11.198) correspond precisely to what one would calculate using the naive integrals of the energy and momentum densities outside the charge shell (which Rohrlich refers to as the Abraham–Lorentz prescription).

The rule advocated above on the basis of the integration theory exemplified in (11.61) on p. 198, namely $d\sigma = \gamma d^3x$, is not so elegant in this respect! The γ factors accumulate here. We have to multiply the Abraham–Lorentz terms by γ^2, even though they already contain their γ factor, according to (3.9). But this is indeed the way we must interpret Rohrlich's definition according to this analysis. And it is corroborated by other accounts of this affair, as we shall see now.

11.5 New Density for the Field Four-Momentum

Let us take a look at what we are in fact advocating here to be the new density for the field four-momentum, if we follow Rohrlich [21] and the prescription $d\sigma = \gamma d^3x$ proposed above. Since

$$T^{\mu\nu} = \begin{pmatrix} -u & -\mathbf{S}/c \\ -\mathbf{S}/c & \mathsf{T} \end{pmatrix} ,$$

where

$$\mathsf{T} := \varepsilon_0 \left[\mathbf{EE} + c^2 \mathbf{BB} - \frac{1}{2}(E^2 + c^2 B^2) \mathbb{I} \right] ,$$

we soon find that (11.198) implies

$$\mathbf{p}_{\text{em}} = \frac{\varepsilon_0}{c} \int \gamma d\sigma \left[\mathbf{E} \times \mathbf{B} + \beta \cdot \mathbf{EE} + c^2 \beta \cdot \mathbf{BB} - \frac{1}{2}\beta(E^2 + c^2 B^2) \right] ,$$

where $\beta := \mathbf{v}/c$. This is similar to what Boyer [23] claims Rohrlich to be using (although with reference to a later paper by Rohrlich), viz.,

$$\mathbf{p}_{em} := \frac{\gamma}{4\pi c} \int d^3x \left[\mathbf{E} \times \mathbf{B} + \mathbf{v} \cdot \mathbf{EE} + \mathbf{v} \cdot \mathbf{BB} - \frac{1}{2}\mathbf{v}(E^2 + B^2) \right] . \quad (11.199)$$

but note the different units in this version and the fact that $c = 1$. If we use Rohrlich's prescription $\gamma d\sigma = d^3x$, we almost get the same thing, but not quite. There is a factor of γ too many in (11.199). With the prescription advocated here, namely $d\sigma = \gamma d^3x$, there is a factor of γ too few in (11.199).

The reader should note that all these discrepancies over the integration measure are not the main argument against Rohrlich's proposal. They are only an incidental problem that illustrates that the integration itself is not such a trivial matter as might be suggested by the little space devoted to it in these papers. The main argument we would like to make against these revisions of the definition of the EM four-momentum is just that it is ad hoc. Let us illustrate this further by considering the paper [24] by Moylan, entitled *An Elementary Account of the Factor of 4/3 in the Electromagnetic Mass*.

This is a naive account, rather than an elementary one, as attested by the claim in the abstract that the resolution of the discrepancy in the title involves the correct, i.e., relativistically covariant, definitions for the momentum and energy of the electromagnetic field. It does have the merit of showing very clearly why the Rohrlich strategy delivers a Lorentz covariant four-momentum for the EM fields around the charge shell, although unfortunately marred by several typographical errors. Here we examine the key part of the article and comment on it in the light of the interpretation we have been making.

We are told that the discrepancy can be resolved by redefining the total energy and the total momentum of the electromagnetic field in a relativistically covariant way. The general definition of the electromagnetic four-momentum in any Lorentz frame is then given as

$$P_\mu = \int_\Sigma T_{\mu\nu} d\sigma^\nu , \quad (11.200)$$

where

$$T_{\mu\nu} = F_{\mu\rho}F^{\rho\nu} + \frac{1}{4}g_{\mu\nu}F_{\rho\sigma}F^{\rho\sigma} . \quad (11.201)$$

This is basically the negative of our own formula (2.32) on p. 11. We have had to adjust Moylan's definition of the energy–momentum tensor, which contained a mistaken use of the Einstein convention for summing over indices. What we have now looks like our own version in (2.32), apart from a factor of $-\varepsilon_0$. The factor of ε_0 is dealt with in Moylan's article by the definitions

$$E_i = \frac{1}{\sqrt{\varepsilon_0}}F_{i0} , \qquad B_i = \frac{1}{2}\sqrt{\mu_0}\varepsilon_{ijk}F_{jk} , \quad (11.202)$$

with the further specification $F_{\mu\nu} = -F_{\nu\mu}$. He takes

$$g_{\mu\nu} = \text{diag}(1, -1, -1, -1) .$$

Recall that $c^2 - 1/\varepsilon_0\mu_0$. A quick check shows that Moylan's EM field tensor is the negative of the one in (2.12).

These are details, showing that our conventions are not so widely removed from each other. The key point is to understand Σ and $d\sigma^v$. We are told that

$$d\sigma^v = \eta^v d^3\sigma , \qquad (11.203)$$

where $d^3\sigma$ is an element of volume on a spacelike hypersurface Σ, with η^v a time-like vector normal to the hyperplane Σ. This vector is expressed in the suggestive form

$$\eta^\mu = \gamma(1,\beta) , \qquad \beta = \mathbf{v}/c . \qquad (11.204)$$

This clearly indicates that the integration is taken over the rest frame HOS, since this is the only one normal to the 4-velocity η^μ of the electron. Then $d^3\sigma$ is an element of volume on the spacelike hyperplane Σ, so it has to correspond to what we called $d^3 x_R$ in Sect. 11.4. This is indeed an invariant volume element because we specify it as the volume element on Σ whatever frame we view from. It coincides with Rohrlich's $d\sigma$.

Looking now at the formula (11.200), it is almost as though we are integrating over the rest frame HOS. However, this is a little misleading. The present view is that $T_{\mu\nu}^R$ must be static for this to work. Then it makes sense to put $d^3\sigma = \gamma dV$, where dV is $d^3 x$ in the relevant coordinate system and its HOS. This is then the rule $d^3 x_R = \gamma d^3 x$ advocated in this book. We can only guess here, because Moylan merely says that $d^3\sigma = dV$ in the rest frame and $d^3\sigma = \gamma dV$ in a general Lorentz frame, without specifying what dV actually is! We thus assume that it is just the coordinate volume measure $d^3 x$ in whatever coordinates we are using.

Note in passing that Moylan imputes his rule $d^3\sigma = \gamma dV$ to the famous book on classical electrodynamics by Jackson [26]. Consulting this source, in particular Sect. 16.5, we do indeed find corroboration of the idea that $d^3\sigma = d^3 x_R$ is the invariant measure, and that the measure $d^3 x$ on the HOS of some relatively moving frame is related to this by $d^3 x_R = \gamma d^3 x$. (But be warned! Jackson uses primed coordinates for his rest frame, and unprimed for the moving frame.)

We should stress the point about the assumption that the energy–momentum tensor is static. We said that it looks as though we are integrating over the rest frame HOS, and in a sense we are. But when we replace $d^3\sigma$ by γdV, the intention is obviously to integrate over the HOS in the moving system! This only works when the energy–momentum tensor is static, as we have shown. The point is that the rest frame and moving frame hyperplanes of simultaneity are different regions of space-time, and if $T_{\mu\nu}^R$ varied with (rest frame) time t_R, we could not successfully change the variables in the integration.

Moylan now produces a long calculation which is intended to show how this 'correct' definition produces something that is indeed Lorentz covariant. Let us exploit this to check the present interpretation of all these goings on. To begin with, using a primed notation in the moving frame,

$$E' = P_0' = \int_{\Sigma} T_{0\nu}' d\sigma^{\nu} . \tag{11.205}$$

We then expand out the energy–momentum tensor in two terms, one going with η^0 and the other with η^i for $i = 1, 2, 3$. We have (putting primes on everything, unlike Moylan),

$$
\int_{\Sigma} T_{0\nu}' d\sigma^{\nu} = -\int_{\Sigma} \left[F_{0i}' F_{0j}' g^{ij} - \frac{1}{4} g_{00} \left(F_{0i}' F'^{0i} + F_{i0}' F'^{i0} + F_{ij}' F'^{ij} \right) \right] \eta^0 d^3\sigma
$$
$$
- \int_{\Sigma} F_{0j}' F_{ik}' g^{jk} \eta^i d^3\sigma , \tag{11.206}
$$

where Latin indices run over $\{1, 2, 3\}$. All the terms here are integrated over the rest frame HOS labelled Σ. In the next step we switch to an integral over the primed HOS. This is signalled by the replacement of Σ by $\mathbb{R}^3$. The two terms in the last formula become (after correcting a crucial typographical error in the sign of the second term)

$$
\int_{\mathbb{R}^3} \left(\frac{1}{2} \varepsilon_0 \mathbf{E}'^2 + \frac{1}{2\mu_0} \mathbf{B}'^2 \right) \gamma^2 dV - \int_{\mathbb{R}^3} \left(\frac{1}{\mu_0 c^2} \mathbf{E}' \times \mathbf{B}' \right) \cdot \mathbf{v} \gamma^2 dV ,
$$

where the measure $d^3\sigma$ on the rest frame HOS has been replaced by γdV, which means what we would call $\gamma d^3x'$. The integrands are functions of the primed fields, but they are constant with changing rest frame time. (To see this, express the primed fields as functions of the unprimed fields, which are independent of the rest frame time.) Note that the other γ factor comes from the 4-velocity as in Rohrlich's formulas (11.197) and (11.198).

The primed fields $\mathbf{E}'$ and $\mathbf{B}'$ are given by

$$\mathbf{E}' = \gamma(\mathbf{E} - c\beta \times \mathbf{B}) - \frac{\gamma^2}{\gamma+1} \beta(\beta \cdot \mathbf{E}) , \tag{11.207}$$

$$\mathbf{B}' = \gamma \left(\mathbf{B} + \frac{1}{c} \beta \times \mathbf{E} \right) - \frac{\gamma^2}{\gamma+1} \beta(\beta \cdot \mathbf{B}) . \tag{11.208}$$

This is a general result for the Lorentz transformation of electric and magnetic fields [2, Chap. 26]. In the present case, $\mathbf{B} = 0$. The point here is not to replace the primed fields by the unprimed fields in the formula, because the aim is in fact to write the 'correct' expression for the field 4-momentum as an integral of the primed fields over the primed HOS. However, using the fact that

$$\mathbf{E}' = \gamma \mathbf{E} - \frac{\gamma^2}{\gamma+1} \beta(\beta \cdot \mathbf{E}) , \qquad \mathbf{B}' = \frac{\gamma}{c} \beta \times \mathbf{E} , \tag{11.209}$$

we can rewrite the expression $(\mathbf{E}' \times \mathbf{B}') \cdot \mathbf{v}$:

$$(\mathbf{E}' \times \mathbf{B}') \cdot \mathbf{v} = (\mathbf{v} \times \mathbf{F}') \cdot \mathbf{B}'$$

$$= \gamma(\mathbf{v} \times \mathbf{E}) \cdot \frac{\gamma}{c}(\boldsymbol{\beta} \times \mathbf{E})$$

$$= \frac{\gamma^2}{c^2}(\mathbf{v} \times \mathbf{E})^2$$

$$= c^2 \mathbf{B}'^2 . \tag{11.210}$$

We deduce finally that

$$E' = \int_{\mathbb{R}^3} \left(\frac{1}{2}\varepsilon_0 \mathbf{E}'^2 - \frac{1}{2\mu_0}\mathbf{B}'^2 \right) \gamma^2 \mathrm{d}V . \tag{11.211}$$

Equation (11.211) is to be compared with (11.193). It agrees entirely. In (11.193), we have γ and in (11.211), we have γ^2, but this is because one γ factor is absorbed into v^i in (11.193).

Note that Moylan does something rather surprising in the last line of his argument. In a last step, he replaces $\gamma \mathrm{d}V$ in (11.211) by his $\mathrm{d}^3 \sigma$. This is presumably due to some embarrassment at finding a factor of γ^2. Recall that, if the above interpretation is correct (and Moylan neglects completely to go into the details), $\mathrm{d}^3 \sigma$ is just a name for $\gamma \mathrm{d}^3 x'$.

It is easy to check that

$$c\mathbf{P}' = P_i' = \gamma\beta \int_{\mathbb{R}^3} \left(\frac{1}{2}\varepsilon_0 \mathbf{E}'^2 - \frac{1}{2\mu_0}\mathbf{B}'^2 \right) \gamma \mathrm{d}V . \tag{11.212}$$

This also agrees entirely with (11.193). Looking at (11.211) and (11.212), we can see just how arbitrary this covariant extension actually is! It agrees with the usual definition in the electron rest frame for the simple reason that the magnetic field is zero there. Elsewhere, the expression (11.211) for the field energy is quite different because the magnetic field energy is subtracted rather than added!

Before leaving Moylan's paper, it is worth considering the line he adopts with regard to Boyer's paper [23], mentioned only in passing. He describes this paper as an interesting and controversial treatment based on Poincaré's approach, and describes Rohrlich's paper [22], to be discussed shortly, as a severe criticism of it. He firmly sides with the author of [22], in favour of what he calls the relativistically covariant approach, which is the one advocated by Rohrlich. Unfortunately, he does not appear to have thought much about what the disagreement really is between the two approaches. The question here is this: is it better to consider the electron as a spatially extended object and take into account the complexity this involves, or is it better to make an ad hoc redefinition of a physical quantity so that one can carry through the point-particle approximation and the resulting mass renormalisation without regard for what may really be going on physically behind such an approximation?

There is another very telling point to make here. If someone gives an account of some physical process in one inertial reference frame, and then gives another account of it in another, relatively moving frame by carrying out a Lorentz transformation, then that person is giving a covariant account of the physical situation. It is not because not all the features of a given description are not separately manifestly covariant that this description is not covariant. The Boyer (or Poincaré) approach is of course Lorentz covariant. The difference in Rohrlich's approach is that something which should not actually be Lorentz covariant is forced to be in a completely ad hoc manner for the purposes of a piece of theory.

11.6 Binding Forces Revisited

The aim in this section will be to use Rohrlich's criticism [22] of Boyer's idea in order to get a more precise picture of how binding forces can be modelled for the charge shell, and also to understand what appears to be a slightly different version of the redefinition approach to the one described in [21].

We ought to note at the outset that Rohrlich's paper, which purports to criticise the ideas described above, actually misses the point on several counts. He opens by saying that Boyer's paper raises some questions that were settled some time earlier, but then provides only one question: whether the EM energy and momentum of the Coulomb field surrounding a charged particle are or are not the components of a four-vector. It is highly unlikely that Boyer ever had any doubts that they would not be. The whole point of involving binding forces is to show how the physics works out in a model where the electron is at least stabilised. In a moment, we shall see another count where Rohrlich clearly misses the point of the collapse model described above, despite the condescending tone of his paper.

As we shall adopt Rohrlich's notation from [22], let us see how he begins this paper. The model of the classical charged particle is a sphere of radius a, mass m, and uniformly distributed surface charge e. As a free object it is a closed system that has a total energy P^0 and momentum $\mathbf{P}$, which transform as components of a four-vector. If the entire particle were expressible by means of a field and an associated energy tensor $\Theta^{\mu\nu}$, such a tensor would necessarily have to satisfy

$$\partial_\alpha \Theta^{\alpha\mu} = 0 ,$$

since the system is closed. The momentum defined by

$$P^\mu = \int_\sigma d^3\sigma_\alpha \Theta^{\alpha\mu}(x)$$

would therefore be independent of the choice of spacelike surface σ.

This much we can understand, as discussed at length earlier (see Sect. 11.2.2). Note, however, that this scenario differs from Boyer's in that Rohrlich considers a closed system. There is, of course, nothing to stop Boyer from treating the stabilising forces as external. According to Rohrlich, the price to pay is a lack of covariance in the various parts.

The particle is not purely electromagnetic, but contains an electromagnetic component (the Coulomb field) and a non-electromagnetic one. We accept the usual assumption that these two components are additive in the energy tensors:

$$\Theta^{\mu\nu} = \Theta_e^{\mu\nu} + \Theta_n^{\mu\nu} .$$

Neither of the two components of $\Theta^{\mu\nu}$ is separately conserved:

$$\partial_\alpha \Theta_e^{\mu\nu} = -\partial_\alpha \Theta_n^{\mu\nu} \neq 0 .$$

This means that the decomposition of P^μ into P_e^μ and P_n^μ given by

$$
\begin{aligned}
P^\mu &= \int_\sigma d^3\sigma_\alpha \Theta_e^{\alpha\mu} + \int_\sigma d^3\sigma_\alpha \Theta_n^{\alpha\mu} \\
&=: P_e^\mu + P_n^\mu ,
\end{aligned}
$$

involves two surface integrals which are not separately independent of σ. But the sum is independent of σ as long as the same σ is chosen in both integrals.

If this trivial observation is the gist of Rohrlich's counterargument, then it is clear that he has missed the point. He is merely restating his own scenario. However, one good thing is that we do get an explicit description of the latter. We are apparently faced with a choice:

- We can allow any inertial observer to use her own HOS σ described by some formula $t = $ constant and calculate things in the way Boyer does.
- Whatever the observer is doing, we can use the electron rest frame HOS $t_R = $ constant (denoted by σ_R) for calculations.

The second choice means that we take the integrals to be

$$P_e^0 := \int_{L\sigma_R} d^3\sigma_\alpha \Theta_e^{\alpha 0} , \qquad P_e^k := \int_{L\sigma_R} d^3\sigma_\alpha \Theta_e^{\alpha k} . \tag{11.213}$$

Rohrlich does not specify what $L\sigma_R$ is supposed to be, although he does say that $\sigma = L\sigma_R$. L is the Lorentz transformation from the rest frame to the observer's frame. Since $\sigma = L\sigma_R$, and since σ is supposed to be the rest frame HOS, which we denoted by σ_R, we shall assume that $L\sigma_R$ is the rest frame HOS as described in the observer's coordinates.

In this paper, Rohrlich takes $d^3\sigma_\alpha = v_\alpha d^3 x_R = \gamma v_\alpha d^3 x$, which seems to confirm that he made a mistake in his earlier paper. Moreover, as we have noted, if this relation is supposed to be a relation between measures on different spacelike hypersurfaces (rest frame HOS_R and observer HOS), it will only serve our purposes here because the integrands turn out to be independent of rest frame time t_R.

We are told that a macroscopic charged sphere would be described by an energy tensor only for its electromagnetic fields, whilst its non-electromagnetic component would be described by a force density $f^\mu(x)$. We are also told that Boyer breaks the latter into a mechanical part leading to a momentum P_m^μ and a part describing the cohesive forces that prevent the charged sphere from expanding. The part f_{coh}^μ is just the force density that provides the cohesive forces (or Poincaré stresses). Rohrlich says that Boyer is wrong to call this an external force. But surely, we can define the system any way we want. We are free to say what is in it and what is not.

Rohrlich states that this force is not physically separable from the rest of the non-electromagnetic components in a classical macroscopic body. According to Rohrlich, without including this cohesive force, the physical charged particle cannot possibly be proven to have a momentum $P^\mu = \int_\sigma d^3\sigma_\alpha\Theta^{\mu\nu}(x)$ that is a four-vector. It may seem surprising to see this, since it would appear to be precisely the point of the Boyer approach to show the relevance of the binding forces. Of course, in his collapsing shell model, the binding forces were only applied when the radius of the shell reached its final value a, at which the collapse comes to a halt. This is why there is momentum transfer from the mechanical momentum of the shell to the EM fields, until the binding forces are switched on.

Indeed this is why Boyer separates the momentum $P^\mu_{\rm m}$ off and treats it as a 4-vector on its own. Then the local conservation law $\partial_\alpha\Theta^{\alpha\mu} = 0$ is reduced to

$$\partial_\alpha\Theta_{\rm e}^{\alpha\mu} + f^\mu_{\rm coh} = 0 . \tag{11.214}$$

This just says that the electromagnetic forces and the cohesive forces balance in the particle. So, of course, $P^\mu_{\rm m}$ is not conserved on its own in this scenario, because the electromagnetic forces were only balanced when the shell reached its radius $r = a$. We have discussed the question of the connection between a quantity being a 4-vector and its being conserved with great care in Sect. 11.2.2 (see in particular p. 204 ff.). When an energy–momentum tensor is conserved, we can define a Lorentz covariant 4-momentum from it by integration over a spacelike hypersurface. Furthermore, this 4-momentum is constant in time [see (11.75) on p. 201].

This presumably explains why Rohrlich wishes to discuss a scenario in which the charged sphere is produced by contraction of an infinite sphere but applying the stabilising $f^\mu_{\rm coh}$ at all times and contracting adiabatically to $r = a$. But it is not easy to see how this idea is born out by the calculations that ensue. (The word 'adiabatically' is often used casually, and it is not always clear exactly what is intended by it.) However, note that Boyer takes $f^\mu_{\rm ext}(t,\mathbf{r}) = -f^\mu_{\rm em}(t,\mathbf{r})\theta(t - t_0)$, and it is the absence of the step function in Rohrlich's analysis which presumably indicates that he balances electromagnetic repulsion by cohesive force throughout his contraction. The problem then is to see exactly where the contraction comes in. As in Boyer's discussion, there is no function $a(t)$ describing the radius of the sphere at time t.

Now Rohrlich is intent on producing a closed system, wherein he may claim that the total momentum is

$$P^\mu = P^\mu_{\rm m} + P^\mu_{\rm e} + P^\mu_{\rm coh} , \tag{11.215}$$

with

$$P^\mu_{\rm e} = \int_\sigma d^3\sigma_\alpha\Theta_{\rm e}^{\alpha\mu} , \qquad P^\mu_{\rm coh} = -\int_{V_4(\sigma)} d^4x\, f^\mu_{\rm coh} . \tag{11.216}$$

This new notation $V_4(\sigma)$ describes the region of spacetime over all times from $-\infty$ up to the spacelike hypersurface σ. What we are calculating to get P^{μ}_{coh} is a force density times a spatial volume element, times a time increment, which is therefore the total momentum imparted by this force field up to the time corresponding to σ. We note that if there existed a tensor $\Theta^{\mu\nu}_{\text{coh}}$ with $f^{\mu}_{\text{coh}} = \partial_\alpha \Theta^{\mu\nu}_{\text{coh}}$, we would have

$$- \int_{V_4(\sigma)} \mathrm{d}^4 x \, f^{\mu}_{\text{coh}} = \int_{\sigma} \mathrm{d}^3 \sigma_\alpha \Theta^{\alpha\mu}_{\text{coh}} \,.$$

Rohrlich takes pains to point out that the σ occurring in the expressions for P^{μ}_{e} and P^{μ}_{coh} must be the same, in case someone might get the idea of using a different σ for each part.

What we shall do in the following is to adapt Rohrlich's argument, which purports to consider an adiabatically collapsing shell, and show that $P^{\mu}_{\text{e}} + P^{\mu}_{\text{coh}} = m_{\text{e}} v^{\mu}$ for a shell that has always been stabilised at radius a by cohesive forces. In other words, m_{e} will add to the mechanical mass, so that we could calculate the mass required for P^{μ} by adding the mechanical and electromagnetic masses. This is surely the essence of Poincaré's original considerations. Once again, Rohrlich seems to have missed the point of the Boyer collapse model, which was to show the physical naturalness of the usual definitions of EM field energy and momentum. What is to be gained by considering an adiabatic collapse?

11.6.1 Model for the Ever-Stable Charged Shell

There are several simple aims here:

- To take Rohrlich's version of Boyer's theory and adapt it from a model for an adiabatically contracting charge shell to a model for a charge shell that has been stable for all time, with electromagnetic repulsion balanced by some unspecified but exactly modelled cohesive forces (the Poincaré stresses).
- To explain explicitly how $P^{\mu}_{\text{coh}} + P^{\mu}_{\text{em}}$ can be a nonzero 4-vector, in fact equal to $m_{\text{em}} v^{\mu}$ for some number m_{em}.
- To compare and contrast the time integration in this approach to the one used by Boyer in his rather different model.

We begin with the electromagnetic fields due to the charge shell, described here by their energy–momentum tensor $\Theta^{\alpha\mu}_{\text{em}}$. In the rest frame, where it is denoted by $\Theta^{\alpha\mu}_{\text{em R}}$, it turns out to be zero within the charge shell and Coulomb outside it. There is a step function factor with step at $r = a$, where a is constant. These fields have energy and momentum defined by an object

$$P^{\mu}_{\text{em}} = \int_{\sigma} \mathrm{d}\sigma_\alpha \Theta^{\alpha\mu}_{\text{em}} \,, \tag{11.217}$$

where σ is a hyperplane of simultaneity in the relevant inertial frame. With the notation established in earlier discussions, we find that, for a system S in which the particle moves with velocity $\mathbf{v}$, the electromagnetic energy and momentum are

$$P_{em}^0 = \int d^3x\, U = \gamma m_e c^2 \left(1 + \frac{1}{3}\frac{v^2}{c^2}\right), \quad \mathbf{P}_{em} = \int d^3x\, \mathbf{S} = \frac{4}{3}\gamma m_{em}\mathbf{v}, \quad (11.218)$$

where

$$m_{em} = \frac{1}{c^2}\int d^3x_R\, U_R = \frac{e^2}{2ac^2}. \quad (11.219)$$

The equation for $\mathbf{P}_{em}$ is just (11.181) on p. 226, already proven there. The expression for P_{em}^0 is proven below. In the rest frame of the charge shell, (11.218) reduces to

$$P_{emR}^\mu = (m_e c, 0, 0, 0). \quad (11.220)$$

The most striking thing about this result is that P_{em}^μ does not transform as a 4-vector in going from one inertial frame to another. Here are the details for the calculation of P_{em}^0.

EM Energy Around a Charge Shell in Motion

We use the formulas (11.178) and (11.179) on p. 225, viz.,

$$\mathbf{E}' = \gamma \mathbf{E} - \frac{\gamma^2}{(\gamma+1)c^2}\mathbf{v}(\mathbf{v}\cdot\mathbf{E}), \quad (11.221)$$

$$\mathbf{B}' = \frac{\gamma}{c^2}\mathbf{v}\times\mathbf{E}. \quad (11.222)$$

Then,

$$\begin{aligned}
P_{em}^0 &= \int d^3x \left[\frac{1}{2}\varepsilon_0|\mathbf{E}'(\mathbf{r})|^2 + \frac{1}{2\mu_0}|\mathbf{B}'(\mathbf{r})|^2\right] \\
&= \int d^3x \left[\frac{1}{2}\varepsilon_0\left|\gamma\mathbf{E} - \frac{\gamma^2}{(\gamma+1)c^2}\mathbf{v}(\mathbf{v}\cdot\mathbf{E})\right|^2 + \frac{1}{2\mu_0}\frac{\gamma^2}{c^4}|\mathbf{v}\times\mathbf{E}|^2\right] \\
&= \frac{1}{2}\varepsilon_0\gamma\int d^3x_R \left[\left|\mathbf{E} - \frac{\gamma}{(\gamma+1)c^2}\mathbf{v}(\mathbf{v}\cdot\mathbf{E})\right|^2 + \frac{1}{c^2}|\mathbf{v}\times\mathbf{E}|^2\right],
\end{aligned}$$

where we have used the fact that $\mathbf{E}$ is constant in the rest frame time t_R (we are considering the static charge shell here) so that we can replace

$$\int_{HOS} d^3x \longrightarrow \int_{HOS(R)} \frac{d^3x_R}{\gamma}.$$

We have also used the relation $c^2 = 1/\mu_0 \varepsilon_0$.

Now note that

$$(\mathbf{v} \times \mathbf{E}) \cdot (\mathbf{v} \times \mathbf{E}) = v^2 E^2 - (\mathbf{v} \cdot \mathbf{E})^2 , \tag{11.223}$$

so

$$\left| \mathbf{E} - \frac{\gamma}{(\gamma+1)c^2} \mathbf{v}(\mathbf{v} \cdot \mathbf{E}) \right|^2 + \frac{1}{c^2} |\mathbf{v} \times \mathbf{E}|^2 \tag{11.224}$$

$$= \left(1 + \frac{v^2}{c^2} \right) E^2 + \frac{(\mathbf{v} \cdot \mathbf{E})^2}{c^2} \left[\frac{\gamma^2}{(\gamma+1)^2} \frac{v^2}{c^2} - \frac{2\gamma}{\gamma+1} - 1 \right] .$$

Since

$$\frac{v^2}{c^2} = \frac{\gamma^2 - 1}{\gamma^2} , \tag{11.225}$$

we find that

$$\frac{\gamma^2}{(\gamma+1)^2} \frac{v^2}{c^2} - \frac{2\gamma}{\gamma+1} - 1 = -2 . \tag{11.226}$$

We now have

$$\begin{aligned}
P_{\text{em}}^0 &= \frac{1}{2} \varepsilon_0 \gamma \int d^3 x_R \left[\left(1 + \frac{v^2}{c^2} \right) E^2 - \frac{2v^2}{c^2} E_x^2 \right] \\
&= \frac{1}{2} \varepsilon_0 \gamma \left[1 + \frac{v^2}{c^2} - \frac{2}{3} \frac{v^2}{c^2} \right] \int d^3 x_R E^2 \\
&= \gamma m_{\text{em}} c^2 \left[1 + \frac{1}{3} \frac{v^2}{c^2} \right] ,
\end{aligned}$$

as claimed.

Looking at (11.218), the main conclusion is that P_{em}^μ is not a 4-vector. This is anathema to Rohrlich. He says that we should not be surprised, because each observer S chooses a different σ, and P_{em}^μ depends on σ. But could this really be a serious argument? The whole point about Lorentz covariance, or the lack of it, is that observers are going to use their own HOS to describe the world. In this case, when observers use their natural σ, relations between observers are no longer governed by the Lorentz transformation. Instructing them to use the same σ, e.g., the rest frame HOS, seems a very cheap ploy indeed for ensuring 'Lorentz covariance'.

Role of the Cohesive Forces

We now turn to the calculation of the energy and momentum P_{coh}^μ imparted to the system by the cohesive forces f_{coh}^μ which we assume to have balanced electromagnetic repulsion within the shell for all time, so that it has constant radius a. Following what was established in Sect. 11.3.3,

$$f^0_{\text{cohR}} = 0 , \qquad \mathbf{f}_{\text{cohR}} = -2\pi\sigma^2 \hat{\mathbf{r}}\delta(r-a) , \tag{11.227}$$

where

$$\sigma = \frac{e}{4\pi a^2} , \tag{11.228}$$

in the rest frame. This is just the negative of the electromagnetic 4-force density exerted by the charge shell on itself, given as (11.162) on p. 223, viz.,

$$f^\mu_{\text{emR}} = \left(0, \frac{e}{4\pi a^2}\delta(r-a)\frac{e\hat{\mathbf{r}}}{2a^2} \right) . \tag{11.229}$$

The minus sign in the second equation of (11.227) tells us that the cohesive forces oppose the electromagnetic ones. In these relations, a is constant. To simplify the notation, we now drop the suffix coh on the cohesive force density. In another inertial frame moving with constant 3-velocity $-\mathbf{v}$ relative to the rest frame,

$$f^0 = \gamma\left(f^0_{\text{R}} + \frac{\mathbf{v}}{c}\cdot\mathbf{f}_{\text{R}} \right) , \qquad \mathbf{f} = \gamma\mathbf{f}^{\parallel}_{\text{R}} + \mathbf{f}^{\perp}_{\text{R}} + \gamma\frac{\mathbf{v}}{c}f^0_{\text{R}} , \tag{11.230}$$

having applied the usual Lorentz transformations. (Note that Rohrlich has the sign of the velocity wrong.) As an aside, we may ask why this cohesive 4-force density should transform as a 4-vector. In fact, in every inertial frame, it must balance the electromagnetic 4-force density exerted by the charge shell on itself, which is a 4-vector, because we have seen [(2.45) on p. 13] that it is given by $f^\mu_{\text{em}} = \Theta^{\alpha\mu}_{\text{em},\alpha}$.

Now the rest frame components f^μ_{R} are functions of the rest frame coordinates, which are in turn functions of the coordinates in the moving frame. We can say

$$P^0_{\text{coh}} = -\int_{V_4(\sigma)} \mathrm{d}^4 x\,\gamma\left[f^0_{\text{R}}(x_{\text{R}}(x)) + \frac{\mathbf{v}}{c}\cdot\mathbf{f}_{\text{R}}(x_{\text{R}}(x)) \right] . \tag{11.231}$$

We can then change variables from x to x_{R}, noting that $\mathrm{d}^4 x \doteq \mathrm{d}^4 x_{\text{R}}$ when $x_{\text{R}} = L^{-1}(x)$, whence

$$P^0_{\text{coh}} = -\int_{V_4(L^{-1}\sigma)} \mathrm{d}^4 x_{\text{R}}\,\gamma\left[f^0_{\text{R}}(x_{\text{R}}) + \frac{\mathbf{v}}{c}\cdot\mathbf{f}_{\text{R}}(x_{\text{R}}) \right] . \tag{11.232}$$

Note that $L^{-1}\sigma$ is the same spacelike hypersurface in spacetime as σ, but it is described in the rest frame coordinates rather than the moving frame coordinates.

We shall carry out the time integral from some large negative cutoff $t^{\text{R}}_{-\infty}$ up to the value of t_{R} that corresponds to the hypersurface σ for the given x_{R}. We have to visualise σ in the $(ct_{\text{R}}, x_{\text{R}})$ plane. The x axis slopes down below the x_{R} axis and points (events) on it satisfy $x^0_{\text{R}} = -vx^1_{\text{R}}/c$ (see Fig. 11.5).

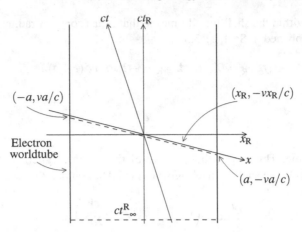

Fig. 11.5 Electron that has always been stabilised by cohesive forces. The electron worldtube is indicated by *bold vertical lines*. The integration is carried out over the region of spacetime between the *dashed lines*

How do the cohesive forces contribute to the 4-momentum in this moving frame? The contribution does not come from the cutoff, which might be viewed physically as a repeat of the Boyer scenario. In this case, it comes from the fact that, for different values of x_R, we include the effect of the cohesive forces up to a different time t_R. It is an artifact of the different notion of simultaneity in the moving frame. More about this later.

Now, bearing in mind that $f_R^0 = 0$, we find

$$
\begin{aligned}
P_{coh}^0 &= -\int_{V_4(L^{-1}\sigma)} \frac{\gamma}{c} d^4 x_R \mathbf{v} \cdot \mathbf{f}_R \\
&= \int_{V_4(L^{-1}\sigma)} \frac{\gamma}{c} 2\pi\sigma^2 d^4 x_R \mathbf{v} \cdot \hat{\mathbf{r}} \delta(r-a) \\
&= \int r^2 \sin\theta \, d\theta \, d\phi \, dr \int_{t_{-\infty}^R}^{-vx_R^1/c^2} dt_R 2\pi\sigma^2 \gamma v \cos\theta \, \delta(r-a) , \quad (11.233)
\end{aligned}
$$

where the polar coordinates cover the spacelike hypersurfaces in the electron rest frame and we have used the fact that $\mathbf{v} \cdot \hat{\mathbf{r}} = v\cos\theta$.

The time integral is easy because the integrand has no time dependence. Note, however, that the factor $\delta(r-a)$ comes from the formula (11.227) for the force, and we might expect a to have a time dependence for the adiabatic contraction Rohrlich advocates. However, in the present case, we turn this into a model for the electron that has always been stable, with constant radius a, so we do not have this problem. This same question is better handled in Boyer's approach, because the force only comes into being instantaneously at some value of t_R, whereafter a no longer changes. Anyway, the time integral merely introduces a factor of

$$-\frac{vx_R}{c^2} - t_{-\infty}^R$$

into the integrand. The term containing $t_{-\infty}^R$ gives zero because the integral of $\sin 2\theta$ over $[0, \pi]$ is zero. We can thus take the limit $t_{-\infty}^R \to -\infty$. We discuss the differences with Boyer's calculation for the collapsing shell below.

We now have

$$\begin{aligned}
P_{coh}^0 &= -2\pi\sigma^2\gamma v \int r^2 \sin\theta \, d\theta \, d\phi \, dr \frac{vx_R}{c^2} \cos\theta \delta(r-a) \\
&= -(2\pi)^2\sigma^2\gamma\frac{v^2}{c^2}a^3 \int_0^\pi \sin\theta \cos^2\theta \, d\theta \\
&= -(2\pi)^2\sigma^2\gamma\frac{v^2}{c^2}a^3 \left[-\frac{1}{3}\cos^3\theta\right]_0^\pi \\
&= -\frac{1}{3}\frac{e^2}{2ac^2}\gamma v^2 \\
&= -\frac{1}{3}m_e\gamma v^2 .
\end{aligned} \tag{11.234}$$

Finally, we calculate

$$\begin{aligned}
\mathbf{P}_{coh} &= -\int_{V_4(\sigma)} d^4x \mathbf{f} \\
&= -\int_{V_4(L^{-1}\sigma)} d^4x_R \left(\gamma\mathbf{f}_R^\parallel + \mathbf{f}_R^\perp + \gamma v f_R^0\right) \\
&= -\int_{V_4(L^{-1}\sigma)} d^4x_R \left(\gamma\mathbf{f}_R^\parallel + \mathbf{f}_R^\perp\right) .
\end{aligned} \tag{11.235}$$

We have dropped the term in f_R^0 because it is zero. Note that Rohrlich seems to have the wrong sign for $\mathbf{v}$ once again and that we have dropped c. The term in $\mathbf{f}_R^\perp$ also gives zero for the following reasons. This term contains the y and z components

$$-2\pi\sigma^2\frac{y_R}{a}\delta(r-a) \quad \text{and} \quad -2\pi\sigma^2\frac{z_R}{a}\delta(r-a) .$$

Consider the second component, and arrange to measure ϕ around the x_R axis and θ the angle with the x_R axis. Then $z_R = r\sin\theta\cos\phi$ and the integral over $\phi \in [0, 2\pi]$ is zero. Hence,

$$\mathbf{P}_{coh} = -\int_{V_4(L^{-1}\sigma)} d^4x_R \gamma\mathbf{f}_R^\parallel . \tag{11.236}$$

But,

$$\mathbf{f}_R^\parallel = -2\pi\sigma^2 \cos\theta \delta(r-a)\frac{\mathbf{v}}{v} , \tag{11.237}$$

so

$$\mathbf{P}_{\text{coh}} = 2\pi\sigma^2 \frac{\mathbf{v}}{v} \int r^2 \sin\theta \, d\theta \, d\phi \, dr \int_{t_{-\infty}^{\text{R}}}^{-vx_{\text{R}}/c^2} \gamma \cos\theta \, \delta(r-a) \, dt_{\text{R}}$$

$$= (2\pi)^2 \sigma^2 a^2 \gamma \frac{\mathbf{v}}{v} \int \sin\theta \, d\theta \left(-\frac{v}{c^2} a \cos\theta - t_{-\infty}^{\text{R}} \right) \cos\theta$$

$$= -\frac{e^2}{4a^2} \frac{\mathbf{v}}{v} \gamma \left[-\frac{1}{3} \cos^3\theta \right]_0^\pi \frac{v}{c^2} a$$

$$= -\frac{e^2}{2ac^2} \frac{\gamma}{3} \mathbf{v}$$

$$= -\frac{1}{3} m_e \gamma \mathbf{v} \, . \tag{11.238}$$

The final result here is therefore

$$P_{\text{coh}}^0 = -\frac{1}{3} m_e \gamma v^2 \, , \qquad \mathbf{P}_{\text{coh}} = -\frac{1}{3} m_e \gamma \mathbf{v} \, . \tag{11.239}$$

We emphasise once again that we do not need to be concerned about a model for the time dependence of a during adiabatic contraction, since we have assumed a charge shell that has been stabilised by Poincaré stresses for all time. Now, in the notation

$$P_{\text{coh}}^\mu = (P_{\text{coh}}^0, \mathbf{P}_{\text{coh}}) \, ,$$

we have shown that

$$P_{\text{coh}}^\mu = -\frac{1}{3} m_e \gamma (v^2, \mathbf{v}) \, . \tag{11.240}$$

It is clear that this quantity is no more a 4-vector than P_{em}^μ. However,

$$P_{\text{em}}^\mu + P_{\text{coh}}^\mu = \gamma m_e (1, \mathbf{v}) = m_e v^\mu \, . \tag{11.241}$$

Note that these results give

$$P_{\text{em R}}^\mu = (m_e c, 0, 0, 0) \, , \qquad P_{\text{coh R}}^\mu = (0, 0, 0, 0) \, , \tag{11.242}$$

in the rest frame. There is no net cohesive force in the rest frame, due to the symmetry of the charge distribution, so the cohesive forces impart no momentum to the system, even though they have always been there, as viewed from this frame.

We are considering a spatially extended electron here, with no intention of taking a point limit $a \to 0$, in contrast to Rohrlich. We studied the fact that the electromagnetic energy–momentum tensor was conserved everywhere outside the charge shell (and indeed inside it, where it was actually identically zero), but not on the shell itself, which meant that it could not be used to define a 4-vector 4-momentum for the electromagnetic fields. We have now included cohesive forces in the model, which balance the non-conservation of the electromagnetic energy–momentum tensor by ensuring

$$f^{\mu}_{\text{coh}} := -\partial_{\alpha} \Theta^{\alpha\mu}_{\text{em}} , \qquad (11.243)$$

and we have found that the 4-momentum they impart to the system, as viewed from a general inertial frame, is

$$P^{\mu}_{\text{coh}} := -\int_{V_4(\sigma)} \mathrm{d}^4 x f^{\mu}_{\text{coh}} = +\int_{V_4(\sigma)} \mathrm{d}^4 x \partial_{\alpha} \Theta^{\alpha\mu}_{\text{em}} , \qquad (11.244)$$

which is not a 4-vector quantity. We are pleased to find, however, that the sum of the 4-momenta from the electromagnetic field and the cohesive forces does transform as a 4-vector. The above proof of this, which comprises quite a lengthy calculation, does not make this result look obvious, and we must ask if there is some quicker way of seeing that we will get a 4-vector quantity by this process.

In this context, it is worth looking at the following mistaken deduction, whereby we seem to show that the sum $P^{\mu}_{\text{em}} + P^{\mu}_{\text{coh}}$ is in fact not just a 4-vector, but actually zero! Suppose we start with

$$f^{\mu}_{\text{coh}} + \partial_{\alpha} \Theta^{\alpha\mu}_{\text{em}} = 0 ,$$

which was just the definition of f^{μ}_{coh}, and then integrate over $V_4(\sigma)$ for some HOS σ in the relevant frame. Why can we not deduce that $P^{\mu}_{\text{em}} + P^{\mu}_{\text{coh}}$ is zero? The obvious answer is that we should obtain a constant of integration on the right-hand side, and this is in fact correct. But it is still not such an inane question. There appears to be something almost circular about this construction, and yet it nevertheless yields a nonzero answer on the right-hand side. For example, it is not obvious that the constant of integration on the right-hand side should have the Lorentz covariant form $m_e v^{\mu}$. It is not obvious why it should not be zero, given that the last displayed relation was the very construction of f^{μ}_{coh} itself, and yet when that relation is integrated over the given region it delivers up a Lorentz covariant object that perfectly includes the electromagnetic 4-momentum for the well known electromagnetic mass m_e given by the rest frame Coulomb energy in the field outside the shell.

The inadmissible step in the proposed deduction is the step that would equate

$$P^{\mu}_{\text{coh}} = \int_{V_4(\sigma)} \mathrm{d}^4 x \partial_{\alpha} \Theta^{\alpha\mu}_{\text{em}} \quad \text{with} \quad -\int_{\sigma} \mathrm{d}\sigma_{\alpha} \Theta^{\alpha\mu}_{\text{em}} .$$

Gauss' theorem leads to this iff the integral of $\Theta^{\alpha\mu}_{\text{em}}$ over the boundaries at infinity gives zero. Let us just consider the proposed deduction in the rest frame where it is at its simplest. In this case,

$$\int_{\sigma_R} \mathrm{d}\sigma^R_{\alpha} \Theta^{\alpha\mu}_{\text{em R}} = (m_e c, 0, 0, 0) ,$$

whereas

$$P^{\mu}_{\text{coh R}} = -\int_{V_4(\sigma_R)} \mathrm{d}^4 x_R f^{\mu}_{\text{coh R}} = 0 ,$$

by symmetry of the cohesive forces in the rest frame. We can see that the integral of $\Theta_{\text{em R}}^{\alpha\mu}$ over the boundaries at infinity does not give zero in this case. Indeed, if we write

$$\int_{V_4(\sigma_R)} d^4 x_R \partial_\alpha \Theta_{\text{em R}}^{\alpha\mu} = \lim_{t_{-\infty} \to -\infty} \int_{t_R=t_{-\infty}}^{t_R=0(\sigma_R)} dt_R d^3 x_R \partial_\alpha \Theta_{\text{em R}}^{\alpha\mu} , \qquad (11.245)$$

we shall find that this becomes, by Gauss' theorem,

$$\int_{\sigma_R} d\sigma_\alpha \Theta_{\text{em}}^{\alpha\mu} - \int_{\sigma(t_R=t_{-\infty})} d\sigma_\alpha \Theta_{\text{em}}^{\alpha\mu} = (m_e c, 0, 0, 0) - (m_e c, 0, 0, 0) ,$$

which is plainly zero. The point is that the surface integral over the HOS at $t_R = t_{-\infty}$ is never zero, no matter how early $t_{-\infty}$ is made.

Now here is a simple explanation for why we obtain a 4-vector for the total energy and momentum. In our earlier discussions (see Sect. 11.2.2), we saw how a conserved energy–momentum tensor could be used to define a Lorentz covariant 4-momentum, and we saw what went wrong in this venture when the energy–momentum tensor was not conserved, for the specific case of the electromagnetic fields outside the charge shell. The problem occurred precisely on the shell surface, where the electromagnetic energy–momentum tensor was not conserved [see (11.122) on p. 210]. By adding something that cancels $\partial_\alpha \Theta_{\text{em}}^{\alpha\mu}$ precisely where the electromagnetic energy–momentum tensor is not conserved, even though what we add is not necessarily itself the divergence of an energy–momentum tensor, this is tantamount to producing an object which can be used to define a Lorentz covariant 4-momentum. A little more work is needed here to show explicitly why this kind of construction does the trick, going back to the integration theory that led to (11.122). But we may leave it at that for our present purposes.

Note that we define the total momentum of the system to be

$$P^\mu = P_{\text{m}}^\mu + P_{\text{em}}^\mu + P_{\text{coh}}^\mu , \qquad (11.246)$$

where $P_{\text{m}}^\mu = m v^\mu$ is a mechanical 4-momentum associated with the particle. Hence, what we have shown is that

$$P^\mu = (m + m_e) v^\mu , \qquad (11.247)$$

which means that the electromagnetic mass merely adds to the mechanical mass. This leaves open the possibility that the mechanical mass is zero, and that all the inertial mass is electromagnetic. This is not the line adopted in this book. Instead we suggest that some other field of force will provide the cohesive forces holding the electron together, whence we would expect this field to make its own contribution to the inertial mass. The total energy–momentum of the electromagnetic field and this other field would be conserved, so that a Lorentz covariant 4-momentum could be defined from it.

Finally, let us try to identify what it is that Rohrlich does not like about this scenario. One possibility is that, since Rohrlich intends to take a point particle limit, this simply makes the idea of cohesive forces rather academic. His approach is to ensure that they always contribute zero to the 4-momentum, and he does this by redefining their contribution to the 4-momentum. He no longer uses a hyperplane of simultaneity of the observer who considers the charge system to be moving, because as we have seen this does not give zero. He uses a HOS in the electron rest frame, because as we have seen this does give zero! The ploy is as simple as that. This ties in with the application of the same ploy for the electromagnetic 4-momentum which then gives $P_{\text{em}}^{\mu} = m_e v^{\mu}$. The sum of the two 4-momenta is the same as before. This is discussed in detail below in Sect. 11.6.3.

11.6.2 Contrast with Collapsing Shell Model

The aim here is to compare the way the time integrations are handled in Boyer's model and in the above, which is basically Poincaré's version of an electron with spatial extension. The point is that these time integrals are rather different in the two cases.

Boyer has the following situation. The cohesive forces are switched on at time $t_R = 0$ in the rest frame, at all points of the shell, just when they can no longer do any work or change the momentum, as viewed in that frame. But when all this is viewed in the coordinates (x, ct) of a frame moving to the left, in which the charge shell appears to be moving at speed v to the right along the x axis, the cohesive forces come into play at different times t for different x positions of bits of the shell, and momentum is transmitted, as far as this observer is concerned.

To do Boyer's calculation, we draw an (x, ct) diagram (see Fig. 11.6). The ct_R axis slants off to the right from the ct axis, whilst the x_R axis climbs from the x axis. The shell is represented in this 2D diagram by two lines parallel to the t_R axis, at equal distances on either side of it. The segment of the x_R axis between points

$$(x, ct) = \left(-\gamma a, -\frac{\gamma v}{c} a\right) \quad \text{and} \quad (x, ct) = \left(\gamma a, \frac{\gamma v}{c} a\right)$$

indicates the region of spacetime in which the cohesive forces become operative. They continue to be operative for all spacetime events above this line segment which are contained within the slanting tube that represents the charge shell. To find the momentum imparted in this frame, we first choose any hyperplane of simultaneity $t = t_*$ for $t_* > \gamma v a/c^2$, this condition ensuring that this observer considers the cohesive forces to have come into play all over the shell, whereupon she will also consider that they balance perfectly and cease to impart any momentum to the shell.

The comments here are based on the calculation on p. 213 and following pages. The cohesive force density is known in the electron rest frame, and Lorentz transformed to the moving frame. We multiply the force density by spatial volume elements and integrate to obtain the total force on the shell, then multiply by time elements and integrate over time to obtain the total momentum imparted by this force. This integration has already been carried out by the time we reach (11.132) on p. 215 and

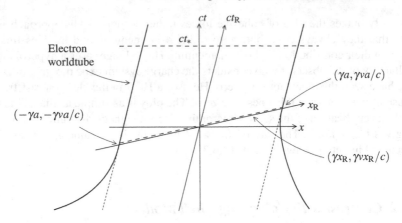

Fig. 11.6 Boyer scenario as viewed in the moving frame. The cohesive forces are switched on at $t_R = 0$. The electron worldtube is indicated by *bold curves*. Integration is carried out over the worldtube and between the *dashed lines*

it introduces the factor

$$\frac{\gamma v}{c^2}(a - x_R) \, .$$

Although we view things in the moving frame, everything gets expressed in terms of rest frame coordinates. We were able to insert the above factor directly, because it represents the time in the moving frame coordinates over which the force applied at x_R remains unbalanced.

Seen from the moving frame, the region of integration in spacetime is a quadrilateral, bounded below by the zero time HOS in the rest frame, at the sides by the world tube of the electron, and at the top by the arbitrary HOS at $t = t_*$ in the moving frame (see Fig. 11.6). For a given value of x_R, the moving frame time t ranges from

$$\frac{\gamma v}{c^2} x_R \quad \text{to} \quad t_* \, . \tag{11.248}$$

The spacetime element thus covered slants in the same way as the electron world tube with respect to the ct axis, being determined by a fixed value of x_R. (It is because the world tube slants like this that x_R is a good variable to use for the integration, even though we are working in the moving frame.) Due to the presence of a temporal step function $\theta(t_R)$ in the expression for the cohesive force density, the range of integration mentioned here can be extended back to $t = -\infty$ with complete impunity. Likewise, the upper bound t_* can be taken as any value later than $\gamma v a/c^2$.

How does the integration range in (11.248) give us the required result, eventually obtained in (11.133) on p. 216? We note that, in going from (11.132) to the calculation leading to (11.133), the upper bound $\gamma v a/c^2$ has dropped out, because it gives zero in the integration over space. The same happens to t_*. We have something like

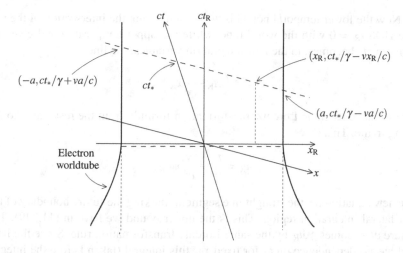

Fig. 11.7 Boyer scenario for collapsing charge shell as viewed in the rest frame. Cohesive forces switched on at $t_R = 0$. Electron worldtube is indicated by *bold curves*. Integration over worldtube between *dashed lines*

$$\int_{x_R=-a}^{a} d^3 x_R \int_{t=\gamma v x_R/c^2}^{t_*} dt \longrightarrow \int_{x_R=-a}^{a} d^3 x_R \left(t_* - \frac{\gamma v}{c^2} x_R \right)$$

$$\longrightarrow -\int_{x_R=-a}^{a} d^3 x_R \frac{\gamma v}{c^2} x_R .$$

In the first step, the integration over t is trivial because the integrand is independent of t. In the second, the term in t_* drops out when the spatial integration is done, by symmetry. This second step is mirrored precisely by the way the term $\gamma v a/c^2$ drops out in going from (11.132) to the calculation leading to (11.133).

In Boyer's own version of the calculation [23], the time integral looks like

$$\int_{t_R=-\infty}^{t_R=t_*/\gamma-v x_R/c^2} . \tag{11.249}$$

Here we have an integration over the rest frame time. We can understand how the lower limit came to be at $-\infty$, because there is a step function at $t_R = 0$ in the cohesive force density, in this scenario. However, the upper limit now seems to mix the all-important term $v x_R/c^2$ with the arbitrary t_*. This is in fact the way the rest frame observer will view what the moving observer is doing.

We see this by drawing a new diagram (see Fig. 11.7), taking (x_R, ct_R) as the orthogonal axes. The world tube of the electron is now represented by vertical lines at equal distances on either side of the ct_R axis, whilst the x axis slopes down below the x_R axis and the ct axis slants off to the left of the ct_R axis as we move forward in time. We are going to change the variables in the last integral from the hybrid pair x_R and t, with slanting elements of spacetime parallel to the ct_R axis in our first diagram, to x_R and t_R, with rectangular elements of spacetime still parallel to the ct_R

axis. Now the lower temporal bound is horizontal, being the intersection of the rest frame HOS $t_R = 0$ with the world tube, while the upper temporal bound given by the line $t = t_*$ becomes, in the new integration variables, the line

$$ct = \gamma\left(ct_R + \frac{v}{c}x_R\right),$$

where we have used the Lorentz transformation formula from the rest frame to the moving frame. This gives

$$t_R = \frac{t_*}{\gamma} - \frac{v}{c^2}x_R$$

as the new equation for the straight line segment marking the future boundary of the quadrilateral integration region. This is the upper bound we have in (11.249). The measure dt becomes γdt_R by the same Lorentz transformation rule. Since the integrand has no dependence on t_R for fixed x_R, this integral (taken before the integral over x_R) naturally gives the same thing as before, viz.,

$$\gamma\left(\frac{t_*}{\gamma} - \frac{v}{c^2}x_R\right).$$

The difference is that we now have the interpretation of the rest frame observer for what the moving frame observer is doing! The rest frame observer needs to explain why the other observer considers that some momentum is imparted to the shell. Her explanation follows the pattern of the integration limits in (11.249): the moving observer is taking into account the cohesive force at $x_R = -a$ from time $t_R = 0$ to time

$$t_R = \frac{t_*}{\gamma} + \frac{v}{c^2}a,$$

but she is only taking into account the cohesive force at $x_R = +a$ from time $t_R = 0$ to time

$$t_R = \frac{t_*}{\gamma} - \frac{v}{c^2}a,$$

an earlier time. This is how the rest frame observer explains why the moving observer gets a different answer.

We now contrast this integration with the one discussed in Sect. 11.6.1. The point is once again to compare the region of integration and see how the same terms survive despite the fact that the approach is somewhat different. Indeed, note that we are not just looking at the same situation in a different way. The physical situation is actually different, even though it leads to the same conclusion about the energy and momentum of the system.

This time the cohesive forces have been there for all time and the charge shell has always been stabilised with radius a. The moving observer chooses some HOS σ in her frame and works out the total energy and momentum P^μ_{coh} imparted to the shell up to the time corresponding to σ in her frame. Although she is working out

$$P^{\mu}_{\text{coh}} = - \int_{V_4(\sigma)} d^4 x f^{\mu}_{\text{coh}}$$

in her frame, and all the quantities here refer to her frame, she finds f^{μ}_{coh} by Lorentz transformation of the rest frame quantity and then changes variables to the rest frame coordinates x_R. In this sense, what we get from the calculation is the rest frame observer's view of what she is doing, wherein we obtain an explanation of how it is that the moving observer finds some energy and momentum imparted to the shell, whereas the rest frame observer does not.

This is illustrated in Fig. 11.5 for the case where σ is taken to be the HOS at $t = 0$ for the moving observer. This was the assumption in the calculation (11.233) on p. 246. The time integral becomes

$$\int_{t^{R}_{-\infty}}^{-vx_R/c^2} dt_R ,$$

but the integrand is constant with respect to t_R so this integral is replaced by a simple factor

$$-\frac{vx_R}{c^2} - t^{R}_{-\infty} .$$

The lower cutoff then drops out by the symmetry of the spatial integrand, leaving only the term $-vx_R/c^2$. What is happening? According to the rest frame observer, the moving observer is looking at the cohesive forces on different parts of the shell up to different times (which the moving observer considers to be the same) and this is why she calculates an overall residue of unbalanced energy and momentum. This is an artifact of the different notion of simultaneity in the moving frame.

So what is different with the Boyer scenario? A glance at Figs. 11.7 and 11.5 shows that they are essentially the same, except that the integration region has been shifted and/or stretched up the rest frame time axis by amounts that do not affect the result of the integration. The reason that these differences do not affect the result is that the spatial part of the integration removes all but one of the terms that arise from the temporal integral. In both cases, the rest frame observer considers that the moving observer is looking at the cohesive forces on different parts of the shell up to different times, so the rest frame observer's final explanation for the existence of a nonzero P^{μ}_{coh} is the same in both cases.

What is shown in Fig. 11.6 is the explanation given by the moving observer. She sees the cohesive forces switched on at different times in different parts of the shell. This is, of course, different to her explanation in the ever-stable scenario, because the cohesive forces are not switched on there, but have always been operative. So how does the moving observer explain the fact that P^{μ}_{coh} is nonzero in her frame in the second scenario? This is the interesting question, which might so easily be overlooked!

The giveaway is the horizontal cutoff at $t_R = t^R_{-\infty}$ in Fig. 11.5. This in turn raises the question of how to calculate the integral over all early times, because it clearly makes a difference whether we cut off at constant rest frame time, or at constant moving frame time! In the second case, it is clear that the integrals giving P^μ_{coh} are going to deliver zero! Physically, this would not be surprising, because it amounts to applying the same cohesive forces non-concurrently to different parts of the shell as viewed by the rest frame observer, but for the same lengths of time. In the moving frame, the picture of this scenario is even simpler!

So why is the cutoff at constant t_R the right way to deal with this? The answer is a purely mathematical one. By cutting off the lower end of the infinite integration region within the world tube with lines slanting at other angles, we could produce any result for the integral! So there is a question of definition: how do we define such an infinite integral over a region of spacetime? The answer is that we cut off the variable covering the infinite dimension, i.e., the rest frame time in this case. To cut off some other variable such as the moving frame time, we must first change the variables of the integration so that one of them is indeed the moving frame time. We would then find that we obtained the same result for P^μ_{coh} and not zero.

Note on the Charge Dumbbell

In Sect. 11.3.6 we noted that the discrepancy or lack of it between energy-derived and momentum-derived EM masses displayed in Table 10.1 for the charge dumbbell could be explained by taking into account the cohesive force required to hold the system at a stable rest frame length d. In the light of the above discussion, the reader will now be able to formulate that exactly.

11.6.3 The Redefinition Approach Revisited

We now come to the approach advocated by Rohrlich, which corresponds to the second choice on p. 240. Hence, whatever the observer is doing, we use the electron rest frame HOS $t_R = $ constant for calculations. This means that, in order to calculate the energy and momentum of the electromagnetic fields, we take the integrals to be

$$P^0_e := \int_{L\sigma_R} d^3\sigma_\alpha \Theta^{\alpha 0}_e , \qquad P^k_e := \int_{L\sigma_R} d^3\sigma_\alpha \Theta^{\alpha k}_e . \qquad (11.250)$$

As mentioned before, we assume that $L\sigma_R$ is Rohrlich's notation for the rest frame HOS σ_R as described using a moving observer's coordinates, where L is the Lorentz transformation from the rest frame to the observer's frame. In this view, we have $d^3\sigma_\alpha = v_\alpha d^3 x_R = \gamma v_\alpha d^3 x$.

Let us examine Rohrlich's demonstration that

$$P_e^0 = \gamma m_e \,, \quad P_e^k = \gamma m_e v^k \,. \tag{11.251}$$

This is the prize. We then have a manifestly Lorentz covariant 4-momentum vector for the electromagnetic field.

Note that the normal to $L\sigma_R$ is the 4-velocity of the electron v^μ. Hence,

$$P_e^\mu = \int_{L\sigma_R} d\sigma \, v_\alpha \Theta_e^{\alpha\mu} \,, \tag{11.252}$$

where $d\sigma = d^3 x_R$, the usual measure on a spacelike hypersurface $t_R = $ constant that would be taken as a HOS in the electron rest frame. Note also that we decree this as part of the definition. There is something hybrid about this: we adopt the components of the electron 4-velocity with respect to the moving frame, but we adopt the natural measure as it would be in the rest frame HOS. This shows just what kind of definition we are dealing with. It is one designed specifically to handle the problem that the old definition was not Lorentz covariant.

Changing the sign of T^{ik} as given by (2.33) on p. 11 and making the replacement $\varepsilon_0 \to 1/4\pi$ to accord with Rohrlich's notation in [22], then using the fact that $v_\alpha = \gamma(c, -\mathbf{v})$, we have

$$v_\alpha \Theta_e^{\alpha 0} = \gamma \begin{pmatrix} c \\ -\mathbf{v} \end{pmatrix} \cdot \begin{pmatrix} U \\ \mathbf{S}/c \end{pmatrix} = \gamma(U - \mathbf{v} \cdot \mathbf{S}) \quad (c = 1) \,, \tag{11.253}$$

and for $k = 1, 2, 3$,

$$
\begin{aligned}
v_\alpha \Theta_e^{\alpha k} &\doteq \gamma \begin{pmatrix} c \\ -\mathbf{v} \end{pmatrix} \cdot (k\text{th column of } \Theta_e^{\alpha k}) \\
&= \frac{1}{4\pi} \gamma \begin{pmatrix} c \\ -\mathbf{v} \end{pmatrix} \cdot \begin{pmatrix} 4\pi \mathbf{S}/c \\ -\mathbf{EE} - c^2 \mathbf{BB} + 4\pi U \mathbb{I} \end{pmatrix} \\
&= \gamma \left[\mathbf{S} + \frac{1}{4\pi} \mathbf{v} \cdot (\mathbf{EE} + c^2 \mathbf{BB} - 4\pi U \mathbb{I}) \right] \,,
\end{aligned} \tag{11.254}
$$

with a reasonably obvious if awkward notation.

If $\mathbf{E}_R$ and $\mathbf{B}_R = 0$ are the rest frame fields, then the fields observed in the frame in which the electron moves with 4-velocity v^μ will be

$$\mathbf{E}^\| = \mathbf{E}_R^\| \,, \quad \mathbf{E}^\perp = \gamma \mathbf{E}_R^\perp \,, \quad \mathbf{B}^\| = 0 \,, \quad \mathbf{B}^\perp = \gamma \mathbf{v} \times \mathbf{E}_R^\perp \,. \tag{11.255}$$

Superscripts $\|$ and $\perp$ refer to vectors parallel and perpendicular to $\mathbf{v}$. These agree with the formulas (11.178) and (11.179) on p. 225, after a little manipulation.

We shall use Rohrlich's conventions in the following. Hence,

$$8\pi U = E^2 + B^2 \,, \quad 4\pi \mathbf{S} = \mathbf{E} \times \mathbf{B} \,, \tag{11.256}$$

and we define

$$m_e := \frac{1}{8\pi} \int d^3 x_R E_R^2 . \tag{11.257}$$

We now have

$$8\pi U = E^2 + B^2 = E_R^{\|2} + \gamma^2 E_R^{\perp 2} + \gamma^2 (\mathbf{v} \times \mathbf{E}_R)^2 , \tag{11.258}$$

and under the integral sign,

$$\begin{aligned}
\int d^3 x_R \gamma U &= \frac{\gamma}{8\pi} \int d^3 x_R (E^2 + B^2) \\
&= \frac{\gamma}{8\pi} \int d^3 x_R \left[\frac{1}{3} E_R^2 + \frac{2}{3} \gamma^2 E_R^2 + \frac{2}{3} \gamma^2 v^2 E_R^2 \right] \\
&= \frac{\gamma}{8\pi} \int d^3 x_R E_R^2 \gamma^2 \left(1 + \frac{1}{3} v^2 \right) ,
\end{aligned} \tag{11.259}$$

after a little manipulation with the useful identity

$$\frac{v^2}{c^2} = \frac{\gamma^2 - 1}{\gamma^2} . \tag{11.260}$$

We also make heavy use of the fact that $\mathbf{E}_R$ is spherically symmetric, which introduces the factors of 1/3 and 2/3, provided we have the integration. Rohrlich takes a short cut in his notation, writing

$$E_R^{\|2} = \frac{1}{3} E_R^2 , \quad E_R^{\perp 2} = \frac{2}{3} E_R^2 .$$

So we now have one of the terms in P_e^0 as given by (11.250) and (11.253), viz.,

$$\int d^3 x_R \gamma U = \gamma^3 m_e \left(1 + \frac{1}{3} v^2 \right) . \tag{11.261}$$

We also have

$$\begin{aligned}
4\pi \mathbf{S} &= (\mathbf{E}_R^{\|} + \gamma \mathbf{E}_R^{\perp}) \times (\mathbf{v} \times \mathbf{E}_R^{\perp}) \gamma \\
&= \gamma \left[(\mathbf{E}_R^{\|} \cdot \mathbf{E}_R^{\perp}) \mathbf{v} - (\mathbf{v} \cdot \mathbf{E}_R^{\|}) \mathbf{E}_R^{\perp} + \gamma E_R^{\perp 2} \mathbf{v} \right] \\
&= \gamma^2 E_R^{\perp 2} \mathbf{v} - \gamma v E_R^{\|} \mathbf{E}_R^{\perp} .
\end{aligned} \tag{11.262}$$

The other term in P_e^0 is now

$$-\int d^3 x_R \gamma \mathbf{v} \cdot \mathbf{S} = -\frac{\gamma^3 v^2}{4\pi} \int d^3 x_R E_R^{\perp 2} = -\frac{4}{3} \gamma^3 v^2 m_e . \tag{11.263}$$

Putting together these results, the total expression for P_e^0 is

$$P_e^0 = \left[\gamma^3 \left(1 + \frac{1}{3} v^2 \right) - \frac{4}{3} \gamma^3 v^2 \right] m_e = \gamma^3 (1 - v^2) m_e = \gamma m_e , \tag{11.264}$$

as required.

Before continuing, it is useful to compare the present expression

$$P_e^0 = \int d^3 x_R \gamma (U - \mathbf{v} \cdot \mathbf{S}) \tag{11.265}$$

with the expression (11.11) on p. 185. In the latter case, we had

$$W = \gamma \int U d\sigma - \frac{\gamma}{c^2} \int \mathbf{S} \cdot \mathbf{v} d\sigma , \tag{11.266}$$

and we concluded there that $d\sigma$ had to be $\gamma d^3 x$, where $d^3 x$ is the measure on a spacelike hypersurface (in fact a HOS) of the moving frame. The reason for using this measure was that we wanted to draw a parallel with the usual Abraham–Lorentz calculation of the energy in the electromagnetic field. In the latter, we integrate the quantity U over the HOS in the moving frame. However, having expressed U in terms of the fields in the electron rest frame, as we have done above, it turns out that the result is independent of the rest frame time t_R and this allows us to rewrite our expression as an integral over the rest frame HOS, provided that we make the replacement $\gamma d^3 x \rightarrow d^3 x_R$.

This does indeed make (11.266) equal to the present expression (11.265). It also confirms once again that Rohrlich was wrong in his 1960 paper when he wrote $\gamma d\sigma = d^3 x$. The present ploy is interesting in that we now bypass any integration over the HOS in the moving frame, by decreeing that the integration has to be over the rest frame HOS with its usual measure $d^3 x_R$, provided of course that the integrand has been suitably converted to a function of the rest frame variables.

We now calculate the 3-momentum of the electromagnetic fields according to Rohrlich's covariant prescription, viz.,

$$P_e^k = \frac{1}{4\pi} \int d^3 x_R \gamma [4\pi \mathbf{S} + \mathbf{v} \cdot (\mathbf{EE} + c^2 \mathbf{BB} - 4\pi U \mathbb{I})] . \tag{11.267}$$

From (11.262), we already have

$$\int d^3 x_R \gamma \mathbf{S} = \frac{4}{3} \gamma^3 m_e \mathbf{v} . \tag{11.268}$$

The second term in (11.262) integrates to zero because of the symmetry of the Coulomb form of the electric field. Now note that

$$\mathbf{v} \cdot (\mathbf{EE} + c^2 \mathbf{BB} - 4\pi U \mathbb{I}) = \mathbf{E}(v E^{\parallel}) - 4\pi U \mathbf{v}$$
$$= (E_R^{\parallel} + \gamma E_R^{\perp}) E_R^{\parallel} v - 4\pi U \mathbf{v}$$
$$= (E_R^{\parallel 2} - 4\pi U) \mathbf{v} + \text{term integrating to } 0 .$$

Hence,

$$\frac{1}{4\pi}\int d^3x_R\,\gamma\mathbf{v}\cdot(\mathbf{EE}+c^2\mathbf{BB}-4\pi U\mathbb{I})=\frac{1}{4\pi}\gamma\int d^3x_R(E_R^{\|2}-4\pi U)\mathbf{v}$$

$$=\frac{2}{3}\gamma m_e\mathbf{v}-m_e\gamma^3\left(1+\frac{1}{3}v^2\right)\mathbf{v}$$

$$=m_e\gamma\mathbf{v}\left[\frac{2}{3}-\gamma^2\left(1+\frac{1}{3}v^2\right)\right]$$

$$=m_e\gamma\mathbf{v}\left[\frac{2}{3}-\gamma^2-\frac{1}{3}(\gamma^2-1)\right]$$

$$=m_e\gamma\mathbf{v}\left(1-\frac{4}{3}\gamma^2\right). \tag{11.269}$$

The full expression for P_e^k is therefore

$$P_e^k=\frac{4}{3}\gamma^3 m_e\mathbf{v}+m_e\gamma\mathbf{v}\left(1-\frac{4}{3}\gamma^2\right)=m_e\gamma\mathbf{v}, \tag{11.270}$$

as required. And so we have yet another prescription for the Lorentz covariant 4-momentum of the electromagnetic fields, and yet another proof that it is Lorentz covariant.

There remains one item here: we calculate the momentum P_{coh}^μ arising due to Rohrlich's cohesive force

$$f_R^0=0, \qquad \mathbf{f}_R=-2\pi\sigma^2\hat{\mathbf{r}}\delta(r-a), \tag{11.271}$$

with

$$\sigma=\frac{e}{4\pi a^2}. \tag{11.272}$$

Naturally, we calculate this using Rohrlich's prescription, whereby f^0 and $\mathbf{f}$ in the moving frame are integrated over the region in the rest frame from $t_R=-\infty$ up to the present rest frame HOS, using the measure $d^4x_R=d^4x$.

We find

$$P_{coh}^0=-\int_{V_4(L\sigma_R)}d^4x f^0$$

$$=-\int_{V_4(\sigma_R)}d^4x_R\,\gamma(f_R^0+\mathbf{v}\cdot\mathbf{f}_R)=0, \tag{11.273}$$

because of the spherical symmetry of $\mathbf{f}_R$. Likewise,

$$\mathbf{P}_{coh}=-\int_{V_4(L\sigma_R)}d^4x\mathbf{f}$$

$$=-\int_{V_4(\sigma_R)}d^4x_R\left(\gamma\mathbf{f}_R^\|+\mathbf{f}_R^\perp+\gamma\mathbf{v}f_R^0\right)=0, \tag{11.274}$$

in a trivial way.

But what could these quantities possibly be? Are we considering an adiabatically collapsing shell of charge? Surely not, because the above cohesive force is the one for the stable electron, i.e., it has no dynamic aspect. So presumably, we are integrating the cohesive force density of the stable electron here and claiming that it contributes no four-momentum to the system. On the other hand, we cannot just integrate these quantities over any region we like and claim that we have found the 4-momentum. Is Rohrlich's region the right one?

11.6.4 Whatever Happened to Lorentz Covariance?

This is a brief conclusion in the form of a summary. The original question was: why do we not get a Lorentz covariant 4-momentum from the electromagnetic fields due to a uniformly charged shell? Ultimately, the answer is: because there are other forces at work that have not been included in the model. Indeed, we need cohesive forces to hold the shell together against the mutual repulsion of the charge elements making it up. The fact that these are needed is equivalent to the fact that the energy–momentum tensor of the electromagnetic fields is not conserved everywhere. To be precise, it is conserved both inside and outside the shell, but not on the shell, exactly where the balancing cohesive forces are required.

Now what goes wrong when an energy–momentum tensor is not conserved? When it is conserved, it can be used to define a Lorentz covariant 4-momentum for the corresponding fields in any hyperplane of simultaneity of whatever inertial frame we are in. (What is more, this 4-momentum remains constant in the time of that inertial frame.) But when the energy–momentum tensor is not conserved, this standard, physically natural definition of a 4-momentum does not yield a Lorentz covariant quantity. This is all there is to the problem.

One approach to show how the cohesive forces can save the day is to define them directly as the forces that would be required to balance electromagnetic repulsive forces on the sphere. Indeed, they are defined by

$$f^\mu_{\text{coh}} := -\partial_\alpha \Theta^{\alpha\mu}_{\text{em}} ,$$

which would be zero if the energy–momentum tensor of the electromagnetic fields were conserved. In whatever frame we examine things, we now find that the cohesive forces impart a 4-momentum P^μ_{coh} to the system. It happens to be zero in the rest frame of the charge shell, but it is not Lorentz covariant. On the other hand, the total 4-momentum due to these forces and the electromagnetic fields (the sum of the two objects P^μ_{coh} and P^μ_{em}) is a 4-vector, and it is equal to

$$P^\mu_{\text{coh}} + P^\mu_{\text{em}} = m_e v^\mu ,$$

where v^μ is the 4-velocity of the shell in the given frame and m_e is the electroma gnetic mass defined from the Coulomb energy in the electric field around the shell in its rest frame (also often referred to as the electrostatic mass).

This is, of course, a minimalist solution to the real physical problem. The idea promoted in this book is that the electron is made up of parts with electric charge and with some other type of charge, source of a second field that acts to hold the thing together against the Coulomb repulsion of the charged elements. The total energy–momentum tensor for the system will contain contributions from both types of field and it will be conserved. The total 4-momentum of the two fields defined in the usual way from the total energy–momentum tensor will therefore be Lorentz covariant. The inertial mass of the system will contain contributions from both the electromagnetic self-force and the self-force arising from the other kind of field.

This is a model in which the electron has spatial structure in any Lorentz frame. If one assumes the electron to be a point charge, or if one wishes to take the point limit, the cohesive forces do seem to be rather superfluous. Presumably, this is what motivated Rohrlich to go to such lengths to defend what is actually a totally artificial ploy for defining a Lorentz covariant 4-momentum for the electromagnetic fields.

Chapter 12
Rigidity in Relativity

In special relativity, one often speaks of rigid rods, looking at them in one inertial frame or another and observing that they do not always have the same length,
despite their rigidity. This chapter is about what happens to the rod as it gets from
one inertial frame to another, i.e., as it accelerates. As we have seen the problem
is not entirely academic. When we try to model extended charge distributions and
their fields, and in particular the forces they exert upon themselves via these electromagnetic effects, when they are accelerating, some hypothesis must be made about
the way the charge distribution shifts around in the relevant spatial hypersurfaces
of Minkowski's spacetime. A notion of rigidity is indeed usually applied and that
is discussed here (Sect. 12.1), in connection with frames of reference adapted to
accelerating observers in the spacetime of special relativity.

The physical legitimacy of adapted frames of reference is discussed in some detail throughout the chapter, but particularly in the context of the Pound–Rebka experiment in Sect. 12.1.10. The immediate aim is to elucidate the roles of what are
usually referred to as the clock and ruler hypotheses. The broader aim is to illustrate
the problems raised by relativity theory for the simple idea of treating particles as
spatially extended objects.

One would also like to consider rigid motions of any material medium in a more
general framework, even in the context of general relativity. The notion of rigidity
can be extended (Sect. 12.2) in a simple but perhaps questionable way. The aim here
will indeed be to cast a critical glance.

12.1 Rigid Rods and Rigid Spheres

12.1.1 A Toy Electron

If one wants to make a model of the electron in which the electric charge is no longer
all concentrated at a mathematical point in any spacelike hypersurface of Minkowski

Lyle, S.N.: *Rigidity in Relativity*. Lect. Notes Phys. **796**, 263–311 (2010)
DOI 10.1007/978-3-642-04785-5_12   © Springer-Verlag Berlin Heidelberg 2010

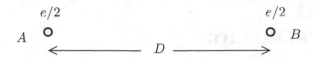

Fig. 12.1 A toy electron at rest in an inertial frame. The system is in equilibrium under the forces between A and B

spacetime, one might get the idea, exploited in the earlier chapters of this book, of dividing the charge into two equal amounts labelled A and B, each one occupying such a point, and spaced apart by some distance D when the system is moving inertially and observed from an inertial frame moving with it. This would certainly be a toy electron (see Fig. 12.1). It would appear to a large extent to defeat the object of giving the electron a spatial extent since there are now two mathematical points of charge instead of just one. But it does nevertheless bring out some of the advantages and some of the difficulties. We shall only be concerned with one of the difficulties here.

The idea of such a model is to allow it to move, then work out the electromagnetic fields due to each point of charge using the Lienard–Wiechert potential, and calculate the force that each charge can thereby exert on the other. One is of course interested in the net force that such a system might exert upon itself when accelerated. In a first approach, one does not worry about the force required to hold the system together. For it should not be forgotten that the two charges are alike and will repel one another. And yet this very question raises another, more urgent one.

For suppose point A, on the left, has a one-dimensional motion given by $x_A(t)$ along the axis from A to B, as observed relative to some inertial frame $\mathscr{I}$ (see Fig. 12.2). What will be the motion $x_B(t)$ of the right-hand end of the system? Let the coordinate speed of A in $\mathscr{I}$ be

$$v_A(t) = \dot{x}_A(t) := \frac{\mathrm{d}x_A}{\mathrm{d}t} .$$

If the system were rigid in the pre-relativistic sense, the speed of B would be

$$v_B(t) = v_A(t) .$$

There is an obvious problem with this: if A and B always have the same coordinate speed, the separation of A and B will always be the same in $\mathscr{I}$, viz., D, whereas we

Fig. 12.2 System in motion in an inertial frame $\mathscr{I}$. If we impose some motion on A, what is the motion of B?

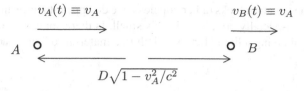

Fig. 12.3 System in uniform motion $v_A(t) \equiv v_A$ in an inertial frame $\mathscr{I}$. We expect $v_B(t) \equiv v_A$ and the length to be contracted as shown

expect lengths of objects to contract in special relativity when viewed from frames in which those objects are moving.

If the system had been set in uniform motion, with A moving at constant coordinate speed v_A in frame $\mathscr{I}$, and if it were rigid in the sense of special relativity, we would expect the separation of A and B to be the contracted length $D\sqrt{1 - v_A^2/c^2}$ as observed from $\mathscr{I}$ (see Fig. 12.3). There would then be no problem for point B to have the same coordinate speed as point A.

The difficulty occurs, of course, when v_A is changing with time, i.e., when there is acceleration, and the system has to adjust all the time. This means that v_B looks as though it might be a very complicated function of time indeed. Perhaps we require some physical assumption about the relaxation time of the system. One immediately wonders how B is supposed to adjust. Indeed, what are the forces on the system? What is accelerating it? Since we started by attributing a motion $x_A(t)$ to A, we might imagine some external force applied to A and ask how the effects of this force might be transmitted to B to make it too accelerate. On the other hand, an external force might equally be applied to both points, as would happen for example if it were due to a force field. In addition, we know that there are mutually repulsive electromagnetic forces between A and B due to their own fields, and we said there had to be binding forces to oppose them, to hold the thing together. How will these react to the changes?

In the pre-relativistic context, there was no problem because the appropriate notion of rigidity automatically delivered the motion of point B given the motion of point A. But this hypothesis was unequivocally pre-relativistic: if the system was accelerated by applying a force to point A, the effects had to propagate instantaneously to point B, and in such a way that the net force on B was always identical. Of course, if the system was accelerated by a uniform field acting simultaneously on both A and B, then one had the advantage of not having to consider the binding forces at all.

So what could be the separation of A and B as observed from $\mathscr{I}$ when $v_A(t)$ changes with time? One idea is that, since A has speed $v_A(t)$, the separation of A and B at time t could just be $D\sqrt{1 - v_A(t)^2/c^2}$, the contracted length for that speed. This might be a good approximation in some cases, but there is an obvious problem with it: because there is some contraction going on, this too will induce a motion of B that ought to be included. And if B does not have the same speed as A, why not use the speed of B to work out the contraction? Or some average of the speeds?

Although the situation looks rather hopeless, we do appear to have an approximation. This looks especially promising if D is small. Is there some way of making D
infinitesimal and taking a limit? Let us switch to a material rod under acceleration.

12.1.2 A Rigid Rod

Of course, we know what happens to a rigid rod when it has uniform motion relative
to an inertial frame $\mathscr{I}$. In other words, we know what we want rigidity to mean in
that context. But can we say how a rigid rod should behave when it accelerates? Can
we still have some kind of rigidity?

Let A and B be the left- and right-hand ends of the rod and consider motion $x_A(t)$
and $x_B(t)$ along the axis from A to B (see Fig. 12.4). Let us first label the particles
in the rod by their distance s to the right of A when the system is stationary in some
inertial frame (see Fig. 12.5). This idea of labelling particles will prove extremely
useful when considering continuous media later on. In the present case, we imagine
the rod as a strictly one-dimensional, continuous row of particles.

Now let A have motion $x_A(t)$ relative to an inertial frame $\mathscr{I}$ (see Fig. 12.6) and
let $X(s,t)$ be a function giving the position of particle s at time t as

$$x_s(t) = x_A(t) + X(s,t) \,,$$

where naturally we require

$$X(0,t) = 0 \,, \qquad X(D,t) = x_B(t) - x_A(t) \,.$$

Let us require the element between s and $s + \delta s$ to have coordinate length

$$\left[1 - \frac{v(s,t)^2}{c^2}\right]^{1/2} \delta s \,, \tag{12.1}$$

where $v(s,t)$ is its instantaneous coordinate velocity, with $v(0,t) = v_A(t)$. This is
precisely the criterion suggested by Rindler [6, pp. 39–40]. We can integrate to find

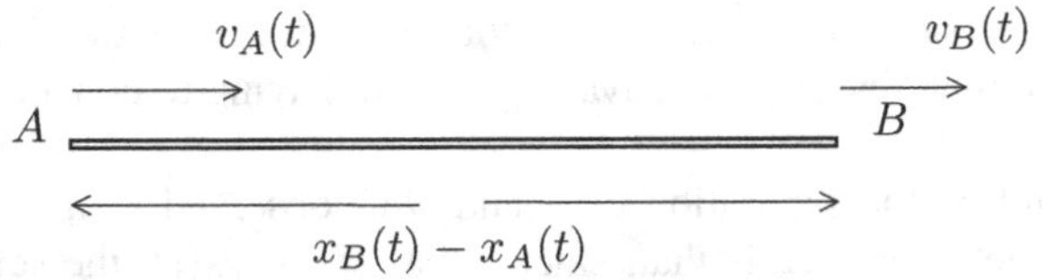

Fig. 12.4 Material rod in motion along its axis in an inertial frame $\mathscr{I}$. The position of the left-hand
end A is given by $x_A(t)$ at time t, and the position of the right-hand end B is given by $x_B(t)$

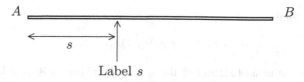

Fig. 12.5 Stationary material rod in an inertial frame $\mathscr{I}$. Labelling the particles in the rod by their distance s from A, so that $s \in [0, D]$

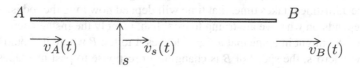

Fig. 12.6 Material rod with arbitrary motion in an inertial frame $\mathscr{I}$

$$X(s,t) = \int_0^s \left[1 - \frac{v(s',t)^2}{c^2} \right]^{1/2} ds' . \tag{12.2}$$

This implies that

$$X_B = \int_0^D \left[1 - \frac{v(s',t)^2}{c^2} \right]^{1/2} ds' . \tag{12.3}$$

Note the highly complex equation this gives for the speed function $v(s,t)$, viz.,

$$v(s,t) = v_A(t) + \frac{\partial X(s,t)}{\partial t} . \tag{12.4}$$

Let us observe carefully that we are not assuming any simple Galilean addition law for velocities here. This is a straightforward differentiation with respect to t of the formula for the coordinate position of atom s at time t, viz., $x_A(t) + X(s,t)$. The partial time derivative of X is not the velocity of s relative to A, that is, it is not the velocity of s measured in a frame moving with A.

Now (12.2) seems to embody the idea of the rod being rigid. For surely this rod could no longer be elastic, in the sense that (12.1) only allows the element δs to relativistically contract for the value of its instantaneous speed, forbidding any other contortions. One could well imagine the rod undergoing a very complex deformation along its length, in which relativistic contraction effects were quite negligible compared with a certain looseness in the molecular bonding, but we are not talking about this. In fact we are seeking a definition of rigidity that does not refer to microscopic structure.

Let us just note what assumption is expressed by the earlier idea that

$$x_B(t) = x_A(t) + D\sqrt{1 - v_A(t)^2/c^2} , \tag{12.5}$$

which would lead to

$$v_B(t) = v_A(t) - \gamma(v_A)v_A(t)\ddot{x}_A(t)D/c^2 , \qquad (12.6)$$

where $\gamma(v_A)$ is the usual function of the speed. Thinking of A as a kind of base point, to which a force is perhaps applied, it says that the relativistic contraction is instantaneous: when A moves at speed v_A, the rod immediately has coordinate length $D/\gamma(v_A)$ for that value of v_A. We are now improving on this, accounting for the fact that, if the adjustment takes time, that time will depend how long the rod is, and that in turn depends on what we are trying to establish, namely the instantaneous length of the rod. We might imagine that a signal leaves A to tell B where it should be, and as it moves across, the speed of B is changing in response to past messages of the same kind. But using (12.2) and (12.4), can we tell when the new signal will get there?

12.1.3 Equation of Motion for Points on the Rod

So far the main equations for the atom labelled s on the rod are (12.2) and (12.4), viz.,

$$X(s,t) = \int_0^s \left[1 - \frac{v(s',t)^2}{c^2}\right]^{1/2} ds' \qquad (12.7)$$

and

$$v(s,t) = v_A(t) + \frac{\partial X(s,t)}{\partial t} . \qquad (12.8)$$

The first implies that

$$\frac{\partial X(s,t)}{\partial s} = \left[1 - \frac{v(s,t)^2}{c^2}\right]^{1/2} . \qquad (12.9)$$

We can write one nonlinear partial differential equation for $X(s,t)$ by eliminating $v(s,t)$ to give

$$c^2 \left(\frac{\partial X}{\partial s}\right)^2 + \left[\frac{\partial X}{\partial t} + v_A(t)\right]^2 = c^2 . \qquad (12.10)$$

This is effectively the equation that we have to solve to find the length of our rod. It is important to see that there is a boundary condition too, viz.,

$$0 = \frac{\partial X(s,t)}{\partial t}\bigg|_{s=0} , \qquad (12.11)$$

because we do require $v(0,t) = v_A(t)$ in conjunction with (12.8).

We shall find a solution to this problem, although not by solving (12.10) directly. Instead we shall follow a circuitous but instructive route and end up guessing the relevant solution.

12.1.4 A Frame for an Accelerating Observer

Let AO be the name for an observer moving with the left-hand end A of the proposed rod. AO is an accelerating observer and it is well known [18] that such a person can find well-adapted coordinates y^μ with the following properties (where the Latin index runs over $\{1,2,3\}$):

- First of all, any curve with all three y^i constant is timelike and any curve with y^0 constant is spacelike.
- At any point along the worldline of AO, the zero coordinate y^0 equals the proper time along that worldline.
- At each point of the worldline of AO, curves with constant y^0 which intersect it are orthogonal to it where they intersect it.
- The metric has the Minkowski form along the worldline of AO.
- The coordinates y^i are Cartesian on every hypersurface of constant y^0.
- The equation for the worldline of AO has the form $y^i = 0$ for $i = 1,2,3$.

Such coordinates could be called semi-Euclidean.

Let us consider a 1D acceleration and temporarily drop the subscript A on the functions $x_A(t)$ and $v_A(t)$ describing the motion of AO in the inertial frame $\mathscr{I}$. The worldline of the accelerating observer is given in inertial coordinates by

$$t = \sigma, \qquad x = x(\sigma), \qquad \frac{dx}{d\sigma} = v(\sigma), \qquad (12.12)$$

$$\frac{d^2x}{d\sigma^2} = a(\sigma), \qquad y(\sigma) = 0 = z(\sigma), \qquad (12.13)$$

using the time t in $\mathscr{I}$ to parametrise. The proper time $\tau(\sigma)$ of AO is given by

$$\frac{d\tau}{d\sigma} = (1 - v^2/c^2)^{1/2}. \qquad (12.14)$$

The coordinates y^μ are constructed on an open neighbourhood of the AO worldline as follows (see Fig. 12.7). For an event (t,x,y,z) not too far from the worldline, there is a unique value of τ and hence also the parameter σ such that the point lies in the hyperplane of simultaneity (HOS) of AO when its proper time is τ. This hyperplane of simultaneity is given by

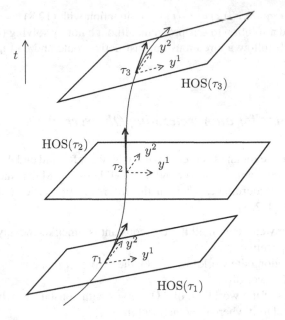

Fig. 12.7 Constructing a semi-Euclidean (SE) frame for an accelerating observer. View from an inertial frame with time coordinate t. The curve is the observer worldline given by (12.12). Three hyperplanes of simultaneity (HOS) are shown at three successive proper times τ_1, τ_2, and τ_3 of the observer. These hyperplanes of simultaneity are borrowed from the instantaneously comoving inertial observer, as are the coordinates y^1, y^2, and y^3 used to coordinatise them. Only two of the latter coordinates can be shown in the spacetime diagram

$$t - \sigma(\tau) = \frac{v(\sigma(\tau))}{c^2}\left[x - x(\sigma(\tau))\right], \qquad (12.15)$$

which solves, for any x and t, to give $\sigma(\tau)(x,t)$.

The semi-Euclidean coordinates attributed to the event (t,x,y,z) are, for the time coordinate y^0, (c times) the proper time τ found from (12.15) and, for the spatial coordinates, the spatial coordinates of this event in an instantaneously comoving inertial frame at proper time τ of AO. In fact, every other event in this instantaneously comoving inertial frame is attributed the same time coordinate $y^0 = c\tau$ and the appropriate spatial coordinates borrowed from this frame. Of course, the HOS of AO at time τ is also the one borrowed from the instantaneously comoving inertial frame.

There is just one detail to get out of the way: there are many different instantaneously comoving inertial frames for a given τ, and there are even many different ways to choose these frames as a smooth function of τ as one moves along the AO worldline, rotating back and forth around various axes in the original inertial frame $\mathscr{I}$ as τ progresses. We choose a sequence with no rotation about any space axis in the instantaneous local rest frame. It can always be done by solving the Fermi–Walker transport equations (see Sect. 12.2.5). The semi-Euclidean coordinates are then given by

$$
\begin{cases}
y^0 = c\tau \,, \\[2mm]
y^1 = \dfrac{[x - x(\sigma)] - v(\sigma)(t - \sigma)}{\sqrt{1 - v^2/c^2}} \,, \\[2mm]
y^2 = y \,, \\[2mm]
y^3 = z \,,
\end{cases}
\qquad (12.16)
$$

where $\sigma = \sigma(t, x)$ as found from (12.15). The inverse transformation, from semi-Euclidean coordinates to inertial coordinates, is given by

$$
\begin{cases}
t = \sigma(y^0) + \dfrac{v(y^0)}{c^2} y^1 \left[1 - \dfrac{v(y^0)^2}{c^2} \right]^{-1/2} \,, \\[3mm]
x = x(y^0) + y^1 \left[1 - \dfrac{v(y^0)^2}{c^2} \right]^{-1/2} \,, \\[3mm]
y = y^2 \,, \\[2mm]
z = y^3 \,,
\end{cases}
\qquad (12.17)
$$

where the function $\sigma(y^0)$ is just the expression relating inertial time to proper time for the accelerating observer, and the functions $x(y^0)$ and $v(y^0)$ should really be written $x(\sigma(y^0))$ and $v(\sigma(y^0))$, respectively.

The above relations are not very enlightening. They are only displayed to show that the idea of such coordinates can be made perfectly concrete. One calculates the metric components in this frame, viz.,

$$
g_{00} = 1/g^{00} = \left[1 + \frac{a(\sigma)[x - x(\sigma)]/c^2}{1 - v(\sigma)^2/c^2} \right]^2 \,,
\qquad (12.18)
$$

where $\sigma = \sigma(t, x)$ as found from (12.15), and

$$
g_{i0} = 0 = g_{0i} \,, \qquad g_{ij} = -\delta_{ij} \,, \qquad i, j \in \{1, 2, 3\} \,,
\qquad (12.19)
$$

and checks the list of requirements for the coordinates to be suitably adapted to the accelerating observer.

Although perfectly concrete, the coordinates are not perfectly explicit: the component g_{00} of the semi-Euclidean metric has been expressed in terms of the original inertial coordinates! This can be remedied as follows. One observes that, with the help of (12.15),

$$y^1 - [x - x(\sigma)]\sqrt{1 - v^2/c^2} \,. \tag{12.20}$$

One calculates the four-acceleration in the inertial frame to be

$$a^\mu := \frac{d^2 x^\mu}{d\tau^2} = a(1 - v^2/c^2)^{-2}\left(\frac{v}{c}, 1, 0, 0\right), \tag{12.21}$$

and transforms this by Lorentz transformation to the inertial frame instantaneously comoving with the observer to find only one nonzero four-acceleration component in that frame, which is called the absolute acceleration of the observer:

$$a_{01} := \text{absolute acceleration} = a(1 - v^2/c^2)^{-3/2}\,. \tag{12.22}$$

The notation a_{01} for the 1-component of the absolute acceleration will appear again on p. 299. One now has the more comforting formula

$$g_{00} = 1/g^{00} = \left[1 + \frac{a_{01}(\sigma)y^1}{c^2}\right]^2. \tag{12.23}$$

To obtain still more explicit formulas, one needs to consider a specific motion $x(\sigma)$ of AO, the classic example being uniform acceleration:

$$x(\sigma) = \frac{c^2}{g}\left[\left(1 + \frac{g^2\sigma^2}{c^2}\right)^{1/2} - 1\right], \qquad t = \sigma, \tag{12.24}$$

where g is some constant with units of acceleration. This does not look like a constant acceleration in the inertial frame:

$$\frac{dx}{d\sigma} = \frac{g\sigma}{(1 + g^2\sigma^2/c^2)^{1/2}}, \qquad \frac{d^2 x}{d\sigma^2} = \frac{g}{(1 + g^2\sigma^2/c^2)^{3/2}}. \tag{12.25}$$

However, the 4-acceleration defined in the inertial frame $\mathscr{I}$ by

$$a^\mu = \frac{d^2 x^\mu}{d\tau^2}, \tag{12.26}$$

where τ is the proper time, has constant magnitude. It turns out that

$$a^2 := a_\mu a^\mu = -g^2,$$

with a suitable convention for the signature of the metric.

In this case, the transformation from inertial to semi-Euclidean coordinates is

$$y^0 = \frac{c^2}{g}\tanh^{-1}\frac{ct}{x + c^2/g}, \tag{12.27}$$

$$y^1 = \left[\left(x + \frac{c^2}{g} \right)^2 - c^2 t^2 \right]^{1/2} - \frac{c^2}{g} \,, \qquad y^2 = y \,, \qquad y^3 = z \,, \qquad (12.28)$$

and the inverse transformation is

$$t = \frac{c}{g} \sinh \frac{g y^0}{c^2} + \frac{y^1}{c} \sinh \frac{g y^0}{c^2} \,, \qquad\qquad (12.29)$$

$$x = \frac{c^2}{g} \left(\cosh \frac{g y^0}{c^2} - 1 \right) + y^1 \cosh \frac{g y^0}{c^2} \,, \qquad y = y^2 \,, \qquad z = y^3 \,. \qquad (12.30)$$

One finds the metric components to be

$$g_{00} = \left(1 + \frac{g y^1}{c^2} \right)^2 \,, \qquad g_{0i} = 0 = g_{i0} \,, \qquad g_{ij} = -\delta_{ij} \,, \qquad (12.31)$$

for $i, j \in \{1, 2, 3\}$, in the semi-Euclidean frame. Interestingly, this metric is static, i.e., g_{00} is independent of y^0. It is the only semi-Euclidean metric that is [4].

It is worth pausing to wonder why AO should adopt such coordinates. It must be comforting to attribute one's own proper time to events that appear simultaneous. But what events are simultaneous with AO? In the above construction, AO borrows the hyperplane of simultaneity of an inertially moving observer, who does not have the same motion at all. AO also borrows the lengths of this inertially moving observer. But if AO were carrying a rigid measuring rod, what lengths would be measured with it?

12.1.5 Lengths Measured by the Rigid Rod

In fact the rigid rod of Sect. 12.1.2 measures the spatial coordinates of AO when this observer uses semi-Euclidean coordinates. Let us prove this for the case of a uniform acceleration g, where formulas are explicit.

We write down the path of a point with some fixed spatial coordinate s along the axis of acceleration (putting the other spatial coordinates equal to zero). The formula we have for the path of the origin of the SE frame as expressed in Minkowski coordinates is

$$x_A(t) = \frac{c^2}{g} \left(\sqrt{1 + \frac{g^2 t^2}{c^2}} - 1 \right) \,, \qquad\qquad (12.32)$$

giving a coordinate velocity

$$v_A(t) = \frac{gt}{\sqrt{1 + \dfrac{g^2 t^2}{c^2}}} . \qquad (12.33)$$

The formula for the path of the point at fixed SE spatial coordinate s from the origin as expressed in Minkowski coordinates is

$$x_s(t) = X(s,t) + x_A(t) = \frac{c^2}{g} \left[\sqrt{\left(1 + \frac{gs}{c^2}\right)^2 + \frac{g^2 t^2}{c^2}} - 1 \right] . \qquad (12.34)$$

We are going to show that the function $X(s,t)$ defined by the last relation actually satisfies our equation of motion (12.10) in the case where the function $x_A(t)$ gives the path of the left-hand end A of the rod, i.e., when the point A is uniformly accelerated by g.

Proof That (12.34) Is a Solution for (12.10)

We begin with the partial derivatives:

$$\frac{\partial X}{\partial t} = \frac{gt}{\sqrt{\left(1 + \dfrac{gs}{c^2}\right)^2 + \dfrac{g^2 t^2}{c^2}}} - v_A(t) , \qquad (12.35)$$

$$\frac{\partial X}{\partial s} = \frac{1 + gs/c^2}{\sqrt{\left(1 + \dfrac{gs}{c^2}\right)^2 + \dfrac{g^2 t^2}{c^2}}} . \qquad (12.36)$$

Hence,

$$\left[\frac{\partial X}{\partial t} + v_A(t) \right]^2 = \frac{g^2 t^2}{\left(1 + \dfrac{gs}{c^2}\right)^2 + \dfrac{g^2 t^2}{c^2}} \qquad (12.37)$$

and

$$c^2 \left(\frac{\partial X}{\partial s} \right)^2 = \frac{c^2 (1 + gs/c^2)^2}{\left(1 + \dfrac{gs}{c^2}\right)^2 + \dfrac{g^2 t^2}{c^2}} . \qquad (12.38)$$

Adding the last two equations together, it is clear that we just get c^2, as required by (12.10). The boundary condition (12.11) on p. 268 is obviously satisfied too. ∎

For a rod with arbitrary 1D acceleration, the formulas are much more involved, due to the lack of explicitness, but the proof is nevertheless straightforward. We begin with the worldline of a point with fixed semi-Euclidean spatial coordinate $y^1 = s$. Using the above formulas for the semi-Euclidean coordinates, this path is given parametrically in Minkowski coordinates by

$$t = \sigma(y^0) + \frac{v_A(y^0)}{c^2} s \left[1 - \frac{v_A(y^0)^2}{c^2}\right]^{-1/2}, \qquad (12.39)$$

$$x = x_A(y^0) + s \left[1 - \frac{v_A(y^0)^2}{c^2}\right]^{-1/2}, \qquad (12.40)$$

where y^0 is considered to be a parameter, which can be eliminated using the first relation to give x as a function of t. We have to be very careful about these formulas, as usual, because $x_A(y^0)$, $v_A(y^0)$ are not what they seem! In fact, $\sigma(y^0)$ gives the Minkowski time as a function of the proper time y^0 of the observer, and $x_A(y^0)$, $v_A(y^0)$ mean

$$x_A\left(\sigma(y^0)\right), \qquad v_A\left(\sigma(y^0)\right),$$

respectively, where $x_A(\sigma)$ and $v_A(\sigma)$ give the Minkowski coordinate path and speed of the observer.

When we use (12.39) to express y^0 in terms of s and t, something we shall only do implicitly, and replace y^0 in (12.40) by the resulting function $y^0(s,t)$, we shall say that the left-hand side of (12.40) is our quantity $X(s,t) + x_A(t)$. In this last expression, $x_A(t)$ really is just the function specifying the Minkowski position of the accelerating observer, who represents the left-hand end A of our rod. Hence,

$$X(s,t) = x_A(y^0) - x_A(t) + s \left[1 - \frac{v_A(y^0)^2}{c^2}\right]^{-1/2}, \qquad (12.41)$$

where $y^0 = y^0(s,t)$ is given by (12.39). Note that $x_A(y^0)$ is quite different from $x_A(t)$.

We now calculate the partial derivatives:

$$\frac{\partial X}{\partial s} = \left\{ v_A(y^0) + \frac{s}{c^2} v_A(y^0)\dot{v}_A(y^0) \left[1 - \frac{v_A(y^0)^2}{c^2}\right]^{-3/2} \right\} \frac{d\sigma}{dy^0} \frac{\partial y^0}{\partial s}$$

$$+ \left[1 - \frac{v_A(y^0)^2}{c^2}\right]^{-1/2}, \qquad (12.42)$$

$$\frac{\partial X}{\partial t} = \left\{ v_A(y^0) + \frac{s}{c^2} v_A(y^0)\dot{v}_A(y^0) \left[1 - \frac{v_A(y^0)^2}{c^2}\right]^{-3/2} \right\} \frac{d\sigma}{dy^0} \frac{\partial y^0}{\partial t} - v_A(t). \quad (12.43)$$

We shall keep the groups

$$\frac{\mathrm{d}\sigma}{\mathrm{d}y^0}\frac{\partial y^0}{\partial s} \quad \text{and} \quad \frac{\mathrm{d}\sigma}{\mathrm{d}y^0}\frac{\partial y^0}{\partial t}.$$

They are obtained by partial differentiation of (12.54). Partial differentiation of (12.39) with respect to t gives

$$1 = \frac{\mathrm{d}\sigma}{\mathrm{d}y^0}\frac{\partial y^0}{\partial t}\left\{1 + \frac{s}{c^2}\dot{v}_A(y^0)\left[1 - \frac{v_A(y^0)^2}{c^2}\right]^{-1/2}\right.$$

$$\left. + \frac{s}{c^2}\dot{v}_A(y^0)\frac{v_A(y^0)^2}{c^2}\left[1 - \frac{v_A(y^0)^2}{c^2}\right]^{-3/2}\right\}.$$

We can immediately improve this, because the last two terms in curly brackets give

$$\frac{s}{c^2}\dot{v}_A(y^0)\left[1 - \frac{v_A(y^0)^2}{c^2}\right]^{-1/2}\left\{1 + \frac{v_A(y^0)^2}{c^2}\left[1 - \frac{v_A(y^0)^2}{c^2}\right]^{-1}\right\}$$

$$= \frac{s}{c^2}\dot{v}_A(y^0)\left[1 - \frac{v_A(y^0)^2}{c^2}\right]^{-3/2}.$$

Hence,

$$\frac{\mathrm{d}\sigma}{\mathrm{d}y^0}\frac{\partial y^0}{\partial t} = \frac{1}{1 + \frac{s}{c^2}\dot{v}_A(y^0)\left[1 - \frac{v_A(y^0)^2}{c^2}\right]^{-3/2}}. \tag{12.44}$$

Differentiating (12.39) with respect to s and carrying out exactly the same manipulations gives

$$\frac{\mathrm{d}\sigma}{\mathrm{d}y^0}\frac{\partial y^0}{\partial s} = \frac{-\frac{v_A(y^0)}{c^2}\left[1 - \frac{v_A(y^0)^2}{c^2}\right]^{-1/2}}{1 + \frac{s}{c^2}\dot{v}_A(y^0)\left[1 - \frac{v_A(y^0)^2}{c^2}\right]^{-3/2}}. \tag{12.45}$$

Now some awkward factors cancel and we have

$$\frac{\partial X}{\partial s} = \left[1 - \frac{v_A(y^0)^2}{c^2}\right]^{-1/2} - v_A(y^0)\frac{v_A(y^0)}{c^2}\left[1 - \frac{v_A(y^0)^2}{c^2}\right]^{-1/2}$$

$$= \left[1 - \frac{v_A(y^0)^2}{c^2}\right]^{1/2}, \tag{12.46}$$

$$\frac{\partial X}{\partial t} = v_A(y^0) - v_A(t). \tag{12.47}$$

It is not difficult to see now that

$$c^2 \left(\frac{\partial X}{\partial s}\right)^2 + \left[\frac{\partial X}{\partial t} + v_A(t)\right]^2 = c^2 \left[1 - \frac{v_A(y^0)^2}{c^2}\right] + v_A(y^0)^2 = c^2 , \qquad (12.48)$$

as required to satisfy (12.10). The boundary condition (12.11) of p. 268 is satisfied too, i.e., we have

$$0 = \left.\frac{\partial X(s,t)}{\partial t}\right|_{s=0} .$$

This follows from (12.47), because we recall that $v_A(y^0)$ really means $v_A(\sigma(y^0))$, and (12.39) tells us that

$$\sigma(y^0(0,t)) = t ,$$

as required.   ∎

So not only have we found the length of our rigid rod when it is accelerating along its own axis, but we discover that any AO with 1D motion could use it to measure semi-Euclidean coordinates along the direction of acceleration. This means that the rigid rod automatically satisfies what is sometimes called the ruler hypothesis, namely, it is at any instant of time ready to measure lengths in an instantaneously comoving inertial frame, since this is precisely the length system used by the semi-Euclidean coordinates.

The accelerating observer would not necessarily have to be holding one end of the rod. It could be lying with one end held fixed at some semi-Euclidean coordinate value $y^1 = s_1$ and the other end would then remain at a constant coordinate value $y^1 = s_2 > s_1$. This is shown by exactly the same kind of analysis as above. In other words, if the rod always manages to occupy precisely this interval on the axis of the SE coordinate system, its length as viewed in the original inertial frame $\mathscr{I}$ will satisfy the rigidity equation (12.10) on p. 268. Hence, a rigid rod whose left-hand end is compelled to follow the worldline $y^1 = s_1$ will always appear to have the same length $s_2 - s_1$ to the SE observer.

Before taking a look at some of the remarkable features of the semi-Euclidean coordinate frame, let us just note in passing that, to first order in the rest length D of the rod, one finds

$$x_B(t) = x_A(t) + \left[1 - \frac{v_A(t)^2}{c^2}\right]^{1/2} D + O(D^2) , \qquad (12.49)$$

which is precisely our original approximation (12.5) back on p. 267.

12.1.6 Properties of a Semi-Euclidean Frame

Since the notion of rigidity expressed by (12.10) fits in so nicely with the semi-Euclidean frame of the associated observer, it is worth summarising some of the features of these frames. We consider any point B with fixed spatial semi-Euclidean coordinates $(y^1, y^2, y^3) = (s, 0, 0)$ and varying y^0. Then we have:

- Viewed from the inertial frame $\mathscr{I}$, B follows an accelerating worldline, but generally with a different acceleration to the observer at the origin of the semi-Euclidean frame. In the case of a uniformly accelerating observer, it turns out that such a point also has uniform acceleration, but a smaller one than the observer at the origin, and ever smaller as s increases [see (12.89) on p. 303].
- Viewed from the inertial frame $\mathscr{I}$, if B is simultaneous with the accelerating observer A at the origin of the semi-Euclidean frame as judged by that observer, i.e., A and B have semi-Euclidean coordinates $(c\tau, 0, 0, 0)$ and $(c\tau, s, 0, 0)$, respectively, for some τ, then they have the same 4-velocity at those two events. Of course, they are not then simultaneous for the original inertial observer with frame $\mathscr{I}$ (except at the coincident origins of the two frames). But since the accelerating observer borrows the hyperplane of simultaneity of an instantaneously comoving inertial observer, the two events in question will be simultaneous for the latter. In other words, when an inertial observer instantaneously comoving with A looks at B, that observer will have the same 4-velocity as B.

The first of these is not difficult to show from the definitions of the semi-Euclidean coordinates. The second follows immediately from the expression

$$\frac{\partial X}{\partial t} = v_A(y^0) - v_A(t) , \tag{12.50}$$

which can be proven for general 1D accelerations of the observer. For then, by (12.8) on p. 268,

$$v_B(t) = v(s, t)$$
$$= v_A(t) + \frac{\partial X}{\partial t}$$
$$= v_A(t) + v_A(y^0) - v_A(t)$$
$$= v_A(y^0) . \tag{12.51}$$

So to find the speed of B at some event on its worldline, we must draw the HOS of the accelerating observer A which contains that event and find the proper time y^0 at which the HOS intersects the worldline of A. The point B has the speed which A had at that proper time. If we draw the worldlines of A and B on the Minkowski diagram, this result about their speeds tells us that any HOS through the A worldline intersects the two worldlines at points where they have the same gradient in the Minkowski (t, x) plane.

12.1.7 Behaviour of a Rigid Rod

From the properties in the last section, we may deduce something about the behaviour of a rigid rod under acceleration, i.e., we may deduce something about what the material points of the rod must do if the rod is to satisfy our rigidity criterion (12.1) on p. 266.

According to the first observation, all points of the rod have different coordinate accelerations and indeed different 4-accelerations relative to the inertial frame $\mathscr{I}$ when the rod is viewed in any hyperplane of simultaneity of AO. It turns out that all points of the rod have different 4-accelerations relative to the inertial frame $\mathscr{I}$ when the rod is viewed in any hyperplane of simultaneity of $\mathscr{I}$. We may deduce that the 4-forces on different material points of the rod must always be different at any instant of time for any inertial observer, and in a specific way that depends on the 4-force at A.

According to the second observation, all points of the rod have the same coordinate velocity and indeed the same 4-velocity relative to the inertial frame $\mathscr{I}$ when the rod is viewed in any hyperplane of simultaneity of AO.

What we have then here is a rather complex system of 4-forces within the rod. We might say that they conspire in such a way that, if AO carries one end of it (labelled A) and uses the semi-Euclidean frame to judge simultaneity, the points of the rod will always have the same speed relative to $\mathscr{I}$. They also conspire in such a way that the rod will always instantaneously have the right length to measure semi-Euclidean coordinate lengths for A, which are also proper lengths for AO in the semi-Euclidean system.

Note, however, that, apart from the first instant when the rod is at rest in $\mathscr{I}$, the rod will never have the relativistically contracted length

$$\left[1 - \frac{v_A(\sigma)^2}{c^2}\right]^{1/2} D \tag{12.52}$$

when observed from $\mathscr{I}$. Its length according to $\mathscr{I}$ will be the quantity $X(D,t)$ defined by inserting $s = D$ in [27]

$$X(s,t) = x_A(y^0) - x_A(t) + s\left[1 - \frac{v_A(y^0)^2}{c^2}\right]^{-1/2}, \tag{12.53}$$

where $y^0 = y^0(s,t)$ is given by

$$t = \sigma(y^0) + \frac{v_A(y^0)}{c^2} s\left[1 - \frac{v_A(y^0)^2}{c^2}\right]^{-1/2}. \tag{12.54}$$

These things are illustrated in Fig. 12.8 for the case of a uniform acceleration of magnitude g, where explicit formulas are possible. Axes t and x are those of the inertial frame $\mathscr{I}$. Of course we have dropped two space dimensions. The rigid rod is the four-dimensional region between the two worldlines [see (12.32) and (12.34)

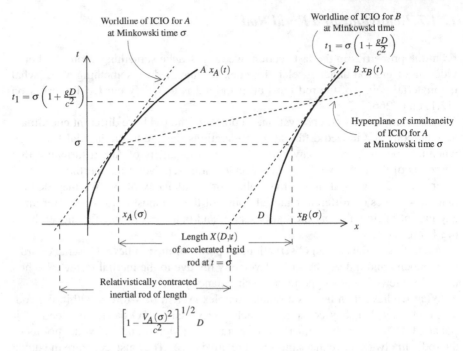

Fig. 12.8 Uniformly accelerating rigid rod. *Slanting dotted axes* are those of the instantaneously comoving inertial observer (ICIO) for Λ at Minkowski time σ or for B at Minkowski time $t_1 = \sigma(1 + gD/c^2)$

on p. 273]

$$x_A(t) = \frac{c^2}{g}\left(\sqrt{1 + \frac{g^2 t^2}{c^2}} - 1\right)$$

and

$$x_B(t) = \frac{c^2}{g}\left[\sqrt{\left(1 + \frac{gD}{c^2}\right)^2 + \frac{g^2 t^2}{c^2}} - 1\right].$$

Sloping dotted axes are those of the instantaneously comoving inertial observer (ICIO) for A at Minkowski time σ. The hyperplane of simultaneity for this ICIO intersects the worldline of B at an event $(t_1, x_B(t_1))$ where that worldline has the same gradient as the worldline of A at the event $(\sigma, x_A(\sigma))$, i.e., the same speed relative to $\mathscr{I}$ as A at the event $(\sigma, x_A(\sigma))$. The Minkowski time t_1 of this event on the worldline of B is found to be

$$t_1 = \sigma\left(1 + \frac{gD}{c^2}\right). \tag{12.55}$$

Note that, because A has the same speed relative to $\mathcal{I}$ at the event $\left(\sigma, x_A(\sigma)\right)$ as B at the event $\left(t_1, x_B(t_1)\right)$, they have the same 4-velocity components relative to $\mathcal{I}$, and hence also relative to the ICIO for A at $\left(\sigma, x_A(\sigma)\right)$. The two 4-velocities (of A and B) are located at different events in spacetime, and the last conclusion follows because the Lorentz transformation from $\mathcal{I}$ to the frame of the ICIO is constant in spacetime. But if we transform the two vectors at different events by a spacetime-dependent transformation, such as the transformation to SE coordinates, we would not expect to end up with the the the same sets of components, and indeed we do not.

Concerning the length of the rod:

- AO always considers the rod to have length D when using the semi-Euclidean system.
- The instantaneously comoving inertial observers with A, or indeed with B, always consider the rod to have length D, but only at the event where they are instantaneously comoving with A or B. As mentioned above, this is precisely what is meant by saying that the rod satisfies the ruler hypothesis.
- The inertial observer with frame $\mathcal{I}$ considers the rod to have length

$$X(D,t) = x_B(t) - x_A(t) = \frac{c^2}{g}\left[\sqrt{\left(1+\frac{gD}{c^2}\right)^2 + \frac{g^2t^2}{c^2}} - \sqrt{1+\frac{g^2t^2}{c^2}}\right]. \quad (12.56)$$

- A rod represented by the 4D region between the two dotted time axes tangent to the worldline of A at Minkowski time σ and the worldline of B at Minkowski time t_1 would have the relativistically contracted length

$$\left[1 - \frac{v_A(\sigma)^2}{c^2}\right]^{1/2} D$$

for $\mathcal{I}$, but there is no such rod here.

12.1.8 Rigid Spheres and Instantaneous Transmission of Motion

We can see what our rod is doing in a spacetime picture, drawn in the inertial frame in which it is originally at rest. It sweeps out a region of spacetime and we are saying that the SE observer is using it to measure length. It would be easy to become euphoric about such calculations, particularly the fact that the rigidity criterion (12.10) is satisfied by an expression like (12.56) set up for rather different reasons. But perhaps we should be asking what we would have to do to get the rod to move like that. Could the SE observer just accelerate the left-hand end and let the rest of the rod adapt somehow to what is happening via its rigidity?

After all we paid no attention to microscopic structure. If we think about the toy electron with its two point charge components, we avoided making any detailed

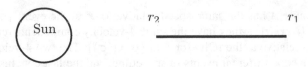

Fig. 12.9 Measuring stick in Schwarzschild spacetime, lying along a fixed radial coordinate interval. The length is given by the usual formula $L = \int_{r_2}^{r_1} dr\,(1 - 2m/r)^{-1/2}$, where r is the Schwarzschild radial coordinate. But is this really the length of a measuring stick? Could a measuring stick really have this motion? Or put another way, what would one have to do to get it to behave in this way?

model of the binding forces. In fact, we appear to have gone a long way without doing any real physics.

To show the generality of the problem, we find this in an elementary course on general relativity (see Fig. 12.9) [28]:

> To get a more quantitative feel for the distortion of the geometry produced by the gravitational field of a star, consider a long stick lying radially in the gravitational field, with its endpoints at the [Schwarzschild] coordinate values $r_1 > r_2$. To compute its length L, we have to evaluate
>
> $$L = \int_{r_2}^{r_1} dr\,(1 - 2m/r)^{-1/2}\,.$$

Since this set of points lies in a hyperplane of simultaneity for the Schwarzschild coordinates, a Schwarzschild observer would call this the proper distance between the two endpoints. But is it really the length of a stick? What would we have to do to get a stick to do this? For example, none of the points of it are in free fall, so they all have some kind of 4-acceleration, and in fact, they all have different 4-accelerations, exactly as we have found for the accelerating rigid rod in a flat spacetime.

Of course, we cannot say whether the measuring stick in the above quote is rigid until we label the material particles in it and extend our definition of rigidity to the curved spacetimes of general relativity. A step is taken in this direction in Sect. 12.2.7. However, it is clear that if real rods do behave like this, there must be some physical reason for it. On the other hand, in pre-relativistic mechanics, rigidity was always an ideal concept, at best a convenient approximation that no one would really have expected to be possible.

One finds the same attitude in calculations of self-force on small charge distributions. This is discussed in the recent book by Yaghjian [7]. Calculations are made for a relativistically rigid spherical shell of charge of radius a, whose center has an arbitrary motion:

> 'Relativistically rigid' refers to the particular model of the electron, proposed originally by Lorentz, that remains spherical in its proper (instantaneous rest) frame, and in an arbitrary inertial frame is contracted in the direction of velocity to an oblate spheroid with minor axis equal to $2a/\gamma$.

This is exactly the kind of rigidity we have been talking about. Like our rod, the sphere always has the same dimensions to the instantaneously comoving inertial observer. Above all, this makes it possible to carry out the self-force calculation. Like so many approximations in physics, it is largely motivated by mathematical convenience. Note, however, that the value of $2a/\gamma$ for the minor axis of the spheroid is only an approximation, as we have been at pains to show [see (12.52) on p. 279].

Yaghjian goes on to say [7]:

> Even a relativistically rigid finite body cannot strictly exist because it would transmit motion instantaneously throughout its finite volume. Nonetheless, one makes the assumption of relativistically 'rigid motion' to avoid the possibility of exciting vibrational modes within the extended model of the electron.

He imputes the last remark to Pauli. But is there really any sense in which motion is transmitted instantaneously? That did seem to be the assumption with pre-relativistic rigidity: if a force was applied to one end of a rod, the same force had to be transmitted instantaneously to all the particles in the rod, so that all particles would always have the same acceleration and the same speed.

Are we assuming something like this in the present case? Viewing from the inertial frame $\mathscr{I}$, imagine the left-hand end A of the rod as being accelerated in some active way, whilst the rest of the rod follows suit in some sense. As the left-hand end moves faster, the other points on the rod pick up speed too. In the case of a uniform acceleration of A, we know that each point of the rigid rod has uniform acceleration, but always lesser, until we come to the right-hand end B, which has the smallest value. However, the end B eventually reaches the speed that A had some time previously. In this view of things, we may be thinking that speed somehow propagates along the rod, with a delay that we ought to be able to calculate.

On p. 278, we showed the following result. If we consider a point $x_A(\sigma)$ on the worldline of A, when it has speed $v_A(\sigma)$, and draw the HOS of the ICIO, this HOS will intersect the worldline $x_B(t)$ of B at a Minkowski time t_1 where $v_B(t_1) = v_A(\sigma)$. In the case of a uniform acceleration g, where we have explicit formulas, we know from (12.55) that

$$t_1 = \sigma \left(1 + \frac{gD}{c^2}\right) .$$

We therefore know how much Minkowski time is required for the speed of A to propagate through to the other end of the rod, if indeed there is propagation, viz., $\sigma gD/c^2$. The Minkowski observer will consider that the signal, if indeed it is one, has propagated from $x_A(\sigma)$ to $x_B(t_1)$, so that it has travelled a distance $x_B(t_1) - x_A(s)$. We can calculate this from the formulas in the last section, and the result is

$$x_B(t_1) - x_A(\sigma) = D \left(1 + \frac{g^2\sigma^2}{c^2}\right)^{1/2} .$$

We know this anyway, because it is the projection onto the x axis of the imaginary inertial rod mentioned earlier.

If we now divide the distance travelled by the putative signal by the time it has taken, as reckoned in the Minkowski frame, we find the value

$$\frac{D(1+g^2\sigma^2/c^2)^{1/2}}{g\sigma D/c^2} = c\left(1 + \frac{c^2}{g^2\sigma^2}\right)^{1/2} > c \tag{12.57}$$

for the speed of propagation. If speed propagates, it does so faster than light.

Is this why our rigid rod is instantaneously ready to measure lengths when accelerated to a new speed? When we look back at the formulation of our equation of motion (12.10) for the material particles making up the rod, it is clear that we never explicitly introduced any delays. We merely hoped that the more sophisticated model would cater for this.

Alternatively, it may be that we should not consider speed as propagating in the rod. After all, the result (12.57) is most catastrophic when $\sigma = 0$, simply because the two ends of the rod happen to have the same speed at the same Minkowski time. The explanation of whatever paradox there seems to be here is more likely to be this. When the motion begins from rest (all points of the rod being at rest when $\sigma = 0$), each one has to instantaneously have the appropriate four-acceleration. This may be a problem in itself, but once that is accomplished, there is no need for the speed to propagate. Each point of the rod is subject to its appropriate four-acceleration and so acquires the required speed locally as it were. The real problem is: how can each point be subject to the appropriate four-acceleration. If a force is applied at one end, it is indeed four-acceleration (or four-force, or just force) that has to take a little time to transmit to the various points of the rod. The rod has to adjust in some way.

So could rigidity be equivalent to instantaneous transmission of four-acceleration? This too looks unnecessary since the rod may have been forever undergoing this motion, at least theoretically. More realistically, one could always wait until the required distribution of four-accelerations has set itself up within the rod and thereafter describe its motion as rigid. Yaghjian says that the assumption of relativistically rigid motion is made to avoid the possibility of exciting vibrational modes within the extended electron. This corresponds to the idea that, in reality, there must always be some time of adjustment after the motion is initiated.

Let us return to the question posed at the beginning of this section, viz., could one just accelerate the left-hand end of a rigid rod and let the rest of the rod adapt somehow to what is happening via its rigidity? We can investigate this idea by applying the result (12.51) on p. 278. We imagine a rod that is stationary in an inertial frame $\mathscr{I}$ and to which an external acceleration is applied to the left-hand end A at some time t_{acc} (reckoned in that frame).

On the Minkowski diagram, the worldline of A is represented by a vertical line which we may take to be the time axis, up to $t = t_{\mathrm{acc}}$, where it begins to curve over to the right. The hyperplanes of simultaneity of an observer moving with A are at first horizontal, but begin to slant upwards after $t = t_{\mathrm{acc}}$, slanting up more and more as A moves faster. This effectively determines what the right-hand end B of the rod will do, if we recall the simple result (12.51) on p. 278. Because the rod is rigid, B always moves with speed equal to the speed of A at the event on the worldline of A that the instantaneously comoving inertial observer moving with A considers to be simultaneous. So the worldline of B is clearly vertical up to the time $t = t_{\mathrm{acc}}$. But what happens next?

As soon as $t > t_{acc}$, the relevant HOS of A through the worldline of B must be one of those that are beginning to slant upwards, no matter how soon after t_{acc} we look at the worldline of B. So B will have the speed of A at a slightly earlier time (as reckoned in the frame $\mathscr{I}$), but nevertheless at some time after t_{acc}, when A was already accelerating. This means that B will have started moving. Its worldline curves over to the right after t_{acc}. It curves over more slowly, but the important thing is that it does so immediately (in the $\mathscr{I}$ reckoning) after the time t_{acc}.

We conclude that the acceleration of B is indeed instantaneous, i.e., simultaneous with the acceleration of A in this frame. And, of course, this means that in some other inertial frame, B will begin to accelerate before A, throwing out the idea that the acceleration of A could be the cause of the acceleration of B. Put another way, if one really were applying an external force only at A, one would not expect B to be able to react for at least the time it takes light to propagate along the length of the rod. This suggests another notion of rigidity, wherein a rigid rod is one in which the speed of sound in the rod is equal to the speed of light [29].

Presumably this shows that one cannot expect any rod to have our kind of rigid motion when tampered with in this way. So rigid motion is not an easy behaviour to achieve, whatever one's medium is made of. However, one could imagine that some other means of accelerating the medium could result in its having rigid motion, e.g., applying different external forces to all particles making up the rod, perhaps by means of a force field like gravity or electromagnetism. Indeed, the accelerations of all particles in the rod are completely determined when it has the rigid motion specified by the criterion (12.1) on p. 266.

The above considerations suggest that one should investigate rigid motion, rather than rigid spheres or rods. One might then have to conclude that rigid motion cannot strictly occur when caused by an external force applied at just one point of the object. This would still leave open the question of motions due to fields of force. We consider the idea of rigid motions of a general medium in Sect. 12.2.

12.1.9 Rigid Electrons and Rigid Atoms

We asked above how each point could be subject to the appropriate four-acceleration when the rigid rod or sphere is made to move. One could envisage an interplay of repulsive and binding forces within these objects conspiring to move each material point in the appropriate way. It does look possible a priori. Indeed, this idea is effectively applied to the spherical shell of charge in the self-force calculations discussed by Yaghjian in his book [7]. The binding forces are precisely those required to keep the shell of charge spherical in its proper frame.

Of course, it could be that small particles like electrons are rigid in this sense. In any case, one has to assume something and this seems to be as convenient an assumption as one could hope for within the framework of relativity theory. Another small particle is the atom. In a pre-quantum model, an electron orbits a nucleus under an electromagnetic attraction so one only has the binding force to worry about.

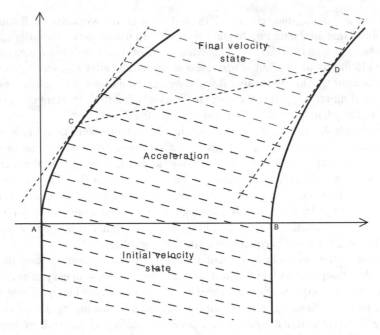

Fig. 12.10 Rigid motion of a rod from one uniform velocity state to another. The rod is represented by the *shaded region* of the spacetime diagram. The initial velocity state continues up to the horizontal line AB, while the acceleration occupies the region up to the slanting line CD

One might, like Bell in his paper *How to Teach Special Relativity* [5], treat the nucleus as an accelerating point charge for which the exact electromagnetic potential (the Lienard–Wiechert potential) is known from Maxwell's theory, then calculate the exact orbit of the electron in this field as the nucleus accelerates. In principle, this would give a perfect description of the way the length of the atom would change in the direction of acceleration.

One might then be forgiven for forgetting the complexity of the rigid rod with its long string of atoms and the unfortunate way they might interfere with one another by being bound together in some manner. One could just suppose that the way measuring sticks contract is just, or should be just, if they are any good for measuring, the way their constituent atoms contract, according to the simple idea of the last paragraph. Would this then give us a different notion of rigidity? It certainly looks likely on the face of it, if one could carry out precise calculations.

However, one might give the following argument for supposing the Bell atom to be at least approximately rigid. Figure 12.10 shows our solution for accelerating a rigid rod from one state of uniform velocity to another. During the acceleration, the length of the rod is always the same for an instantaneously comoving inertial observer. This is precisely what one expects for the radius of the Bell atom if the acceleration is slow enough, i.e., if the electron gets in plenty of revolutions around the nucleus before the acceleration ever has time to change very much. In an accurate calculation, one would expect to require some adjustment time, but to a certain level of approximation, the atom will behave rigidly. This is discussed further at the end of Sect. 12.2.7.

The aim of the second part of this discussion (Sect. 12.2) will be to consider rigid motion in a more general context, for any material medium. We are trying here to avoid awkward questions about how the motion is brought about, or how the medium manages to get into or remain in a state of rigid motion.

12.1.10 A Note on the Pound–Rebka Experiment

This is just a simplified overview to identify some hidden assumptions. We imagine an emitter E at the bottom of a tower and a receiver R at the top (in fact, iron nuclei emitting and absorbing gamma rays). The latter is going to detect the gravitational redshift predicted by general relativity. We may suppose that the gravitational field is perfectly uniform here, by which we mean that there are coordinates relative to which the metric takes the form (12.31) on p. 273. (This assumption in itself is worth examining much more closely [4].) Of course, this spacetime is flat, i.e., there is a global inertial frame, often called the freely falling frame. By the strong equivalence principle, electromagnetic effects relating to E and R can be examined using the ordinary Maxwell equations, or ordinary quantum electrodynamics, in the freely falling frame.

In his book [14, Sect. 5.7], Brown mentions the need to assume the clock hypothesis, which declares that a clock worthy of the name will measure the proper time along whatever worldline it happens to be following. This could be taken as the definition of an ideal clock, or a hypothesis, to be tested, that some particular putative clock approximates to ideality. Let us understand this in the context of the Pound–Rebka experiment. E emits waves that leave at precise intervals, but relative to what time scale? The proper time associated with its worldline? If the iron nuclei used as emitter and receiver satisfy the clock hypothesis, we would answer affirmatively there.

As an aside, which is nevertheless quite relevant to the general ethos of this book, we may well ask why this should be. What is the physical explanation? It is suggested here that the Bell approach may show that this is just a good approximation [5], but that detailed calculations with the relevant theories in special relativity (this spacetime is flat) or minimally extended from flat spacetime in general relativity where necessary [12], would give a better answer. So we are suggesting that the clock 'hypothesis' is necessary insofar as one needs to know when the waves or photons are emitted, but that one could also prove that this is a good approximation, so that the only assumption needed is the assumption that one has good theories for the emission process.

But there is already an interesting problem with identifying the emitter and receiver worldlines in a global inertial frame when the tower is uniformly accelerating. In fact, there are at least two obvious possibilities:

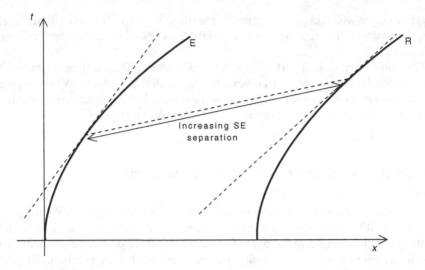

Fig. 12.11 Case 1. The emitter E follows the usual hyperbola in spacetime of a uniformly accelerating point and the receiver R likewise, with the same uniform acceleration. With a ruler satisfying the ruler hypothesis, E considers R to recede

1. The emitter E follows the usual hyperbola in spacetime of a uniformly accelerating point and the receiver R likewise, with the same uniform acceleration, i.e., with the same hyperbola but shifted along the space axis (see Fig. 12.11). If E uses semi-Euclidean (Rindler) coordinates, i.e., rigid rulers as described in this chapter, then R recedes from it.
2. The emitter follows the usual hyperbola of a uniformly accelerating point and the receiver likewise, but a different one, viz., the hyperbola of a point at fixed semi-Euclidean distance from E (see Fig. 12.12). If that distance is fixed, it must have a lower uniform acceleration.

In general relativity, which is what we are doing here (even though Brown is considering a case where one is still trying to do special relativity but finding that the results of the Pound–Rebka experiment create a problem with the notion of inertial frame), the emitter and receiver are accelerating because they are *not* being allowed to fall freely. But should they have the same acceleration for some reason related to the fact that the gravitional field is uniform, as in (1) above (it has zero curvature, but then that does not tell us how strong the field is, only that there are no tidal effects, hence no variation in it); or should they have constant spatial separation as in 2, if indeed that is what we should mean by separation (ruler hypothesis)?

If the receiver is supported by the roof of the tower, then it is indeed the structure of the tower that determines the motion of the receiver. In fact it is usually assumed that the tower is rigid, i.e., case 2 above. If that is so, it is important to see that one is assuming that the emitter and receiver have different accelerations, i.e., they are being supported differently against the uniform gravitational field. This might look surprising when one considers that the gravitational field is supposed to be uniform.

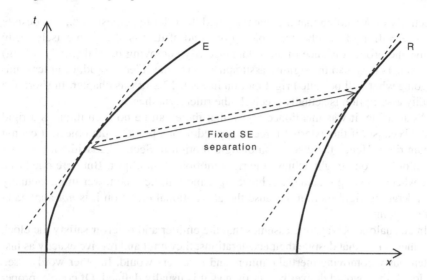

Fig. 12.12 Case 2. The emitter follows the usual hyperbola of a uniformly accelerating point and the receiver likewise, but a different one, viz., the hyperbola of a point at fixed semi-Euclidean distance from E

But it just illustrates the fact that supporting something in a gravitational field in general relativity introduces effects that are quite different from the gravitational field, viz., a supporting force which causes an acceleration (while freely falling objects have no acceleration). In fact, supporting something is a rather arbitrary thing to do in a certain sense. The notion of supporting is just specified by saying that a thing is not allowed to move relative to some coordinates one happens to be using, and coordinates are not fundamental in general relativity. A classic case would be an object held at some fixed values of the usual coordinates for a Schwarzschild spacetime.

So what coordinates are being used in the Pound–Rebka experiment and how would they be set up? Presumably one wants to say that the distance between the emitter and receiver is constant. But what distance is this? Suppose one measures distances up the tower using a ruler held by an observer sitting with the emitter. If it satisfies the ruler hypothesis, then it measures semi-Euclidean (Rindler) spatial coordinates. In case 1 above, where the receiver is supported in such a way that it has the same uniform acceleration as the emitter, the receiver would be measured to recede according to such measurements. But if the tower is rigid (in the usually accepted sense, discussed in this chapter) and the receiver is fixed relative to a point of the tower, then the receiver would be considered to remain at a fixed distance from the emitter.

The point about mentioning this is just to say that, just as one discusses the clock hypothesis in this context, there is a similar consideration of the ruler hypothesis. If one did measure the emitter–receiver separation with a ruler and wanted to say that this gave the semi-Euclidean spatial coordinate (because the constancy of the separation would allow us to do the redshift calculation in the usual way), one would effectively be assuming that the ruler satisfied the ruler hypothesis, i.e., that despite the acceleration of the observer holding one end of it, it is always precisely ready to give the proper distance of an instantaneously comoving inertial (freely falling) observer. This is also the rigidity assumption, viz., the ruler is rigid, or at least undergoing what will be called rigid motion in Sect. 12.2 of this chapter. In short, the rigidity assumption is actually precisely the ruler hypothesis.

As an aside, it was mentioned above that there can be no such thing as a rigid object because, if the external force is applied at one point of the object, it cannot remain rigid. Hence the discussion of rigid motion in Sect. 12.2, without consideration of how one might achieve the rigid motion of an object. But here one has a case where one might actually achieve rigid motion, i.e., the tower might actually be undergoing rigid motion, because the gravitational effect on it is *not* applied at just one point.

In an analogous way, one assumes that the emitter and receiver satisfy the clock hypothesis, i.e., that despite their accelerations, they emit and receive exactly as instantaneously comoving inertial emitters and receivers would. In other words, used as clocks, they would deliver proper time as it is usually defined. Of course, proper time is perfectly well defined in a mathematical sense along arbitrary worldlines in special relativity, without the need to mention any clocks. The only hypothesis one needs there in a context like this is the hypothesis that what one is actually hoping to use as a clock does read proper time. Of course, if it did not, it would not be regarded as a clock. What we would like to add here is Bell's idea that one should be able to show theoretically that any particular device is or is not a clock, or is a good or bad approximation to a clock, and this entirely within special relativity if one is using special relativity. The only extra assumption in the latter case is that one's theories about how the clock is working (e.g., electromagnetism for an electron going round an atom, or quantum electrodynamics for a better model) are actually valid theories in that context.

In this context nothing is really different in general relativity, except that one has to add the strong equivalence principle in order to be able to apply non-gravitational bits of physics when the spacetime is curved. In a certain sense one can consider special relativity as a special case of general relativity, viewing special relativity as general relativity with no gravitational effects (and saying, of course, that special relativity treats gravity very differently when there is any gravity). Moreover, general relativity adds nothing as far as acceleration is concerned. One can perfectly well consider accelerating test particles in special relativity, as in general relativity. But if some process is occurring in the particle, e.g., an electron orbiting a central nucleus, we do not know a priori whether that process is going just as it would for an instantaneously comoving inertial particle of the same kind, insofar as the two processes could be compared. It seems unlikely, but presumably a detailed calculation with the relevant theories would allow one to estimate the discrepancy. Presumably it would also show that the discrepancy is very small for most things we use as clocks, and the scale of accuracy on which no physical process fits with the theoretical proper time would be the one where we would have to admit that general relativity was beginning to fail.

Why do some people claim that special relativity cannot treat accelerated motions? Perhaps they are thinking, not of accelerated test particles, but accelerated observers. The view here is that one has exactly the same problem with accelerating observers in general relativity as in special relativity. If an observer is uniformly accelerating in special relativity, what coordinates would this observer set up? Everyone seems to use the semi-Euclidean (Rindler) coordinates as though there were something special about them. Of course, the observer remains at the spatial origin of those coordinates, the time coordinate is the proper time of the observer, and other obvious things like that. But are those the coordinates the observer would set up? If we are thinking about using clocks and rulers to set them up in a real world, it would seem that we do not actually know. The clock and ruler hypotheses merely assert that they would be in that context. Whether our actual physical clocks and rulers would fit the bill is another matter.

But would general relativity help here? Of course there are nice coordinates for any timelike worldline, in which the worldline remains at the spatial origin and the time coordinate is the proper time, etc. But are those the coordinates that an observer following that worldline would set up using clocks and rulers? It would seem that we are in exactly the same situation as in the last paragraph.

Both of the above cases 1 and 2 lead to redshift. The point here is just to see that the receiver with the same uniform acceleration as the observer will still detect a redshift (case 1), since the other case is the standard one. Consider the situation in the local inertial frame, which happens to be global for a uniform gravitational field. In the spacetime diagram, we have two identically shaped curves, curving over to the right, translates of one another along the space axis. The one on the left is the emitter and the one on the right is the receiver. A signal from the emitter leaves it when the emitter has a certain speed and arrives at the receiver when the receiver has a higher speed. In this freely-falling frame view, the redshift is just a Doppler shift. (To apply this analysis in the general relativistic case considered here, where we have a uniform gravitational field, we are of course also assuming the strong equivalence principle.) The only difference in case 2 is that the shape of the receiver worldline in the spacetime diagram for a freely falling observer is different from the shape of the emitter worldline, because it curves over more slowly (lower proper acceleration).

The redshift calculation for case 2 can be found in [4, Sect. 15.6] along with a critical discussion of the way semi-Euclidean coordinate systems are interpreted. Case 1 here is straightforward in the freely falling frame. The worldlines of E and R are (see Fig. 12.13)

$$x_E(t) = \frac{c^2}{g}\left[\left(1+\frac{g^2t^2}{c^2}\right)^{1/2} - 1\right], \tag{12.58}$$

as in (12.24) on p. 272, and $x_R(t) = x_E(t) + \kappa$, for some constant κ. The two worldlines have the same shape, because they have the same uniform acceleration, by hypothesis 1. Then

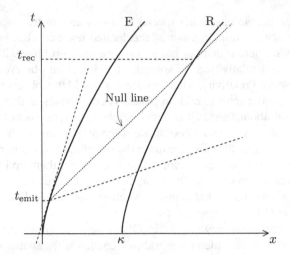

Fig. 12.13 Calculating the redshift in case 1. The emitter E emits a signal at time t_{emit} which is received by the receiver R at time t_{rec}. The two worldlines have the same shape. The receiver worldline is identical to the emitter worldline but shifted a distance κ along the space axis. By the time the signal reaches the receiver, the receiver worldline has curved over due to the increasing speed of the receiver

$$v_E(t) = \frac{gt}{(1+g^2t^2/c^2)^{1/2}} = v_R(t) \,. \tag{12.59}$$

To get the redshift, imagine a light signal sent from the worldline of E at time t_{emit} and find the time of reception t_{rec} on the worldline of R. Find the apparent relative velocity

$$v := v_R(t_{rec}) - v_E(t_{emit}) = v_E(t_{rec}) - v_E(t_{emit}) \,, \tag{12.60}$$

and plug it into the usual special relativistic formula for the Doppler shift. If z is the redshift, then

$$1+z = \frac{(1+v/c)^{1/2}}{(1-v/c)^{1/2}} \,. \tag{12.61}$$

Interestingly, the result is much less elegant than in the standard case 2. However, it is easy to show that

$$z \approx g\kappa/c^2 \,, \tag{12.62}$$

for small κ, i.e., the same result as for case 2. On the other hand, for large κ, the two results will obviously differ. What is not obvious is whether experimental accuracy could yet distinguish the two cases.

12.2 Rigid Motion

This section is based on the discussion of rigid motions in B.S. DeWitt's Stanford lectures on relativity [15]. These lectures (to be published by Springer) deal at some length with the problem of continuous media in the context of curved spacetime. The ideas below extend naturally to general relativity, but we consider a flat spacetime for the presentation below.

12.2.1 General Motion of a Continuous Medium

The component particles of the medium are labeled by three parameters ξ^i, $i = 1, 2, 3$, and the worldline of particle ξ is given by four functions $x^\mu(\xi, \tau)$, $\mu = 0, 1, 2, 3$, where τ is its proper time. In general relativity, the x^μ may be arbitrary coordinates in curved spacetime, but here we assume them to be standard coordinates of some inertial frame.

If $\xi^i + \delta\xi^i$ are the labels of a neighbouring particle, its worldline is given by the functions

$$x^\mu(\xi + \delta\xi, \tau) = x^\mu(\xi, \tau) + x^\mu{}_{,i}(\xi, \tau)\delta\xi^i ,$$

where the comma followed by a Latin index denotes partial differentiation with respect to the corresponding ξ (DeWitt's notation). Note that the quantity $x^\mu{}_{,i}(\xi, \tau)\delta\xi^i$, representing the difference between the two sets of worldline functions, is formally a 4-vector, being basically an infinitesimal coordinate difference. However, it is not generally orthogonal to the worldline of ξ. In other words, it does not lie in the hyperplane of simultaneity of either particle.

To get such a vector one applies the projection tensor onto the instantaneous hyperplane of simultaneity:

$$P^{\mu\nu} = \eta^{\mu\nu} + \dot{x}^\mu\dot{x}^\nu ,$$

where the dot denotes partial differentiation with respect to τ, and we note that in general relativity the projection tensor takes the form

$$P^{\mu\nu} = g^{\mu\nu} + \dot{x}^\mu\dot{x}^\nu ,$$

with $g^{\mu\nu}$ the metric tensor of the curved spacetime. The result is

$$\delta x^\mu := P^\mu{}_\nu x^\nu{}_{,i}(\xi, \tau)\delta\xi^i . \tag{12.63}$$

One finds that application of the projection tensor corresponds to a simple proper-time shift of amount

$$\delta\tau = \eta_{\mu\nu}\dot{x}^\mu\dot{x}^\nu{}_{,i}\delta\xi^i ,$$

so that

$$\delta x^{\mu} = x^{\mu}(\xi + \delta\xi, \tau + \delta\tau) - x^{\mu}(\xi, \tau) .$$

Indeed,

$$x^{\mu}(\xi + \delta\xi, \tau + \delta\tau) = x^{\mu}(\xi, \tau) + x^{\mu}_{,i}\delta\xi^{i} + \dot{x}^{\mu}\delta\tau ,$$

and feeding in the proposed expression for $\delta\tau$, we do obtain precisely δx^{μ} as defined above.

What can we conclude from this analysis? The two particles ξ and $\xi + \delta\xi$ appear, in the instantaneous rest frame of either, to be separated by a distance δs given by

$$(\delta s)^{2} = (\delta x)^{2} = \gamma_{ij}\delta\xi^{i}\delta\xi^{j} , \tag{12.64}$$

where

$$\gamma_{ij} = P_{\mu\nu}x^{\mu}_{,i}x^{\nu}_{,j} . \tag{12.65}$$

DeWitt calls the quantity γ_{ij} the proper metric of the medium.

12.2.2 Rigid Motion of a Continuous Medium

At this point, one can introduce a notion of rigidity. One says that the medium undergoes rigid motion if and only if its proper metric is independent of τ. This is therefore expressed by

$$\dot{\gamma}_{ij} = 0 . \tag{12.66}$$

Under rigid motion the instantaneous separation distance between any pair of neighbouring particles is constant in time, as they would see it. Note that this criterion is independent of the coordinates used because γ_{ij} is a scalar.

Let us see whether this coincides with the notion of rigidity discussed earlier, i.e., whether the rigid rod of Sect. 12.1.2 is DeWitt rigid, or put differently, whether the rod described in Sect. 12.1.2 is undergoing rigid motion according to the criterion (12.66). DeWitt's ξ correspond to s in Sect. 12.1.2 (see p. 266). In a given inertial frame, particle s has motion described by $X(s,t)$, where

$$\frac{\partial X}{\partial s} = \frac{1}{\gamma} , \qquad \gamma = \gamma(v(s,t)) ,$$

and

$$v(s,t) = v_{A}(t) + \frac{\partial X}{\partial t} ,$$

where $v_A(t)$ is the speed of the end of the rod. Suppose we now change to a frame moving instantaneously at speed $v(s,t)$ and measure the distance between particle s and particle $s + \delta s$ as viewed in this frame. Will it be constant in this model, as required for DeWitt rigid motion? In the original frame where both particles are moving, we have separation

$$X(s + \delta s, t) - X(s,t) = \frac{\partial X}{\partial s} \delta s = \frac{\delta s}{\gamma} \,.$$

In the new frame moving at speed $v(s,t)$, this has length

$$\gamma \frac{\delta s}{\gamma} = \delta s = \text{constant}.$$

This is what DeWitt rigid motion requires.

12.2.3 Rate of Strain Tensor

The aim here is to express the rigid motion condition $\dot{\gamma}_{ij} = 0$ in terms of derivatives with respect to the coordinates x^μ by introducing the relativistic analog of the rate of strain tensor in ordinary continuum mechanics.

The non-relativistic strain tensor can be defined by

$$e_{ij} := \frac{1}{2}\left(\frac{\partial u_j}{\partial x_i} + \frac{\partial u_i}{\partial x_j}\right),$$

where $u_i(x)$ are the components of the displacement vector of the medium, describing the motion of the point originally at x when the material is deformed. One also defines the antisymmetric tensor

$$\omega_{ij} := \frac{1}{2}\left(\frac{\partial u_j}{\partial x_i} - \frac{\partial u_i}{\partial x_j}\right),$$

which describes the rotation occurring when the material is deformed. Clearly,

$$e_{ij} - \omega_{ij} = \frac{\partial u_i}{\partial x_j},$$

and hence, if all distortions are small,

$$\Delta u_i = (e_{ij} - \omega_{ij})\Delta x_j \,.$$

We can consider that e_{ij} describes non-rotational distortions, i.e., stretching, compression, and shear.

In the present discussion, u_i is replaced by a velocity field v_i and we have a rate of strain tensor. The non-relativistic rate of strain tensor is

$$r_{ij} = v_{i,j} + v_{j,i} , \qquad (12.67)$$

where v_i is a 3-velocity field and the differentiation is with respect to ordinary Cartesian coordinates. Let us look for a moment at this tensor. The nonrelativistic condition for rigid motion is

$$r_{ij} = 0 \qquad \text{everywhere}.$$

This equation implies

$$0 = r_{ij,k} = v_{i,jk} + v_{j,ik} , \qquad (12.68)$$

$$0 = r_{jk,i} = v_{j,ki} + v_{k,ji} . \qquad (12.69)$$

Subtracting (12.69) from (12.68) and commuting the partial derivatives, we find

$$v_{i,jk} - v_{k,ji} = 0 , \qquad (12.70)$$

which, upon permutation of the indices j and k, yields also

$$v_{i,kj} - v_{j,ki} = 0 . \qquad (12.71)$$

Adding (12.68) and (12.71), we obtain

$$v_{i,jk} = 0 ,$$

which has the general solution

$$v_i = -\omega_{ij} x_j + \beta_i , \qquad (12.72)$$

where ω_{ij} and β_i are functions of time only. The condition $r_{ij} = 0$ constrains ω_{ij} to be antisymmetric, i.e.,

$$\omega_{ij} = -\omega_{ji} ,$$

and nonrelativistic rigid motion is seen to be, at each instant, a uniform rotation with angular velocity

$$\omega_i = \frac{1}{2} \varepsilon_{ijk} \omega_{jk}$$

about the coordinate origin, superimposed upon a uniform translation with velocity β_i. Because the coordinate origin may be located arbitrarily at each instant, rigid motion may alternatively be described as one in which an arbitrary particle in the medium moves in an arbitrary way while at the same time the medium as a whole rotates about this point in an arbitrary (but uniform) way. Such a motion has six degrees of freedom.

Note that when r_{ij} is zero, we can also deduce that $v_{i,i} = 0$, i.e., $\operatorname{div} v = 0$, which is the condition for an incompressible fluid. This is evidently a weaker condition than rigidity.

Let us see how this generalises to special relativity. We return to the continuous medium in which particles are labelled by ξ^i, $i = 1, 2, 3$. Just as the coordinates x^μ are functions of the ξ^i and τ, so the ξ^i and τ can be regarded as functions of the x^μ, at least in the region of spacetime occupied by the medium. Following DeWitt [15], we write

$$u^\mu := \dot{x}^\mu \,, \qquad u^2 = -1 \,, \qquad P_{\mu\nu} = \eta_{\mu\nu} + u_\mu u_\nu \,.$$

If f is an arbitrary function in the region occupied by the medium then

$$f_{,\mu} = f_{,i}\xi^i_{,\mu} + \dot{f}\tau_{,\mu} \,,$$

where the comma followed by a Greek index μ denotes partial differentiation with respect to the coordinate x^μ. We also have

$$\dot{x} \cdot \ddot{x} = 0 \quad \text{or} \quad u \cdot \dot{u} = 0 \,,$$

since $u^2 = -1$, and

$$u_\mu u^\mu_{,\nu} = 0 \,, \qquad \dot{u}_\mu = u_{\mu,\nu} u^\nu \,, \qquad u_\mu u^\mu_{,i} = 0 \,,$$

$$x^\mu_{,i}\xi^i_{,\nu} + \dot{x}^\mu \tau_{,\nu} = \delta^\mu_{\ \nu} \,,$$

$$\xi^i_{,\mu} x^\mu_{,j} = \delta^i_j \,, \qquad \xi^i_{,\mu}\dot{x}^\mu = 0 \,,$$

$$\tau_{,\mu} x^\mu_{,i} = 0 \,, \qquad \tau_{,\mu}\dot{x}^\mu = 1 \,,$$

$$P_{\mu\nu}\dot{x}^\nu_{,i} = P_{\mu\nu} u^\nu_{,i} = u_{\mu,i} \,.$$

We now define the rate of strain tensor for the medium:

$$r_{\mu\nu} := \dot{\gamma}_{ij}\xi^i{}_{,\mu}\xi^j{}_{,\nu}$$

$$= \left(\dot{P}_{\sigma\tau}x^\sigma{}_{,i}x^\tau{}_{,j} + P_{\sigma\tau}\dot{x}^\sigma{}_{,i}x^\tau{}_{,j} + P_{\sigma\tau}x^\sigma{}_{,i}\dot{x}^\tau{}_{,j}\right)\xi^i{}_{,\mu}\xi^j{}_{,\nu}$$

$$= (\dot{u}_\sigma u_\tau + u_\sigma\dot{u}_\tau)(\delta^\sigma{}_\mu - u^\sigma\tau_{,\mu})(\delta^\tau{}_\nu - u^\tau\tau_{,\nu})$$

$$+ u_{\tau,i}\xi^i{}_{,\mu}(\delta^\tau{}_\nu - u^\tau\tau_{,\nu}) + (\delta^\sigma{}_\mu - u^\sigma\tau_{,\mu})u_{\sigma,j}\xi^j{}_{,\nu}$$

$$= \dot{u}_\mu u_\nu + u_\mu\dot{u}_\nu + \dot{u}_\mu\tau_{,\nu} + \tau_{,\mu}\dot{u}_\nu$$

$$+ u_{\nu,\mu} - \dot{u}_\nu\tau_{,\mu} + u_{\mu,\nu} - \dot{u}_\mu\tau_{,\nu}$$

$$= u_{\mu,\sigma}u^\sigma u_\nu + u_\mu u^\sigma u_{\nu,\sigma} + u_{\nu,\mu} + u_{\mu,\nu}$$

$$= P_\mu{}^\sigma P_\nu{}^\tau(u_{\sigma,\tau} + u_{\tau,\sigma}) .$$

This is to be compared with (12.67) to justify calling it the rate of strain tensor. At any event x^μ, it lies entirely in the instantaneous hyperplane of simultaneity of the particle ξ^i that happens to coincide with that event.

Note in passing that this generalises to curved spacetimes. We define

$$r_{\mu\nu} := \dot{\gamma}_{ij}\xi^i{}_{,\mu}\xi^j{}_{,\nu} , \tag{12.73}$$

as before, noting that it is a tensor, since γ_{ij}, $\dot{\gamma}_{ij}$, ξ^i and ξ^j are scalars under change of coordinates. At any x, there are coordinates such that $g_{\mu\nu,\sigma}\big|_x = 0$, whence covariant derivatives with respect to the Levi-Civita connection are just coordinate derivatives at x, and it follows immediately that

$$r_{\mu\nu} = P_\mu{}^\sigma P_\nu{}^\tau(u_{\sigma;\tau} + u_{\tau;\sigma}) , \tag{12.74}$$

where semi-colons denote covariant derivatives and $P^{\mu\nu}$ is given by

$$P^{\mu\nu} = g^{\mu\nu} + \dot{x}^\mu\dot{x}^\nu ,$$

for metric $g^{\mu\nu}$.

Returning now to the context of special relativity, the result

$$r_{\mu\nu} := \dot{\gamma}_{ij}\xi^i{}_{,\mu}\xi^j{}_{,\nu} = P_\mu{}^\sigma P_\nu{}^\tau(u_{\sigma,\tau} + u_{\tau,\sigma}) \tag{12.75}$$

expresses the rate of strain tensor in terms of coordinate derivatives of the four-velocity field of the medium. We now characterise relativistic rigid motion by

$$r_{\mu\nu} = 0 , \qquad \dot{\gamma}_{ij} = 0 . \tag{12.76}$$

Once again, we observe that the criterion for rigid motion, viz., $r_{\mu\nu} = 0$, is independent of the coordinates, because $r_{\mu\nu}$ is a tensor, even in a curved spacetime.

12.2.4 Examples of Rigid Motion

The next problem is to find some examples. We choose an arbitrary particle in the medium and let it be the origin of the labels ξ^i. The problem here is to choose these labels smoothly throughout the medium. Let the worldline $x^\mu(0, \tau)$ of the point $\xi^i = 0$ be arbitrary (but timelike). We now introduce a local rest frame for the particle, characterized by an orthonormal triad $n_i^{\ \mu}(\tau)$:

$$n_i \cdot n_j = \delta_{ij} , \qquad n_i \cdot u_0 = 0 , \qquad u_0^2 = -1 , \qquad u_0^{\ \mu} := \dot{x}^\mu(0, \tau) .$$

We now assume that the worldlines of all the other particles of the medium can be given by

$$x^\mu(\xi, \tau) = x^\mu(0, \sigma) + \xi^i n_i^{\ \mu}(\sigma) , \tag{12.77}$$

where σ is a certain function of the ξ^i and τ to be determined. On the left, τ is the proper time of the particle labelled by ξ. To achieve a relation of this type, given τ and ξ, we must find the unique proper time σ of the particle $\xi = 0$ such that the point $x^\mu(\xi, \tau)$ is simultaneous with the event $x^\mu(0, \sigma)$ in the instantaneous rest frame of the particle $\xi = 0$. Then the label ξ^i for our particle is defined by the above relation. There is indeed an assumption here, namely that these ξ^i really do label particles. That is, if we look at events with the same ξ^i but varying τ, we are assuming that we do follow a single particle. It is unlikely that all motions of the medium could be expressed like this, but we can obtain some rigid motions, as we shall discover.

To determine the function $\sigma(\xi^i, \tau)$, write

$$u^\mu = \dot{x}^\mu(\xi, \tau) = (u_0^{\ \mu} + \xi^i \dot{n}_i^{\ \mu}) \dot{\sigma} ,$$

all arguments being suppressed in the final expression. Here and in what follows, it is to be understood that dots over u_0 and the n_i denote differentiation with respect to σ, while the dot over σ denotes differentiation with respect to τ.

In order to proceed further, one must expand $\dot{n}_i$ in terms of the orthonormal tetrad u_0, n_i :

$$\dot{n}_i^{\ \mu} = a_{0i} u_0^{\ \mu} + \Omega_{ij} n_j^{\ \mu} . \tag{12.78}$$

The coefficients a_{0i} are determined, from the identity

$$\dot{n}_i \cdot u_0 + n_i \cdot \dot{u}_0 = 0 ,$$

to be just the components of the absolute acceleration of the particle $\xi = 0$ in its local rest frame [see an example in (12.22) on p. 272]:

$$a_{0i} = n_i \cdot \dot{u}_0 , \tag{12.79}$$

and the identity

$$\dot{n}_i \cdot n_j + n_i \cdot \dot{n}_j = 0$$

tells us that Ω_{ij} is antisymmetric:

$$\Omega_{ij} = -\Omega_{ji} .$$

We now have

$$u^\mu = \left[(1 + \xi^i a_{0i}) u_0{}^\mu + \xi^i \Omega_{ij} n_j{}^\mu \right] \dot{\sigma} .$$

But

$$-1 = u^2 = -\left[(1 + \xi^i a_{0i})^2 - \xi^i \xi^j \Omega_{ik} \Omega_{jk} \right] \dot{\sigma}^2 ,$$

whence

$$\dot{\sigma} = \left[(1 + \xi^i a_{0i})^2 - \xi^i \xi^j \Omega_{ik} \Omega_{jk} \right]^{-1/2} . \tag{12.80}$$

The right hand side of this equation is a function solely of σ and the ξ^i. Therefore the equation may be integrated along each worldline $\xi = $ const., subject to the boundary condition

$$\sigma(\xi, 0) = 0 .$$

We shall, in particular, have the necessary condition

$$\sigma(0, \tau) = \tau .$$

Note that the medium must be confined to regions where

$$(1 + \xi^i a_{0i})^2 > \xi^i \Omega_{ik} \xi^j \Omega_{jk} \quad (\geq 0) . \tag{12.81}$$

Otherwise, some of its component particles will be moving faster than light.

We can now calculate the proper metric of the medium. We have

$$n_i \cdot u = -\Omega_{ij} \xi^j \dot{\sigma} , \tag{12.82}$$

$$x^\mu{}_{,i} = n_i{}^\mu + (u_0{}^\mu + \xi^j \dot{n}_j{}^\mu) \sigma_{,i} = n_i{}^\mu + u^\mu \dot{\sigma}^{-1} \sigma_{,i} ,$$

$$u_\mu x^\mu{}_{,i} = -\Omega_{ij} \xi^j \dot{\sigma} - \dot{\sigma}^{-1} \sigma_{,i} ,$$

$$\gamma_{ij} = P_{\mu\nu}x^{\mu}{}_{,i}x^{\nu}{}_{,j}$$

$$= \delta_{ij} - \Omega_{ik}\xi^k\sigma_{,j} - \Omega_{jk}\xi^k\sigma_{,i} - \dot{\sigma}^{-2}\sigma_{,i}\sigma_{,j}$$

$$+\left(\Omega_{ik}\xi^k\dot{\sigma} + \dot{\sigma}^{-1}\sigma_{,i}\right)\left(\Omega_{jl}\xi^l\dot{\sigma} + \dot{\sigma}^{-1}\sigma_{,j}\right)$$

$$= \delta_{ij} + \dot{\sigma}^2\Omega_{ik}\Omega_{jl}\xi^k\xi^l$$

$$= \delta_{ij} + \frac{\Omega_{ik}\Omega_{jl}\xi^k\xi^l}{\left(1 + \xi^m a_{0m}\right)^2 - \xi^n\xi^r\Omega_{ns}\Omega_{rs}}, \qquad (12.83)$$

using the above expression (12.80) for $\dot{\sigma}$.

From this expression we see that there are two ways in which the motion of the medium can be rigid:

- All the Ω_{ij} are zero.
- All the Ω_{ij} and all the a_{0i} are constants, independent of σ.

In the second case the motion is one of a six-parameter family, with the Ω_{ij} and the a_{0i} as parameters. DeWitt refers to these special motions as superhelical motions. One example, constant rotation about a fixed axis, is discussed in Sect. 12.2.6. Let us first consider the case where all the Ω_{ij} are zero.

12.2.5 Rigid Motion Without Rotation

Saying that the Ω_{ij} are all zero amounts to saying that the triad $n_i{}^{\mu}$ is Fermi–Walker transported along the worldline of the particle $\xi = 0$. Let us see briefly what this means.

If $u_0(\sigma)$ is the 4-velocity of the worldline, the equation for Fermi–Walker transport of a contravector A^{μ} along the worldline is

$$\dot{A}^{\mu} = (A \cdot u_0)u_0 - (A \cdot u_0)\dot{u}_0 . \qquad (12.84)$$

This preserves inner products, i.e., if A and B are FW transported along the worldline, then $A \cdot B$ is constant along the worldline. Furthermore, the tangent vector u_0 to the worldline is itself FW transported along the worldline, and if the worldline is a spacetime geodesic (a straight line in Minkowski coordinates), then FW transport is the same as parallel transport.

Now recall that the Ω_{ij} were defined by

$$\dot{n}_i{}^{\mu} = a_{0i}u_0{}^{\mu} + \Omega_{ij}n_j{}^{\mu} . \qquad (12.85)$$

When $\Omega_{ij} = 0$, this becomes

$$\dot{n}_i{}^{\mu} = a_{0i}u_0{}^{\mu} . \qquad (12.86)$$

This is indeed the Fermi–Walker transport equation for $n_i{}^\mu$, found by inserting $A = n_i$ into (12.84), because we insist on $n_i \cdot u_0 = 0$ and we have $a_{0i} = n_i \cdot \dot{u}_0$ [see (12.79) on p. 299].

In fact, the orientation in spacetime of the local rest frame triad $n_i{}^\mu$ cannot be kept constant along a worldline unless that worldline is straight (we are referring to flat spacetimes here). Under Fermi–Walker transport, however, the triad remains as constantly oriented, or as rotationless, as possible, in the following sense: at each instant of time σ, the triad is subjected to a pure Lorentz boost without rotation in the instantaneous hyperplane of simultaneity. (On a closed orbit, this process can still lead to spatial rotation of axes upon return to the same space coordinates, an effect known as Thomas precession.) For a general non-Fermi–Walker transported triad, the Ω_{ij} are the components of the angular velocity tensor that describes the instantaneous rate of rotation of the triad in the instantaneous hyperplane of simultaneity.

Of course, given any triad $n_i{}^\mu$ at one point on the worldline, it is always possible to Fermi–Walker transport it to other points by solving (12.84). We are then saying that motions that can be given by (12.77), viz.,

$$x^\mu(\xi,\tau) = x^\mu(0,\sigma) + \xi^i n_i{}^\mu(\sigma) , \qquad (12.87)$$

where the ξ^i are assumed to label material particles in the medium, are rigid in the sense of the criterion given above.

Furthermore, the proper geometry of the medium given by the proper metric γ_{ij} in (12.65) on p. 294 is then flat, i.e.,

$$\gamma_{ij} = \delta_{ij} .$$

We also note that (σ, ξ^i) are the semi-Euclidean coordinates for an observer with worldline $x^\mu(0,\sigma)$, moving with the base particle $\xi = 0$. This generalises the construction of Sect. 12.1.4 to the case of a general 3D acceleration.

What we are doing here is to label the particle ξ^i by its spatial coordinates ξ^i in the semi-Euclidean system moving with the particle $\xi = 0$. Geometrically, we have the worldline of the arbitrarily chosen particle O at the origin, viz., $x^\mu(0,\sigma)$, with σ its proper time. We have another worldline $x^\mu(\xi^i,\tau)$ of a particle P labelled by ξ, with proper time τ. For given τ, we seek σ such that $x^\mu(\xi^i,\tau)$ is in the hyperplane of simultaneity of O at its proper time σ. Then (ξ^i) is the position of P in the tetrad moving with O. Indeed, $\{\xi^i\}$ are the space coordinates of P relative to O in that frame.

As attested by (12.82) on p. 300, we also have

$$n_i \cdot u = 0 , \qquad (12.88)$$

so that the instantaneous hyperplane of simultaneity of the particle at $\xi = 0$ is an instantaneous hyperplane of simultaneity for all the other particles of the medium as well, and the triad $n_i{}^\mu$ serves to define a rotationless rest frame for the whole medium. In other words, the coordinate system defined by the particle labels ξ^i

may itself be regarded as being Fermi–Walker transported, and all the particles of the medium have a common designator of simultaneity in the parameter σ. In the semi-Euclidean system, σ is taken to be the time coordinate.

Put another way, (12.88) says that the $n_i(\sigma)$ are in fact orthogonal to the worldline of the particle labelled by ξ^i at the value of τ corresponding to σ. This happens because $u(\xi, \tau) = u_0(0, \sigma)$. In words, the 4-velocity of particle ξ at its proper time τ is the same as the 4-velocity of the base particle when it is simultaneous with the latter in the reckoning of the base particle (quite a remarkable thing).

Because σ is not generally equal to τ, however, it is not possible for the particles to have a common synchronization of standard clocks. The relation between σ and τ is given by (12.80) on p. 300 as

$$\dot{\sigma} = \left(1 + \xi^i a_{0i}\right)^{-1}.$$

We can thus find the absolute acceleration a_i of an arbitrary particle in terms of a_{0i} and the ξ^i:

$$a_i = n_i \cdot \dot{u} = n_i \cdot \frac{\partial u}{\partial \sigma} \dot{\sigma} = \dot{\sigma} n_i \cdot \frac{\partial}{\partial \sigma} \left[\left(1 + \xi^j a_{0j}\right) u_0 \dot{\sigma}\right]$$

$$= \dot{\sigma} n_i \cdot \dot{u}_0$$

$$= \frac{a_{0i}}{1 + \xi^j a_{0j}}. \tag{12.89}$$

Here we have used the fact that $u = \left(1 + \xi^j a_{0j}\right) u_0 \dot{\sigma} = u_0$. We see that, although the motion is rigid and rotationless in the sense described above, not all parts of the medium are subject to the same acceleration.

It is important to note that, when we find ξ^i and σ, they constitute semi-Euclidean coordinates (adapted to $\xi = 0$) for the point $x^\mu(\xi, \tau)$ whether or not that point follows a particle for fixed ξ. What we have here are material particles that follow all these points with fixed ξ, for a whole 3D range of values of ξ.

In these coordinates, the metric tensor takes the form

$$g_{00} = \left.\frac{\partial x^\mu}{\partial \sigma}\right|_\xi \left.\frac{\partial x^\nu}{\partial \sigma}\right|_\xi \eta_{\mu\nu} = u^2 \dot{\sigma}^{-2} = -\left(1 + \xi^i a_{0i}\right)^2,$$

$$g_{i0} = g_{0i} = \left.\frac{\partial x^\mu}{\partial \xi^i}\right|_\sigma \left.\frac{\partial x^\nu}{\partial \sigma}\right|_\xi \eta_{\mu\nu} = (n_i \cdot u) \dot{\sigma}^{-1} = 0,$$

$$g_{ij} = \left.\frac{\partial x^\mu}{\partial \xi^i}\right|_\sigma \left.\frac{\partial x^\nu}{\partial \xi^j}\right|_\sigma \eta_{\mu\nu} = n_i \cdot n_j = \delta_{ij},$$

which has a simple diagonal structure. We note that this metric becomes static, i.e., time-independent, with the parameter σ playing the role of time, in the special case in which the absolute acceleration of each particle is constant. This should be compared with (12.18) and (12.19) on p. 271, and also (12.23) on p. 272 (but note that the sign convention for the metric has been reversed).

We conclude that this rigid motion possesses only the three degrees of freedom that the particle $\xi = 0$ itself possesses. The base particle $\xi = 0$ can move any way it wants, but the rest of the medium must then follow in a well defined way.

12.2.6 Rigid Rotation

The simplest example of a medium undergoing rigid rotation is obtained by choosing

$$a_{0i} = 0, \qquad \Omega_{12} = \omega, \qquad \Omega_{23} = 0 = \Omega_{31}.$$

The worldline of the particle at $\xi = 0$ is then straight, but the worldlines of all the other particles are helices of constant pitch. We have

$$\dot\sigma = \left\{ 1 - \omega^2 \left[(\xi^1)^2 + (\xi^2)^2 \right] \right\}^{-1/2}$$

and the proper metric of the medium takes the form

$$(\gamma_{ij}) = \begin{pmatrix} 1 + (\dot\sigma\omega\xi^2)^2 & -(\dot\sigma\omega)^2\xi^1\xi^2 & 0 \\ -(\dot\sigma\omega)^2\xi^1\xi^2 & 1 + (\dot\sigma\omega\xi^1)^2 & 0 \\ 0 & 0 & 1 \end{pmatrix}.$$

Relabelling the particles by means of three new coordinates r, θ, z given by

$$\xi^1 = r\cos\theta, \qquad \xi^2 = r\sin\theta, \qquad \xi^3 = z, \tag{12.90}$$

the proper metric of the rotating medium takes the form

$$\mathrm{diag}\left(1, \frac{r^2}{1 - \omega^2 r^2}, 1 \right).$$

Indeed, we have

$$\dot\sigma^2 = \frac{1}{1 - \omega^2 r^2},$$

whence

$$\gamma_{rr} = \frac{\partial \xi^i}{\partial r} \frac{\partial \xi^j}{\partial r} \gamma_{ij}$$

$$= \cos^2 \theta \left[1 + (\dot\sigma \varpi r)^2 \sin^2 \theta\right] - 2(\dot\sigma \varpi r)^2 \sin^2 \theta \cos^2 \theta$$

$$+ \sin^2 \theta \left[1 + (\dot\sigma \varpi r)^2 \cos^2 \theta\right]$$

$$= 1,$$

$$\gamma_{r\theta} = \gamma_{\theta r} = \frac{\partial \xi^i}{\partial r} \frac{\partial \xi^j}{\partial \theta} \gamma_{ij}$$

$$= -r \sin \theta \cos \theta \left[1 + (\dot\sigma \varpi r)^2 \sin^2 \theta\right] - r(\dot\sigma \varpi r)^2 \sin \theta \cos^3 \theta$$

$$+ r(\dot\sigma \varpi r)^2 \sin^3 \theta \cos \theta + r \sin \theta \cos \theta \left[1 + (\dot\sigma \varpi r)^2 \cos^2 \theta\right]$$

$$= 0,$$

$$\gamma_{rz} = \gamma_{zr} = \frac{\partial \xi^i}{\partial r} \frac{\partial \xi^j}{\partial z} \gamma_{ij} = 0,$$

$$\gamma_{\theta\theta} = \frac{\partial \xi^i}{\partial \theta} \frac{\partial \xi^j}{\partial \theta} \gamma_{ij}$$

$$= r^2 \sin^2 \theta \left[1 + (\dot\sigma \varpi r)^2 \sin^2 \theta\right] + 2r^2 (\dot\sigma \varpi r)^2 \sin^2 \theta \cos^2 \theta$$

$$+ r^2 \cos^2 \theta \left[1 + (\dot\sigma \varpi r)^2 \cos^2 \theta\right]$$

$$= r^2 \left[1 + (\dot\sigma \varpi r)^2\right] = r^2 \left(1 + \frac{\omega^2 r^2}{1 - \omega^2 r^2}\right) = \frac{r^2}{1 - \omega^2 r^2},$$

$$\gamma_{\theta z} = \gamma_{z\theta} = \frac{\partial \xi^i}{\partial \theta} \frac{\partial \xi^j}{\partial z} \gamma_{ij} = 0, \qquad \gamma_{zz} = \frac{\partial \xi^i}{\partial z} \frac{\partial \xi^j}{\partial z} \gamma_{ij} = 1.$$

In terms of these coordinates the proper distance δs between two particles separated by displacements δr, $\delta \theta$, and δz therefore takes the form

$$\delta s^2 = (\delta r)^2 + \frac{r^2}{1 - \omega^2 r^2} (\delta \theta)^2 + (\delta z)^2.$$

We are merely applying (12.64) on p. 294 for the new particle labels. This gives the distance of one particle as reckoned in the instantaneous rest frame of the neighbouring particle. The second term on the right of this equation may be understood as arising from relativistic contraction. At first sight, it may look odd to find that, when a disk of radius r is set spinning with angular frequency ω about its axis, so that radial distances are unaffected by relativistic contraction, distances in the direction of rotation contract in such a way that the circumference of the disk gets reduced to the value $2\pi R\sqrt{1 - \omega^2 R^2}$. It appears to contradict the Euclidean nature of the ordinary 3-space that the disk inhabits! DeWitt describes this as follows [15]:

> What in fact happens is that, when set in rotation, the disk must suffer a strain that arises for kinematic reasons quite apart from any strains it suffers on account of centrifugal forces. In particular, it must undergo a stretching of amount $(1 - \omega^2 r^2)^{-1/2}$ in the direction of rotation, to compensate the Lorentz contraction factor $(1 - \omega^2 r^2)^{1/2}$ that appears when the disk is viewed in the inertial rest frame of its axis, thereby maintaining the Euclidean nature of 3-space. It is this stretching factor that appears in the proper metric of the medium.

Let us try to put this more explicitly. Suppose A and B are two neighbouring particles at distance R from the centre and with labels θ and $\theta + \delta\theta$. When the disk is not rotating, the proper distance between them as reckoned by either in its instantaneously comoving inertial frame (ICIF) is $R\delta\theta$. When the disk is rotating, the expression for γ_{ij} tells us that the proper distance between them in the new ICIF will increase to $R\delta\theta/(1 - \omega^2 R^2)^{1/2}$. Seen by an inertial observer moving with the centre of the disk, this separation will thus be $R\delta\theta$, as before, and there will be no contradiction with the edicts of Euclidean geometry. This shows that the matter between A and B is stretched in the sense of occupying a greater proper distance as judged in an ICIF moving with either A or B.

There is a direct parallel with the two accelerating rockets mentioned at the beginning of Bell's well known paper *How to Teach Special Relativity* [5]. The separation of A and B seen by an inertial observer moving with the centre of the disk is unchanged when the rotation gets under way, so their proper separation is greater, leading to a strain which DeWitt claims to be due to kinematic reasons. If the disk could somehow be made of a very fragile material already stretched to its limit in the inertial frame moving with the center of the disk, it would shatter under rotation, just as the fragile thread joining Bell's two accelerating rockets was doomed to break.

The above discussion does assume that θ labels the material particles! And this follows from the relations in (12.90) and the fact that ξ^1, ξ^2, ξ^3 label the particles. It would be easy to miss this point. There remains therefore the question as to whether any association of material particles could have, or is likely to have this motion.

We note that the medium must be confined to regions where $r < \omega^{-1}$ and that its motion will not be rigid if ω varies with time. There are no degrees of freedom in this kind of (superhelical) motion: once the medium gets into superhelical motion, it must remain frozen into it if it wants to stay rigid. We note also that the proper geometry of the medium is not flat, i.e., $\gamma_{ij} \neq \delta_{ij}$.

12.2.7 Rigid Motion in Schwarzschild Spacetime

As an example in a curved spacetime, let us show that a medium in which particles are labelled by Schwarzschild coordinates $\xi := (r, \theta, \phi)$ is in rigid motion. The metric, displayed as a matrix, is

$$(g_{\mu\nu}) = \begin{pmatrix} -c^2 B(r) & 0 & 0 & 0 \\ 0 & B(r)^{-1} & 0 & 0 \\ 0 & 0 & r^2 & 0 \\ 0 & 0 & 0 & r^2 \sin^2\theta \end{pmatrix}, \qquad B(r) := 1 - \frac{A}{r},$$

where $A := 2GM/c^2$ is the usual constant. The motion we have in mind is described by the four functions

$$x^0(\xi, \tau) = t, \quad x^1(\xi, \tau) = r, \quad x^2(\xi, \tau) = \theta, \quad x^3(\xi, \tau) = \phi,$$

following the general scheme set out at the beginning of Sect. 12.2.1. It is then a very simple matter indeed to show that [27]

$$\gamma_{ij} = g_{ij}, \qquad i, j \in \{1, 2, 3\},$$

and also

$$\dot{\gamma}_{ij} = 0.$$

This is all rather obvious and it is easy to see how to obtain a host of rigid motions in curved spacetimes where the metric has a static form. Alternatively, one can calculate the rate of strain tensor $r_{\mu\nu}$ of (12.74) on p. 298, and it is a trivial matter to show that $r_{\mu\nu} = 0$. So any medium in which the particles could be labelled by the Schwarzschild space coordinates is undergoing a rigid motion according to this criterion.

Let us think back briefly to the short quotation from a standard textbook presentation on p. 282. In fact it is interesting to see how that account continues with regard to the related question of proper distance [28]:

Note that the [increment in the] proper radius R of a two-sphere [centered on the singularity], obtained from the spatial line element by setting $\theta = \text{const.}$, $\phi = \text{const.}$, is

$$dR = (1 - 2m/r)^{-1/2} dr > dr. \tag{12.91}$$

In other words, the proper distance between spheres of radius r and radius $r + dr$ is $dR > dr$, and hence larger than in flat space.

It is intriguing to wonder what the last comment means. For this is not really a comparison with any spheres in flat space. The coordinate interval dr need not be at a point where the spacetime is even approximately flat. The so-called proper distance is something that is related to the coordinate r in this way. In fact, the quoted relation (12.91) is telling us how to understand the coordinates.

As an aside, we have the same kind of pedagogical difficulty in the following, still in the context of the Schwarzschild metric [28]:

Let us consider proper time for a stationary observer, i.e., an observer at rest at fixed values of r, θ, ϕ. Proper time is related to coordinate time by

$$d\tau = (1 - 2m/r)^{1/2} dt < dt. \tag{12.92}$$

Thus clocks go slower in a gravitational field.

But they go slower than what? Of course, this is a neat inequality and very simple. But does it really tell us that the clock is going slower than the same clock in flat spacetime? It does not seem so. dt is a coordinate change at a place where $r \neq \infty$ and spacetime is not flat. The above relation tells us how to understand the coordinate t at the relevant point, provided that we understand how to interpret proper time as given by the metric.

Now a rod permanently occupying $[r_1, r_2]$ would be undergoing rigid motion, and so would a rod permanently occupying $[r'_1, r'_2]$. But we do not yet know whether there is some motion of the points making up the rod occupying $[r_1, r_2]$ that could serve as a transition of the same rod from the unprimed to the primed state. It seems likely that one could find a DeWitt rigid motion making the transition from $[r_1, r_2]$ to $[r'_1, r'_2]$ if and only if the proper lengths (rather than the coordinate lengths) are the same, and indeed it is not difficult to give a heuristic argument. It is worth drawing the analogy with a rod in Minkowski (flat) spacetime when it is accelerated from one state of constant velocity to another, as illustrated by the 4D region shaded in Fig. 12.10 (see p. 286). In fact, we have a similar problem here to the one discussed in Minkowski spacetime: we may know, or assume, that the proper length of a rod will be different for a given inertial observer $\mathscr{I}$ when it moves at different constant velocities relative to $\mathscr{I}$, but we do not have a theory for what it will look like in the transition between the two velocity states.

In the usual special relativistic discussion, rigid means suitably contracted in one uniform velocity state as compared with the other, but we do not usually try to say what rigid means during the transition between the states. In Fig. 12.10, the rod, initially in one velocity state, then under acceleration, then in the final velocity state, is represented by the shaded 4D region. The proposal in this chapter is just one proposal, i.e., we have found a possible solution for the motion during the transition, but it is not based on any microscopic theory of the atomic structure of the rod.

Rigid motion is a natural enough notion, but what of a microscopic theory? There is a clear parallel with the discussion of the acceleration of an atom in Minkowski spacetime, as discussed by Bell [5] and mentioned in relation to Fig. 12.10 on p. 286. When Bell's (pre-quantum) atom is accelerated slowly enough, we expect it to contract in exactly the way proposed for Fig. 12.10, i.e., so that it always has the same radius to the instantaneously comoving inertial observer. Slowly enough just means that many periods of the electron orbit fit in before the acceleration has changed the velocity very much.

What about a Bell atom in Schwarzschild spacetime? In fact, a version of the equivalence principle shows that, if an atom has radius r_{atom} in flat spacetime, then when held fixed at the value R of the Schwarzschild radial coordinate in such a way that the plane of the electron orbit contains the Schwarzschild radial direction, it will have Schwarzschild coordinate radius [12]

$$\left(1 - \frac{A}{R}\right)^{1/2} r_{\text{atom}} \tag{12.93}$$

as viewed in the hyperplane of simultaneity of the Schwarzschild observer; whence its proper radius will still be r_{atom} in the hyperplane of simultaneity of the SO. This is seen as follows. One finds coordinates $\{z^\mu\}$ at R such that an atomic nucleus fixed at R is at the origin $z^\mu = 0$ and instantaneously has speed zero relative to these coordinates, and such that

$$g_{\mu\nu}^z\big|_{z=0} = \eta_{\mu\nu}, \qquad \Gamma_{z\nu\sigma}^\mu\big|_{z=0} = 0.$$

Assuming that the electron orbits at small enough radius and with short enough period relative to the curvature, a standard rather strong version of the equivalence principle says that it will behave relative to these coordinates like an atom in flat spacetime for a certain number of orbits, e.g., following a circular orbit with radius r_{atom} and the same period as in flat spacetime too. When we transform this orbit to the Schwarzschild coordinate description, we find the Schwarzschild coordinate radius (12.93).

By this kind of argument, the strong equivalence principle shows that a thing like an atom, or a rod made of a row of such atoms (without worrying about how binding forces affect it), always measures proper length in whatever hyperplane of simultaneity it is observed, provided that it is instantaneously stationary there relative to the relevant coordinates, where proper length is the quantity usually obtained from the metric, and usually just assumed without further discussion to represent the lengths of such real (if ideal) physical objects. One might say that, wherever it is, whatever it is doing, this kind of atom or rod always measures proper lengths if used correctly. The last proviso just refers to the fact that the atom must be instantaneously stationary relative to suitable coordinates.

Is there a link with rigidity as we have been describing it? Are these Bell atoms rigid? It looks as though they are. Such an atom can sit at constant Schwarzschild space coordinates and have constant coordinate radius. Moved elsewhere, if moved slowly enough, its Schwarzschild coordinate radius changes in such a way that its proper radius in the hyperplane of simultaneity of the SO is roughly constant, just like the above infinitesimal rod subjected to an approximate DeWitt rigid motion.

As an aside, the Schwarzschild coordinates arise in a purely mathematical way in many presentations of this metric, by solving Einstein's equations, and no attempt is made to associate some physical counterpart with them. Although the notation may be suggestive, the discussion after the solution should perhaps address the question: how do we now relate these coordinates to what we measure? Furthermore, one should perhaps also ask why clock readings and measuring stick readings correspond the way they do to our coordinates, bearing in mind what a measuring stick must do to lie quietly between the points r_1 and r_2 with all its atoms under different 4-accelerations. What principle of the theory are we applying?

The above idea of a rigid rod (measuring stick) in Schwarzschild spacetime is thus that we can support it in the gravitational field in such a way that the 4D region it sweeps out crosses any Schwarzschild plane of simultaneity in the fixed coordinate interval from r_1 to r_2. The term 'support' covers up for some complex continuum of different 4-forces on its various atoms. Perhaps we can just hold one end of it at r_1, say, and let the internal forces within the rod do the rest naturally. According to the above analyses, the material of such a rod would indeed be undergoing rigid motion if all particles in the rod could be labelled by constant Schwarzschild space coordinates. This would then be a rigid rod, quite analogous to the one discussed for an accelerating observer.

It seems that our measuring sticks have to be like this for the theory to have a practical application, and the principle hiding away here is (a version of) the equivalence principle.

12.3 Conclusion

Rigid rods are commonly referred to in the special theory of relativity. In a certain sense they hardly need to be rigid. If one is moving inertially with a rigid rod, it has length L, and if one then changes to another inertial motion, it has another length L' which is shorter than L. Nothing is required of the rod here.

One does not even have to be present in this scenario. In the paradigm provided by the general theory of relativity, one just has to adopt coordinates. It does not matter what the observer is doing, only what coordinates he or she may adopt. In this view, the new length of the rod is just an illusion, not caused by anything real. The proper length depends on the choice of spacelike hypersurface used to intersect the essentially 4D spacetime region occupied by the rod.

This is not the view described by Bell [5]. The relationship between observer and rod, or between coordinate frame and rod, in which one moves relative to the other, can be achieved in another way, namely, by accelerating the rod. The observer does not have to do anything. The rod is accelerated and one would like to say that as a consequence the particles making it up adjust their relationship to one another in such a way that the rod becomes shorter as judged by the observer. There is even a theory for this: Maxwell's theory of electromagnetism.

In this view, rigidity would just be something like the assumption that there are no transient oscillatory effects during or after acceleration, or that such effects can be neglected. It would seem that the notions of rigidity discussed here are an approximation of this kind.

Put another way, rigidity is a constraint on the motion of (the particles within) a measuring stick that allows one to say exactly what is happening to it when it makes the transition from one uniform velocity state to another (in flat spacetime). Bell was considering just such a transition in his paper [5], but using the microphysical theory provided by Maxwell, which is presumably more realistic. The aim here is to draw attention to the fact that the rigidity constraint is artificial, and show that the standard, often uncritically interpreted semi-Euclidean coordinate system adapted to the worldline of an accelerating observer (in flat spacetime) fundamentally uses this constraint, and hence remind us that we ought to be wary of non-inertial coordinate systems (see also [4]).

The distinguishing feature of special relativity, when it is considered as a special case of general relativity, is that there are preferred frames of reference adapted to observers with inertial motion. However, even in special relativity, there are no preferred frames of reference adapted to accelerating observers. If they know the theory, they may as well adopt inertial coordinates (relative to which they accelerate, of course). One may nevertheless wonder what such people would measure with a measuring stick, or with the kind of (pre-quantum, non-radiating) atom described by Bell [5]. If the acceleration is not too great, one expects the Bell atom to adjust rather quickly, whereas the rigid rod described in Sect. 12.1 (accelerated along its axis) adjusts immediately for any acceleration to measure proper length in the instantaneously comoving inertial frame of the observer, i.e., in the spacelike hypersurface borrowed from an instantaneously comoving inertial observer. In other words, the rigid rod satisfies what is usually known as the ruler hypothesis.

Something like the clock and ruler hypotheses are necessary to interpret the Pound–Rebka experiment. This is used as an example to illustrate the idea that one should be wary of naive interpretations of appealing coordinate systems, which often involve assumptions of this kind in a covert way.

In a curved spacetime, one expects to find a rigid motion of a rod between two states if and only if the two states correspond to the same proper length relative to suitably adapted frames. Once again, an atom of the type described by Bell would provide an approximation.

Chapter 13
Mass in Elementary Particle Physics

The aim in this chapter is to examine the way inertial mass is handled in elementary particle physics. Only a superficial knowledge of this subject and the related quantum theory will be needed. The discussion is very much based on the superb account in [10], but another excellent reference here is [32], which fills in literally *all* the details concerning the various symmetry considerations prevalent in particle physics.

13.1 Energy and Mass

Everyone knows that $E = mc^2$. Literally everyone. This seems to be telling us that inertial mass is just another form of energy. And indeed we convert it to energy to run factories and make bombs, with typical human vigour. It seems a very neat and practical result, but where did it come from? In his refreshing and idiosyncratic book *Theoretical Concepts in Physics*, Longair provides the following fast track [30].

We have discovered the special theory of relativity, and conceive of it as a study in invariance. That is to say, we set out to build mathematical objects that are invariant under Lorentz transformations. So we have a Minkowski spacetime with its metric, and we come to consider the problem of dynamics. We have seen the beauty of the four-velocity, expressed as $U := (\gamma c, \gamma \mathbf{u})$ in some inertial frame, where $\mathbf{u}$ is the coordinate three-velocity in that inertial frame.

Now if we wish to get something like what we used to call momentum, the obvious thing to do is to look at the four-component object

$$P := m_0 U ,\qquad (13.1)$$

where m_0 is the mass. Actually, we are already forced to call m_0 the rest mass, because with hindsight we know that the inertial mass, or the inertia, of a particle increases as it moves faster. This is the very thing we hope to understand by the theory we are developing right now. The point about (13.1) is that P will be a four-

Lyle, S.N.: *Mass in Elementary Particle Physics*. Lect. Notes Phys. **796**, 313–378 (2010)
DOI 10.1007/978-3-642-04785-5_13   © Springer-Verlag Berlin Heidelberg 2010

vector provided that m_0 is a constant. This statement in turn means something about how to represent P when we look at it from another inertial frame of reference. It means that m_0 will be the same and the components $(\gamma' c, \gamma' \mathbf{u}')$ will be got from the components $(\gamma c, \gamma \mathbf{u})$ by a suitable Lorentz transformation.

Another good thing about (13.1) is that at least some of it corresponds approximately to something we used to call momentum. The three last components look just like the usual three-momentum when $v \ll c$, because then $\gamma \approx 1$. But keep an eye on that first component, tagging along so discretely.

One senses a certain unease in Longair's account which is fully justified. After making this definition on what one might describe as purely theoretical grounds, he immediately reminds us that there is a proviso for calling P a suitable four-momentum: this must agree with experiment. One is as yet somewhat at a loss to know how an experiment could come in at this early stage, but the warning is well taken. Definitions are fine, but experiment is the final judge of their usefulness.

So we have

$$P := (\gamma m_0 c, \gamma m_0 \mathbf{u}) , \tag{13.2}$$

and we call $\mathbf{p} := \gamma m_0 \mathbf{u}$ the relativistic three-momentum (γ times the Newtonian three-momentum) and γm_0 the relativistic inertial mass. Once again, we are reminded that we have not proved that γm_0 is really the right measure of inertia. We can certainly call it the inertial mass, but that may not necessarily be a good name for it. We need to check that a force on a moving particle will give it less three-acceleration than the same force when it was stationary, and in the proportions provided by the theory. We need to do experiments.

But first we require a force four-vector. By analogy with Newtonian dynamics, we make the hypothesis that the four-vector generalisation of Newton's second law of motion should be

$$F = \frac{dP}{d\tau} , \tag{13.3}$$

where τ is the proper time of the particle, generally used to parametrise its worldline. The right-hand side is a four-vector, because τ is an invariant. Longair asks how we should relate the force we measure in the laboratory with F, as though F were somehow defined by (13.3), but it seems better to treat this relation as a hypothesis and just ask what F should be. For example, for a charge e moving in a magnetic field, what are the four components of F?

One of the questions in the air here concerns a possible conservation law, and Longair notes that, in a collision between two particles which initially have four-momenta P_1 and P_2, it might turn out that

$$P_1 + P_2 = P_3 + P_4 ,$$

where P_3 and P_4 are their final four-momenta. One would call this conservation of four-momentum. Experimentally, one would have to check componentwise, i.e., check that

$$\mathbf{p}_1 + \mathbf{p}_2 = \mathbf{p}_3 + \mathbf{p}_4 \,, \qquad m_1 + m_2 = m_3 + m_4 \,,$$

where the first relation is about the *relativistic* three-momenta, not the Newtonian ones, so it contains factors of γ, and the second is about relativistic inertial masses, also containing γ factors.

We shall be particularly interested in that second relation, but what about the first? We make the suggestion that

$$\mathbf{f} = \frac{d\mathbf{p}}{dt} = \frac{d}{dt}(\gamma m_0 \mathbf{u}) \,, \tag{13.4}$$

with t the coordinate time in the given inertial frame, might be the normal three-force measured in the laboratory. This is the crucial hypothesis. Longair asks whether this definition is good enough in relativity. However, it is not a definition. It may certainly look like one on the face of it, but when one replaces $\mathbf{f}$ by $e\mathbf{u} \times \mathbf{B}$, for example, where $\mathbf{B}$ is a magnetic field and e the electric charge of the particle, one is making a prediction about experimental results.

A better question here is perhaps: does it accurately predict experimental results? And indeed Longair immediately asks this question, although once again one senses a certain embarrassment that there is nothing like a mathematical proof. He provides a heuristic argument regarding Newton's third law (of action and reaction), but even that has to appeal to experiment. In conclusion to this brief and searching paragraph (what makes Longair's book interesting is that it really is searching), the author reminds us that relativistic dynamics cannot come out of pure thought. The best we can do is make it logically self-consistent, and also consistent with experiment (up to present levels of accuracy).

But of course, what gets left out of the textbooks of any age are precisely the pointers that guided the early explorers through the dark territory between the previous theory and the new theory. Indeed, the textbooks often neglect even to say what problem those explorers were trying to solve. They simply deal in logical self-consistency and consistency with experiment, giving the distinct impression that what came before, including the struggle to arrive at this new theory, can be promptly forgotten.

But this is not the place to go into historical detail (something much more difficult to do accurately than many scientists would imagine). Let us just say that, if one had calculated the momentum in the EM fields of a charged particle moving with constant velocity, and if one had had the idea that the momentum of the charge might simply be the momentum of its EM fields, and that its inertial mass might just be what we called the electromagnetic mass in Chap. 3, then one would be trying to engineer a factor of γ into the mass and the momentum [see, for example, (3.24) on p. 37]. This was one of the pointers, according to Feynman in [2, Sect. 28.3] (which is not an ordinary textbook).

For the moment, let us see what the hypothesis (13.4) delivers, following Longair's account. If A is the four-acceleration $dU/d\tau$, then as in (13.3), we would like to extend the hypothesis to

$$F = m_0 A ,\tag{13.5}$$

which is comforting for those who seek continuity with past ideas. But note that $F = (\gamma f_4, \gamma \mathbf{f})$ for some f_4 that we have not specified. If $\mathbf{f}$ is going to be things like $e\mathbf{u} \times \mathbf{B}$ in an electromagnetic context, what should f_4 be? It looks as though this remains open, but it does not if we make the hypothesis that $F = m_0 A$. Let us see why.

Since $U^2 = c^2$ is constant (U^2 means the usual scalar product of U with itself in the Lorentz pseudometric), it follows that

$$U \cdot \frac{dU}{d\tau} = 0 ,$$

which in turn means that $U \cdot A = 0$. So we also have $F \cdot U = 0$ for any F that satisfies our equation of motion (13.3) or (13.5). But then

$$F \cdot U = 0 \quad \Longleftrightarrow \quad \gamma f_4 \gamma c = \gamma^2 \mathbf{f} \cdot \mathbf{u}$$

$$\Longleftrightarrow \quad f_4 = \frac{\mathbf{f} \cdot \mathbf{u}}{c} ,$$

so what we must choose for f_4 is completely determined by our proposed law of motion (13.5).

However, (13.4) is not the only content of this law. The first component of (13.5) tells us that

$$\gamma f_4 = \frac{d(m_0 \gamma c)}{d\tau} ,$$

and since $dt/d\tau = \gamma$, this means that

$$f_4 = c \frac{dm}{dt} ,$$

where $m := m_0 \gamma$ is the putative relativistic inertial mass. So if we have $f_4 = \mathbf{f} \cdot \mathbf{u}/c$, we must have

$$\boxed{\frac{d(mc^2)}{dt} = \mathbf{f} \cdot \mathbf{u}}\tag{13.6}$$

The quantity $\mathbf{f} \cdot \mathbf{u}$ is the rate at which work is done on the particle by the force $\mathbf{f}$, i.e., the rate of increase of energy of the particle, so mc^2 is identified with the total energy of the particle. Longair describes this as the formal proof that $E = mc^2$.

But nobody would claim that, from here, one could immediately make a hydrogen bomb. That 'experiment' is several large steps away from this 'proof'. Indeed, we ask in this chapter: what does the above argument actually prove? Longair concludes from (13.6) that there is a certain amount of inertial mass associated with the energy produced when work is done. That is indeed what is being laid down by (13.3) [or its alternative version (13.5)], because (13.4) replaces the constant inertial mass m_0 of Newtonian theory by the function $m := m_0\gamma(u)$ of the speed u, and this means precisely that a force on a moving particle will give it less three-acceleration than the same force when it was stationary.

We turn the above reasoning – what Longair calls a study in invariance – into a bold hypothesis. It does not matter what the form of the energy is. It could be electrostatic, magnetic, kinetic, elastic, or any other, in Longair's own words. All energies are the same thing as inertial mass. Then reading the equation backwards, still in Longair's own words, we conclude that inertial mass is energy, and that nuclear explosions are vivid demonstrations of this identity. Hopefully, the reader will agree that this is hypothesis, the stuff of science, and not proof.

It is hypothesis, and with a good dose of definition in it too. The m on the left-hand side of (13.6) is a very different thing to m_0, because we absorbed γ into it. Since $u < c$, we can expand the function $\gamma(u)$ as a series in powers of u/c to obtain

$$m = m_0\gamma(u) = m_0 \left(1 + \frac{1}{2}\frac{u^2}{c^2} + \frac{3}{8}\frac{u^4}{c^4} + \cdots \right),$$

or put another way,

$$mc^2 = m_0c^2 + \frac{1}{2}m_0u^2 + \frac{3}{8}m_0\frac{u^4}{c^2} + \cdots. \tag{13.7}$$

The first term here is optimistically called rest mass energy, although it seems unlikely that the pioneers really expected to get energy out of it in the early days, and the second term is the old kinetic energy. So we have simply fed the kinetic energy into our mc^2 on the left-hand side of (13.6). There are of course the remaining terms, which are part of the boldness of the hypothesis. But the real innovation here seems to be (13.4), because it says that everything on the right-hand side of (13.7) contributes to the inertia of the particle.

One of the themes of this book is that we do not understand inertia, i.e., that there is something to be understood and that we have not found it yet. And here is the special theory of relativistic dynamics telling us that m is the measure of the particle's inertia, and not m_0. There must be a message here for those who seek to understand why particles resist being accelerated, and in different proportions depending on their nature and circumstances. But the usual attitude is one of instrumentalism, as Popper called it [31]. The theory is simply required to be empirically adequate, in the sense of being well-supported by evidence or by the measure of its predictive success.

Having established (13.3) as the best, possibly even the only way to play the Lorentz invariance game, we then simply apply it and note that it works. Here, surely, is a missed opportunity. We did not know why Newton's law had the form $\mathbf{F} = m\mathbf{a}$, and now that we find exactly how we have to adjust this to get it in line with the principles of relativity, we do not ask why the dynamical law still has the form (13.3), even though we have learnt something quite significant about what contributes to inertia.

13.2 Bound State Particles

There are two kinds of particle, although both kinds are often referred to out of habit as elementary particles. Some particles, like electrons, neutrinos, and quarks, are generally considered not to be made up of anything smaller (although not always), so they are truly elementary, if that is the case. They are effectively treated as point particles, even in quantum field theory, and this leads to a need for renormalisation, as we have seen in the pre-quantum context. Naturally, in the pre-quantum view, it would be difficult to model them as extended in space and yet not made up of anything smaller, but in quantum theory, there are wave functions, and this changes our conceptions.

The other kind of particle is the bound state, which includes things like the proton, currently conceived of as a bound state of three quarks, or the π mesons, currently taken to be quark–antiquark bound states. This section is about these. It is based on the superb textbook [10] by Griffiths. We shall skate over things, examining only what is relevant to inertia.

13.2.1 Generalities

We can pick up from where we left off in Sect. 13.1. The total energy, hence inertia, of a composite system is a sum of three contributions:

- The rest energy of the constituents.
- The kinetic energy of the constituents.
- The potential energy of the configuration.

The latter two are usually comparable in size, as confirmed by the famous virial theorem, which has its classical and quantum counterparts. There is a good rule of thumb here. When the binding energy, as given by the potential energy of the configuration, is much less than the rest energies of the constituents, so too will be their kinetic energies. In this case, one can often assume that the motions of the constituents will be non-relativistic (speeds much less than c). But when the mass of the composite structure is significantly greater than the sum of the rest masses of the constituents, one expects the kinetic energies also to be significant, whence the motions of the constituents are likely to be relativistic.

It is easier to analyse bound states when the constituents have non-relativistic speeds, because then one can apply the simpler theory of non-relativistic quantum mechanics. The classic example in QM textbooks is the hydrogen atom, considered as made up of a proton and an electron. But any baryon or meson made up of heavy quarks like c, b, or t may also succumb to this treatment. The problems come with the light quarks, u, d, and s. Then only quantum field theory can give realistic results. But therein lies another difficulty, because quantum field theory usually assumes that the particles it deals with are free before and after some brief interaction, and that is not at all what we have in mind when we consider bound states.

   The binding energy of a hydrogen atom is 13.6 eV, a small figure compared with
the rest energy of an electron, which stands at 511 000 eV, so the system is non-
relativistic, and it is well known that excellent models can be made by solving a
non-relativistic Schrödinger equation, and then tweaking the results to account for
relativistic corrections, spin–orbit coupling, and minor perturbations of this kind.
A first rate overview of this standard material can be found in [10], and the main
results are listed qualitatively in the next section. We also have a very good theory
of charmonium, a $c\bar{c}$ bound state, often called the ψ meson, and bottomonium, the
$b\bar{b}$ bound state, or Υ meson. But the excited states of $u\bar{u}$ or $d\bar{d}$, easier to produce in
the accelerator, are much more difficult to treat theoretically.
   Looking at what contributes to the total energy of the bound state particle, we
recall the idea in the last section that these three features will also contribute to
the inertia of this entity. Hopefully, the reader will realise that there is a big step
from the dynamical requirements of symmetry under Lorentz transformations as
outlined there to the bold hypothesis that the kinetic energy and binding energy of
the subparticles have to be added (suitably divided by c^2) to the inertial mass of the
whole. Some may feel that the symmetry route to this hypothesis is not persuasive,
but of course, the key thing for others may merely be to reach the hypothesis and
see if it works.

13.2.2 From Hydrogen to Positronium

Let us just see for the record what we know about the hydrogen atom, a bound state
of a proton and an electron:

- In a first, non-relativistic approximation, we find that there are energy levels

$$E_n = -\frac{\alpha^2 mc^2}{2n^2} = -\frac{13.6}{n^2} \text{ eV} , \quad n \in \mathbb{Z}^+ , \tag{13.8}$$

  where $\alpha = e^2/\hbar c \approx 1/137$ is the fine structure constant.
- Each of the levels except $n = 1$ is degenerate, meaning that there are many states
  with energy E_n. For given n, there are n states labelled by integers l ranging from
  0 to $n - 1$, where this label specifies the orbital angular momentum, and for each
  value of l there are $2l + 1$ states labelled by integers m_l ranging from $-l$ to $+l$,
  where this label specifies the component of the orbital angular momentum in
  some previously chosen direction, so the degeneracy of the nth level is

$$\sum_{l=0}^{n-1} (2l+1) = n^2 .$$

All the rest of the theory of the hydrogen atom is concerned with showing that
there is in reality no degeneracy.

- There is a relativistic correction, obtained by recalling that the kinetic energy of a particle with three-momentum $\mathbf{p}$ is not $\mathbf{p}^2/2m$ but

$$T_{\text{rel}} = \frac{\mathbf{p}^2}{2m} - \frac{\mathbf{p}^4}{8m^3c^2} + \cdots ,$$

whence we add a lowest order relativistic correction

$$\Delta H_{\text{rel}} = -\frac{1}{8m^3c^2}\mathbf{p}^4$$

to the main Hamiltonian and apply perturbation theory. The states specified by quantum numbers n and l now have energies shifted by an amount

$$\Delta E_{\text{rel}} = -\alpha^4 mc^2 \frac{1}{4n^3}\left(\frac{2n}{l+1/2} - \frac{3}{2}\right) . \tag{13.9}$$

This is a very small correction. The Bohr levels E_n go as α^2, while these shifts go as α^4. This is why α is called the fine structure constant, and here we are talking about the fine structure of the hydrogen atomic spectrum. A lot of degeneracy has already gone from the level labelled by n, because we now have a whole range of energies close to E_n, one for each value of l.

- The spin $\mathbf{S}$ of the electron models the fact that it constitutes a magnetic dipole with dipole moment

$$\mu = -\frac{e}{mc}\mathbf{S} ,$$

while in the electron rest frame, the proton appears to be circling it, and hence creates a magnetic field $\mathbf{B}$. (There are other ways of looking at this.) The result is an interaction energy

$$W = -\mu \cdot \mathbf{B} ,$$

and this is called the spin–orbit interaction. We obtain another perturbation to the main Hamiltonian, viz.,

$$\Delta H_{\text{so}} = \frac{e^2}{2m^2c^2r^3}\mathbf{L}\cdot\mathbf{S} ,$$

where $\mathbf{L}$ is the electron orbital angular momentum. This leads to a further fine structure shift in the energy level of a state labelled by n and l, given by

$$\Delta E_{\text{so}} = \alpha^4 mc^2 \frac{j(j+1) - l(l+1) - 3/4}{4n^3 l(l+1/2)(l+1)} , \tag{13.10}$$

where j is a quantum number corresponding to the total angular momentum of the electron given by $\mathbf{J} = \mathbf{L} + \mathbf{S}$.

- Taking the two fine structure adjustments (13.9) and (13.10) together, we obtain the surprisingly neat result

$$\Delta E_{\text{fs}} = -\alpha^4 mc^2 \frac{1}{4n^3} \left(\frac{2n}{j+1/2} - \frac{3}{2} \right) . \tag{13.11}$$

These are very small adjustments, but they change the inertial mass of the hydrogen atom in the corresponding state.

- When the EM field itself is quantised, using the sophisticated theory known as quantum electrodynamics (QED), one has to make another adjustment to the energy levels of the various states. This is the Lamb shift:

$$\Delta E_{\text{Lamb}} = \begin{cases} \alpha^5 mc^2 \dfrac{1}{4n^3} k(n,0) , & l = 0 , \\[2ex] \alpha^5 mc^2 \dfrac{1}{4n^3} \left[k(n,l) \pm \dfrac{1}{\pi(j+1/2)(l+1/2)} \right] , & l \neq 0, \ j = l \pm 1/2 , \end{cases}$$

where $k(n,l)$ is a slowly varying function of n for $l = 0$, with values in the range $[12.7, 13.2]$, and a very small, slowly varying function of n and l (with value less than 0.05) when $l \neq 0$. Qualitatively, the key thing here is that this removes a great deal more of the remaining degeneracy in the energy levels. When we considered only the fine structure adjustments, we found that the formula (13.11) depended only on j, not l, so some states with the different l values still shared the same energy, provided they had the same j value. The classic example is $2S_{1/2}$ with $n = 2, l = 0, j = 1/2$, and $2P_{1/2}$ with $n = 2, l = 1, j = 1/2$. The Lamb shift shows that these two states actually have slightly different energies.

- The last adjustment we shall mention here is called hyperfine structure. It arises when we take into account the spin of the proton, which interacts not only with the electron orbital motion, but also directly with the electron spin. Being concerned with spin, these are of course magnetic effects once again. For states with $l = 0$, where there is only a spin–spin coupling, it turns out that

$$\Delta E_{\text{hf}} = \frac{8\pi\gamma_{\text{p}} e^2}{3mm_{\text{p}}c^2} (\mathbf{S}_{\text{p}} \cdot \mathbf{S}_{\text{e}}) |\psi_{n00}(0)|^2 , \tag{13.12}$$

where $\gamma_{\text{p}} \approx 2.79$ is a numerical factor in the expression for the proton spin dipole moment, m_{p} is the proton mass, and ψ_{n00} is the electron wave function, evaluated here at the origin, where the proton is located. It turns out that

$$|\psi_{n00}(0)|^2 = \frac{1}{\pi n^3 a^3} . \tag{13.13}$$

If

$$\mathbf{F} := \mathbf{L} + \mathbf{S}_{\text{e}} + \mathbf{S}_{\text{p}} = \mathbf{J} + \mathbf{S}_{\text{p}}$$

is the total angular momentum of the atom, with quantum numbers f and m_f, then we finally obtain for the case $l = 0$,

$$\Delta E_{\text{hf}} = \frac{m}{m_p} \alpha^4 mc^2 \frac{4\gamma_p}{3n^3} \left[f(f+1) - \frac{3}{2} \right] . \qquad (13.14)$$

It is the prefactor of m/m_p that makes this such a small adjustment, roughly a factor of a thousand less than the fine structure shifts. When $l = 0$, f can only be 0 in the singlet state where the spins of the electron and proton are oppositely aligned, or 1 in the triplet state where the spins of the electron and proton are parallel. So each $l = 0$ level splits into two levels, with the singlet pushed down and the triplet pushed up. Transitions between these two closely spaced levels give the famous 21 cm line in astronomy. When $l > 0$, there is no spin–spin coupling, only the nuclear spin–electron orbit coupling, and the energy shift turns out to be

$$\Delta E_{\text{hf}} = \frac{m}{m_p} \alpha^4 mc^2 \frac{\gamma_p}{2n^3} \frac{f(f+1) - j(j+1) - 3/4}{j(j+1)(l+1/2)} .$$

This supersedes the last formula because, although calculated in a rather different way, it actually gives (13.14) when $l = 0$ (and hence $j = 1/2$). The proton spin is 1/2, so f can only be $j + 1/2$ or $j - 1/2$, and it turns out that the last formula can be written

$$\Delta E_{\text{hf}} = \frac{m}{m_p} \alpha^4 mc^2 \frac{\gamma_p}{2n^3} \frac{\pm 1}{(f+1/2)(l+1/2)} , \qquad \text{for } f = j \pm 1/2 .$$

Each level characterised by particular values of n, l, and j is split into two.

The upshot of the above results is that there is never really any degeneracy, because in the real world there is always some interaction that concerns one of any two different states in a different way to the other. And each interaction within the bound state system slightly alters its inertial mass by application of the rule of thumb that energy is mass. The key word here is 'slightly', in the case of the hydrogen atom. The energy levels are all close to the ground state, compared with the rest energy of the whole system, so excited states are still considered to be excited states of the same thing, viz., the hydrogen atom.

Another interesting system, taking us a step toward the panoply of bound state particles in elementary particle physics, is positronium, a bound state e^+e^- of an electron and a positron. It provides a model for quarkonium, i.e., quark–antiquark bound states to discussed in the next section. In contrast to the hydrogen atom, where the much heavier nucleus was treated as being essentially stationary, with the electron in orbit around it, the two particles in positronium have equal masses.

The two-body problem is converted into a one-body problem by introducing the reduced mass

$$m_{\text{red}} := \frac{m_1 m_2}{m_1 + m_2} ,$$

which in this case is equal to $m_e/2$. If the two-body Hamiltonian has the form

$$H = \frac{\mathbf{p}_1^2}{2m_1} + \frac{\mathbf{p}_2^2}{2m_2} + V(\mathbf{r}_1, \mathbf{r}_2), \qquad (13.15)$$

where the potential V is a function only of $r = |\mathbf{r}_2 - \mathbf{r}_1|$, and if we work in the center of mass frame where $\mathbf{p}_1 = -\mathbf{p}_2 = \mathbf{p}$, then

$$H = \frac{\mathbf{p}^2}{2m_{\text{red}}} + V(r),$$

which looks like the Hamiltonian for a single particle of momentum $\mathbf{p}$ and mass m_{red}. The coordinate r is then just the distance between the two components.

In a first approximation, the Hamiltonian for positronium does have the form (13.15), with $V(r) = -e^2/r$. We call this the unperturbed Hamiltonian, because we intend to treat later adjustments merely as small perturbations. At this stage, we get the unperturbed (first approximation) energy levels for positronium by substituting $m \to m/2$ in the Bohr formula (13.8) for the hydrogen energy levels (dropping the subscript on the electron mass):

$$E_n^{\text{pos}} = \frac{1}{2}E_n = -\alpha^2 mc^2 \frac{1}{4n^2}, \quad n \in \mathbb{Z}^+.$$

The ground state binding energy is thus 6.8 eV, half the value for hydrogen. Even the wave functions for the energy eigenstates are functionally the same, with a small adjustment in one of the fixed parameters (in fact, the Bohr radius).

It is interesting to see what happens to the perturbations discussed above for the hydrogen atom, i.e., terms leading to higher approximation:

- The relativistic correction to the Hamiltonian is now

$$\Delta H_{\text{rel}} = -\frac{1}{8m_1^3 c^2}\mathbf{p}_1^4 - \frac{1}{8m_2^3 c^2}\mathbf{p}_2^4 = -\frac{1}{4m^3 c^2}\mathbf{p}^4,$$

  which looks like a factor of 2 bigger than in the hydrogen case, except that first order perturbation theory requires us to find the expectation value in the positronium state ψ_{nlm}, because this gives the shift in the energy level labelled by n. It turns out that the expectation value of $\mathbf{p}^4$ in the hydrogenic state ψ_{nlm} goes as $(mc)^4$, whence for positronium is is reduced by a factor of $(1/2)^4$. Finally, the relativistic correction for positronium is one eighth what it is for hydrogen.

- Regarding spin effects, viz., spin–orbit and spin–spin coupling between electron and positron, there is now no distinction between fine structure and hyperfine structure, because the factor of m/m_p for hydrogen is replaced by unity in the present case. So all such perturbations go as $\alpha^4 mc^2$ and can be classified as fine structure.

- There is a new perturbation due to the fact that there is no stationary nucleus and the potential is in reality non-static, whence the finite propagation time for the EM field must be taken into account. Classical electrodynamics suggests the following adjustment to the Hamiltonian:

$$\Delta H_{\text{ret}} = -\frac{e^2}{2m^2c^2}\frac{1}{r}\left[\mathbf{p}^2 + (\mathbf{p}\cdot\hat{\mathbf{r}})^2\right],$$

which also eventually leads to an adjustment going as $\alpha^4 mc^2$.

- The fine structure adjustments taken together amount to

$$\Delta E_{\text{fs}}^{\text{pos}} = \alpha^4 mc^2 \frac{1}{2n^3}\left[\frac{11}{32n} - \frac{l+\varepsilon/2}{2l+1}\right],$$

where ε is 0 for the singlet spin combination and

$$\varepsilon = \begin{cases} -\dfrac{3l+4}{(l+1)(2l+3)}, & \text{for } j = l+1, \\[2ex] \dfrac{1}{l(l+1)}, & \text{for } j = l, \\[2ex] \dfrac{3l-1}{l(2l-1)}, & \text{for } j = l-1, \end{cases}$$

for the triplet. Note that, since the electron and positron spins contribute in the same way, they are combined to give $\mathbf{S} := \mathbf{S}_1 + \mathbf{S}_2$, and the symbol $\mathbf{J}$ can then be used for the total angular momentum

$$\mathbf{J} := \mathbf{L} + \mathbf{S} = \mathbf{L} + \mathbf{S}_1 + \mathbf{S}_2.$$

- There is still a Lamb shift correction going as $\alpha^5 mc^2$, but all degeneracy has already been removed at the fine structure level in this case, so as Griffiths points out, the Lamb shift loses some of its interest.

However, in the case of positronium, there is a completely new perturbation to the energy levels of the system due to the fact that e^+ and e^- can 'temporarily' annihilate to produce a virtual photon. This is a QED correction with the Feynman diagram shown in Fig. 13.1.

It is not expected to contribute anything when $l > 0$, because the wave function is then zero at the origin, i.e., when there is no separation between electron and positron. We understand this disappearance of the wave function for zero separation intuitively as having something to do with the centrifugal force that must be pushing the two components apart when $l \neq 0$. On the other hand, when $l = 0$, it transpires that this effect contributes something going as $|\psi(0)|^2$, where ψ is the wave function. This is explained intuitively in terms of the idea that the two components must meet in space in order to annihilate.

Furthermore, the virtual photon has spin 1 like any other photon, and QED would forbid a spin singlet combination of the electron and positron to annihilate 'temporarily' in this way. So this particular perturbation only affects the zero angular momentum spin triplet configurations. These are all rather feeble attempts to put words to what the Feynman diagram appears to show. Basically, one just has to carry out

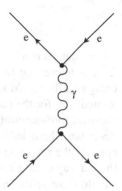

Fig. 13.1 Feynman diagram for electron–positron pair annihilation, which affects the positronium spectrum of energy levels. γ is a virtual photon

the QED calculation, and it is comforting to find that one does have a vestige of intuition about it. Anyway, the final result of that calculation is a positive shift in the energy of the $l = 0$ triplet states by the amount

$$\Delta E_{\text{ann}} = \alpha^4 mc^2 \frac{1}{4n^3} \qquad (l = 0, \, s = 1). \qquad (13.16)$$

By convention, positronium energy eigenstates are labeled $n^{2s+1}l_j$, where l is given in spectroscopic notation, i.e., S for $l = 0$, P for $l = 1$, D for $l = 2$, and so on, and s is the total spin, equal to 0 for the singlet and 1 for the triplet.

Like hydrogen, positronium makes transitions between its energy eigenstates, emitting or absorbing photons of the corresponding wavelength. But in contrast to hydrogen, positronium can simply annihilate itself to form two or more photons. (One photon is not possible, by conservation of momentum, because there is no rest frame for the photon.) When in an energy eigenstate specified by quantum numbers l and s, positronium is in a charge conjugation eigenstate with eigenvalue $(-1)^{l+s}$, whereas n photons are in a charge conjugation eigenstate with eigenvalue $C = (-1)^N$. Since QED preserves charge conjugation eigenstates, there is a selection rule for annihilation into photons, namely,

$$(-1)^{l+s} = (-1)^N,$$

where N is here the number of resulting photons. One can check that, to first order in perturbation theory, the decay is forbidden for $l > 0$. Intuitively, this comes back to the idea that the positron and electron wave functions have to overlap in space, and they do not unless $l = 0$. However, to higher orders, and hence in much less probable situations, positronium can decay directly from a state with $l > 0$. It transpires that it would be much more likely to cascade down to an S state first, then decay. Anyway, if l is equal to zero, the above selection rule is telling us that the spin singlet $s = 0$ must decay to an even number of photons, generally two, while the spin triplet $s = 1$ has to decay to an odd number of photons, and hence at least three, since just one photon is out for other reasons. One can estimate the lifetime τ of positronium by lowest order QED calculations. It turns out that

$$\tau = \frac{2\hbar}{\alpha^5 mc^2} = 1.25 \times 10^{-10} \text{ s}.$$

All the details can be found in [10].

It is interesting to reflect that all these perturbations with their corresponding energy shifts should also shift the way the system as a whole reacts to being accelerated. But for the basic EM potential $-e^2/r$, we have already seen that there is a classical (pre-quantum) explanation for this, namely the self-force. The point about describing the details of the hydrogen and positronium bound states here is to encourage the reader to consider the other energy shifts in this light. For example, is there a way of understanding why the one due to the Feynman diagram of Fig. 13.1 should increase the inertia of positronium?

13.3 Quark Bound States

The vast majority of elementary particles are not elementary. The mesons, such as the π mesons, are quark–antiquark bound states, while the baryons, such as the neutron and proton, are bound states of three quarks. We shall now consider these, and in particular, the way their inertial masses can be estimated.

We begin with a comment on the masses of the quarks themselves. Then, in Sect. 13.3.2, we discuss what Griffiths refers to as quarkonium, viz., states $q_1 \bar{q}_2$, where q_1 and q_2 are heavy quarks c, b, or t, and the bar denotes the antiparticle. Such particles are of course mesons. These models are inspired by the theory of positronium. Section 13.3.3 is a very brief introduction to the notion of multiplet, all-important for rationalising the vast array of particles churned out of our accelerators.

In Sect. 13.3.4, the light quark mesons are treated separately, because intrinsically relativistic, and Sect. 13.3.5 is about baryons, which are states of the form $q_1 q_2 q_3$, where q_1, q_2, and q_3 are any three quarks. The ultimate aim here will be the discussion of the meson masses in Sect. 13.3.4 and the baryon masses in Sect. 13.3.5. This discussion will necessarily skate over details, so is only intended to illustrate how this aspect of inertia is approached in particle physics.

Sections 13.3.6–13.3.8 describe some attempts to relate the masses of different baryons, or different mesons. They illustrate the new kind of thinking one has in the quantum theoretical context, which uses symmetry considerations and the relativistic rule-of-thumb $E = mc^2$, only paying lip service to any classical idea of a bound state comprising several subparticles.

We will barely mention quantum chromodynamics (QCD), which is a very sophisticated quantum field theory for modelling the strong force. Perhaps this is the best place to outline the way it works, at least qualitatively. A good reference to begin with is [10], while full up-to-date details can be found in [33]. In this model, quarks are sources of the strong force through something known misleadingly as colour, which plays the role of the charge in electrodynamics. The basic process is

$$\text{quark} \longrightarrow \text{quark} + \text{gluon},$$

the analogue of $e^- \longrightarrow e^- + \gamma$ in quantum electrodynamics (QED). The gluon is a massless intermediate vector boson, like the photon. However, in contrast to QED with its single kind of charge, there are three colours, called red, green, and blue. Furthermore, the gluons have colour attributes. Any given quark has a colour state, in the sense of quantum mechanics, and the set of all colour states constitutes a 3D vector space upon which one can define SU(3) transformations, i.e., unitary transformations with unit determinant. The theory is then constructed in such a way that the dynamics is invariant under such transformations by building everything up from an SU(3) invariant Lagrangian.

Just as QED construes the electromagnetic interaction as being due to exchange of virtual photons, so QCD construes the strong interaction as being due to exchange of virtual gluons. So any strongly bound state is conceived of as being full of virtual gluons. There are eight different possible colour attributes for a gluon, a way of saying that the gluon state space carries an 8D irreducible representation of those SU(3) transformations mentioned above.

Now one of the new laws of nature coming with this theory is that all naturally occurring particles are colourless or white, i.e., the total amount of each colour is zero or the three colours are present in equal amounts (bearing in mind that an antiquark carries anticolour). This guarantees that the theory will not sport bound states of two quarks or of four quarks. The only combinations satisfying the law have the form $q\bar{q}$, which are mesons, qqq, which are baryons, and $\bar{q}\bar{q}\bar{q}$, which are antibaryons, together with packages containing several such combinations, of course. We shall say a little more about this law later (see p. 351), but the reader should perhaps be asking how such a rule could be explained physically.

Returning to the strong force mediators, the gluons, it turns out that they can actually interact amongst themselves, in the sense that there are vertices in Feynman diagrams which involve only gluons. Bound states of interacting gluons known as glueballs are therefore a possibility, occurring in colourless combinations to satisfy the law mentioned above. Since gluons have a colour attribute, one would not expect to find them as isolated particles, in contrast to photons. According to Griffiths, there is evidence, from deep inelastic scattering experiments, that roughly half the momentum of a proton is carried by electrically neutral constituents that we may presume to be gluons. There is even a possibility that glueballs may have been observed.

In the present context, this momentum contribution must be something like the strong analogue of the EM field momentum discussed in Chap. 5. If quantum chromodynamics had a classical formulation like electrodynamics, one would presumably find a colour field momentum that corresponded to this. So ultimately, this large fraction of the proton momentum in the form of gluons is another manifestation of the self-force idea, but complicated by the fact that it arises from a non-Abelian gauge theory.

It could be interesting to try to formulate this in a more concrete way. One of the obstacles would be the lack of a clear ontology which thwarts most attempts to obtain a physical understanding of quantum theory. For instance, the gluons here are presumably virtual, a notion that only gets a meaning from terms in a Feynman graph expansion.

Table 13.1 What we know about quark masses. The intrinsic masses are taken from the new edition of [10], which no longer tabulates a distinction between the mass in mesons and the mass in baryons, but simply quotes an average effective value for the mass in these bound states. The third and fourth columns are thus taken from the old edition, with an adjustment for the top quark. The intrinsic masses of the light quarks have gone down considerably between the two editions. The values come with a disclaimer: they are model dependent. All the most up-to-date data can be found at the website of the Particle Data Group [11]

	Quark flavour	Intrinsic mass [MeV/c^2]	Mass in mesons [MeV/c^2]	Mass in baryons [MeV/c^2]
Light quarks	u	2	310	363
	d	5	310	363
	s	95	483	538
Heavy quarks	c	1 300	1 550	1 550
	b	4 200	4 700	4 700
	t	174 000	177 000	177 000

13.3.1 Quark Masses

We do not know the quark masses. In a way, this is the whole problem in what follows, and it is the interesting part of the discussion. The mass of a quark bound state will be made up of different ingredients, and we can only measure the result. We can only measure the mass of the resulting meson or baryon. It is like trying to guess what went into a cake only when it comes out of the oven. Except that here, there are many cakes, each containing different combinations of the ingredients, and in particular six ingredients: the masses of the six different quarks u, d, s, c, b, and t. The difficulty is that we can never catch the baker in the act. We never see a quark on its own.

So in a way, the really ultimate aim of all this may not be the discussion of the meson and baryon masses, but what we can actually deduce about the quark masses. Note that the Standard Model of particle physics does not fix their values, nor even give a hint as to what they should be. (And fortunately, the Standard Model contains many other parameters it cannot explain, so we should not run out of things to do in the near future.) Consequently, let us say that all the rest of this chapter is an attempt to answer the question: what are the quark masses?

For the moment, it is worth quickly summarising what Griffiths has to say about it [10, Chap. 4]. Table 13.1 does this. There are reasons for thinking that the u and d quarks are intrinsically light. The word 'intrinsically' refers to what we would find if we could ever get them out of a bound state. (Unfortunately, Griffiths also uses the term 'bare mass', now enslaved by renormalisation theorists, for whom the bare mass of any point particle is always infinite.) The present view is that the intrinsic masses of the u and d are well below 10 MeV/c^2 [11].

But in a hadron, i.e., any particle partaking in strong interactions, which means the mesons and baryons, the quarks have a higher effective mass. The precise value depends on the context. It seems to be higher in baryons than in mesons. Why is there an effective mass and an intrinsic mass? One reason is presumably the motion of the quark within its bound state. If it has kinetic energy, its relativistic mass will be higher than its rest mass. This is not mentioned by Griffiths. Instead he gives the analogy of an object moving through a viscous liquid: its effective inertia is greater than its true inertia.

This analogy is sometimes given to explain how the truly elementary particles get their mass via the Higgs mechanism in the Standard Model, something we discuss later (see Sect. 13.4). The analogy may not be perfect. There is a difference between this kind of inertia and the real thing, which is that the kind of inertia in a viscous liquid is velocity dependent. On the other hand, relativistic mass is velocity dependent.

There is another reason why a quark mass might be different in different contexts, apart from its kinetic energy in the given bound state. This is the simple fact that the quark may itself be a bound state. This is something that does not seem to be much discussed, but it is clear that, if the quark is itself a bound state, built up of various ingredients, the way it is bound together will be affected by the neighbourhood it finds itself in.

One model to explain the effective quark mass is the MIT bag model, in which the hadron is treated as a bag that confines the quarks [33, Sect. 3.3]. Although confined, they are considered to be able to move around like free particles, and one calculates a bag pressure. Clearly, this notion of pressure is related to the idea of the quarks having kinetic energy within the bag. The model is entirely phenomenological and, although it proves surprisingly useful in many situations, it is beset with theoretical shortfallings too. One expects the full theory of quantum chromodynamics (QCD) to deliver all the answers. This is the currently favoured theory of the strong force. The main difficulty with it is that it is very hard to do calculations, although a promising approach has been lattice QCD [33].

Finally, we note that the effective masses of the u and d in hadrons are much closer in relative terms than their intrinsic masses. The latter would not contribute much to considerations of hadron masses, and furthermore, the effective masses could be taken as equal to a reasonable approximation. However, the s quark has a much higher intrinsic mass, and much higher effective mass in the hadrons. It is thought that the strong force may treat all flavours equally, but the different mass of the s quark breaks the flavour SU(3) symmetry, a point discussed further in Sect. 13.3.3.

13.3.2 Quarkonium

Quarks have electrical charge, so there will be EM forces in the $q_1\bar{q}_2$ bound state. But quarks are also sources for the strong force, i.e., they have colour, which is the strong force equivalent for charge. They exchange not only photons, but also gluons. The full quantum field theory for the strong force is quantum chromodynamics (QCD). The book by Griffiths is a good introduction to this, because it shows right away how to do calculations using the Feynman rules [10], but we do not need to go into such detail here.

Assuming that we can treat this system as non-relativistic, due to the supposed high masses of the constituent quarks, we nevertheless face the problem of what (strong force) potential to stick into our Schrödinger equation. Further, when it comes to considering spin coupling effects, we do not know the strong force analog of the magnetic field. Some bold assumptions can be made, however.

There are two striking features of the strong force, related to the fact that the number α_s standing in as coupling constant in QCD (the analog of the fine structure constant α in electrodynamics) is not constant at all:

- At short distances, less than the dimensions of the proton, this running coupling is very small, implying that the strong force is actually quite weak. Griffiths rather nicely describes the quarks as rattling around within the proton without interacting very much. This is confirmed by deep inelastic scattering experiments, and goes by the name of *asymptotic freedom*, i.e., the coupling tends to zero on small length scales.
- At greater distance, the running coupling becomes very big. This is related to the fact that we have never managed to isolate a single quark, by breaking up a meson or baryon, for example, and it is referred to as *quark confinement*.

Short distance behaviour, as within our quarkonium bound state, is still likely to be dominated by one-gluon exchange, just as EM effects within such a system would be dominated by one-photon exchange. In addition, both the gluon and the photon are massless spin one particles, and one would expect (this is the bold hypothesis) the strong interactions to be identical in form to the EM interactions, apart from the different coupling constant α_s in the place of α, and some colour factors, simple numerical factors arising from the accounting procedure we call QCD.

So we guess a strong force potential of the form

$$V(r) = -\frac{4}{3}\frac{\alpha_s \hbar c}{r} + F_0 r \,, \tag{13.17}$$

where the factor of 4/3 is a colour factor, the first term is of Coulomb form, and F_0 is a very large constant that engineers quark confinement. Of course, F_0 is supposed to go to zero on very short length scales, like α_s, so this picture comes with a pinch of salt.

Now the light quark mesons like the pions are intrinsically relativistic because the binding energies they involve (a few hundred MeV) are so much greater than the masses of the constituent quarks u, d, and s. Although we made relativistic corrections for hydrogen and positronium, we did assume them to feature merely as perturbations. In the present case we are hoping that bound states of the more massive quarks c, b, and t are accessible to the same procedure. There is nevertheless a proviso. The binding energy E is still expected to represent a substantial fraction of the total energy of the system, implying that different energy levels will be rather widely spaced. It is thus standard practice in particle physics to consider these different energy levels as different mesons, with masses

$$M = m_1 + m_2 + E/c^2, \tag{13.18}$$

where m_1 and m_2 are the masses of the constituent quarks.

It is not really obvious that the heavy quark mesons will be non-relativistic. For a Coulomb potential, one finds that $E/mc^2 \sim \alpha^2$ [consult (13.8) on p. 319]. This makes the binding energy a fixed fraction of the rest energy of the constituents, whatever their mass. But then in QCD, if the strong force had a purely Coulomb style potential, the heavy quark mesons would be just as relativistic as their light quark cousins. However, the bound state of two heavy particles is generally smaller than the bound state of two light particles, because the Bohr radius of a bound state goes as $1/m$. One sees this in the case of the hydrogen atom, for which the Bohr radius is given by

$$a = \frac{\hbar^2}{me^2}, \tag{13.19}$$

where m is the electron mass in that case. For positronium, the Bohr radius of the system is $a_{pos} = 2a$, because we feed in the reduced mass in the place of m. What does this smaller radius of the bound system imply? There are two points:

- Light quark mesons will be more sensitive to the confining term in the potential, which is linear in (13.17). But it turns out that binding energies go more like $m^{-1/3}$ than m for such a potential.
- Asymptotic freedom corresponds to the idea that α_s is smaller at short range, so the strong coupling that operates for the closer heavy quark pair will be smaller in the analog $E/mc^2 \sim \alpha_s^2$ of the relation that triggered this discussion.

So we shall go ahead with this assumption.

The first candidate here is charmonium, a non-relativistic bound state $c\bar{c}$ of a charm quark and an anti-charm quark. The famous ψ meson, also known as J/ψ, discovered in 1974, is in fact the 1^3S_1 state of charmonium. It was produced by e^+e^- annihilation via a virtual photon at SLAC, hence has spin 1, as indicated by the superscript 3 here. It has more than 3 times the mass of the proton and an unusually long lifetime of 10^{-20} s, compared with a standard of 10^{-23} s for the typical hadron.

By comparison with positronium, one would expect there to be a spin 0 state 1^1S_0 (the subscript zero indicates a total angular momentum $j = 0$) with lower mass. One would also expect six $n = 2$ states. Shortly after the discovery of J/ψ, the ψ' meson was found, corresponding to 2^3S_1. Note that this too has spin 1, and was found by e^+e^- annihilation, but at a higher beam energy. A whole range of such mesons are now known, raising a problem of nomenclature. Griffiths describes the following system:

- Singlet S states, i.e., spin 0 and $l = 0$, are denoted by η_c.
- Triplet S states, i.e., spin 1 and $l = 0$, are denoted by ψ.
- Triplet P states, i.e., spin 1 and $l = 1$, hence $j = 0$, 1, or 2, are denoted by χ_0, χ_1, or χ_2.

- The value of n, originally denoted by putting $n - 1$ primes on the symbol, is now usually denoted by putting the energy equivalent of the mass of the particle in brackets (in MeV).

Obviously, the above list continues. Examples are: $\psi = \psi(3097)$ for $n = 1$, $\psi' = \psi(3686)$ for $n = 2$, $\psi'' = \psi(4040)$ for $n = 3$, $\psi''' = \psi(4160)$ for $n = 4$, and so on.

There is a very good correlation between the charmonium and positronium states, implying that the model is working well. What is interesting here is to note the wide spacing of the levels compared with hydrogen or positronium, for example. The gap between the two $n = 1$ levels, called hyperfine splitting in hydrogen, is 10^{11} times greater in charmonium than in positronium. But despite the enormous scale change, the ordering of the energy levels, and their relative spacing for a given value of n, are strikingly similar. However, rather than considering them as levels, or excited states, of the same thing, one tends to think of them as different things. After all, inertial mass is usually considered a defining property of a particle.

Griffiths quotes results from numerical solutions of the Schrödinger equation for the potential (13.17) with its Coulomb style and linear terms. F_0 is chosen to fit the data and a value of around 900 MeV/fm is found. (But note that other potentials can also be fitted to the data.) Since the potential corresponds to a force model, we can make a connection between the idea that the mass of the bound state should be given by a relation like (13.18), which includes the binding energy E, and the classical (pre-quantum) idea that this binding energy contributes (negatively) to the inertial mass of the system precisely because it is the measure of a force the system exerts on itself to assist acceleration, i.e., in the same direction as the acceleration, if one tries to accelerate it. In this case, the self-force is a strong force, making much bigger contributions (reductions here) to the inertia.

What are m_1 and m_2 in the relation (13.18)? They should be roughly the inertial masses of the component quark and antiquark taken alone, adjusted for the appropriate speeds of those components within the system, in the usual relativistic way. Even if this system is considered to be non-relativistic, so that the quark and antiquark are moving rather slowly compared with c, they are considered to be massive particles in the sense of having a high rest mass, so they should have a significant kinetic energy. Indeed, by the virial theorem, their kinetic energy is considered to be on a par with the binding energy. Presumably, this is part of the explanation for the different effective masses of the quarks within different bound systems. Note that if we take the effective value of the charm quark mass as 1550 MeV/c^2, as given in Table 13.1, twice this falls almost directly on the mass of the ψ meson as given three paragraphs ago. This suggests that the effective masses quoted by Griffiths include the maximum absolute value of the binding energy already.

A word should be said about the long lifetimes of the $n = 1$ and $n = 2$ charmonium particles (but see [10] for more detail on that). This is explained by the OZI rule, which suppresses their strong decay to pions, because the only possible Feynman diagrams for such decays can be cut in two by slicing only internal gluon lines. For $n = 3$ and higher, the charmonium states are massive enough to be able to decay to charmed D mesons, viz., D^0 and $\overline{D}^0$ with mass 1865 MeV/c^2, or $D^{\pm}$ with mass 1869 MeV/c^2. These are bound states of the charm quark with the light quarks, viz., $D^+ = c\overline{d}$, $D^- = d\overline{c}$, $D^0 = c\overline{u}$, and $\overline{D}^0 = u\overline{c}$. The $n = 1$ and $n = 2$ states of charmonium are considered to be bound, while for $n \geq 3$, they are considered to be quasi-bound, because their lifetimes are so much shorter.

There are also bottomonium $b\overline{b}$ bound states. Here, even the $n = 3$ states are considered to be bound states, because the B mesons, viz., $B^+ = u\overline{b}$, $B^- = b\overline{u}$, $B^0 = b\overline{d}$, and $\overline{B}^0 = d\overline{b}$, to which bottomonium might decay, are massive enough to limit this process. The level spacings in the charmonium ψ and bottomonium Υ systems turn out to be not just qualitatively, but also quantitatively similar, despite the fact that the bottom quark is several times more massive than the charm quark. As noted above, for a Coulomb style potential, level spacings are expected to be proportional to the mass, while for a linear potential, they are expected to go as $m^{-1/3}$. So if the potential (13.17) is a good approximation, the almost equal spacings in the two systems have to be due to some kind of conspiracy between the two terms in the potential or, as Griffiths puts it, an accidental feature of the value found for F_0, adjusted to fit the ψ data.

If it could be made, the toponium system $t\overline{t}$ should be more sensitive to the short range, Coulomb part of the potential, because the top quark is so much more massive, and the Bohr radius is inversely proportional to mass [see (13.19)]. One would therefore expect the spacings to be different there. However, according to Griffiths, the top quark is too shortlived to form bound states. The reaction producing them is $u + \overline{u} \longrightarrow t + \overline{t}$, with the top and anti-top decaying immediately. Their presence is merely hypothesis for explaining the ensuing decay products.

13.3.3 Multiplets

Particle accelerators churn out scores of different particles. These are classified by arranging them into multiplets, i.e., sets of particles which have something in common with one another. And what they have in common is basically their inertial mass. However, masses are not exactly constant within any multiplet. So one of the main activities in particle physics is to fit observed particles into these multiplets, and another is to explain why they do not quite fit.

The explanation for the existence of any such groupings of particles always comes back to the idea of a symmetry group, and the aim in this short section is just to glimpse the connection between the multiplet groupings, the related symmetry, and the notion of inertial mass as it is handled in particle physics today. A very good reference for getting all the details about that is [32].

Historically, the first multiplet contained just two particles, the proton and neutron. They have masses $m_p = 938.28$ MeV/c^2 and $m_n = 939.57$ MeV/c^2. The idea is to consider p and n as two states of the same particle, called the nucleon:

$$\psi_p = \psi(\mathbf{r}, t, s, \tau = +1), \qquad \psi_n = \psi(\mathbf{r}, t, s, \tau = -1),$$

where s is the spin coordinate and τ is called the isospin coordinate. The wave function of the nucleon is represented by a two-component column vector

$$\psi = \begin{pmatrix} u_1(\mathbf{r},t,s) \\ u_2(\mathbf{r},t,s) \end{pmatrix} ,$$

where $|u_1(\mathbf{r},t,s)|^2$ gives the probability density for a proton at position $\mathbf{r}$ and at time t, with spin s, while $|u_2(\mathbf{r},t,s)|^2$ does the analogous job for the neutron. These so-called isospinors carry a representation of the group SU(2) (unitary, complex-valued 2×2 matrices with unit determinant), just as spinors do. By this we mean that, for any $A \in$ SU(2), $A\psi$ is another isospinor, representing some other nucleon (or the same nucleon from some other angle). And the big idea here is that all $A \in$ SU(2) commute with the Hamiltonian for the strong interaction.

We still have to see why this is a big idea. The following is an attempt to expose this as explicitly as possible. It is absolutely standard theory, but not always easy to get so quickly from textbooks, even though it is qualitatively the key feature, because textbooks have to build up to things. So here, we are standing on the shoulders of giant textbooks like [32].

Isospin is theoretically a carbon copy of spin. Just as we had a spin vector $\mathbf{S}$ for spin 1/2 particles whose components were multiples by $\hbar/2$ of the Pauli spin matrices σ_i, $i = 1, 2, 3$, so we have an isospin vector $\mathbf{T}$ whose components are multiples by 1/2 of these matrices. The matrix components of $\mathbf{T}$ generate the matrices of the isospin SU(2) group, by which we mean that exponentials of linear combinations of them deliver SU(2) matrices. The most general isospin SU(2) matrix thus has the form

$$\begin{aligned} U_{\text{iso}}(\varepsilon) &= U_{\text{iso}}(\varepsilon_1, \varepsilon_2, \varepsilon_3) \\ &= \exp(-i\varepsilon_i T_i) \\ &= \exp\left(-\frac{i}{2}\varepsilon n_i \sigma_i\right) \\ &= \mathbb{I}\cos(\varepsilon/2) - in_i\sigma_i\sin(\varepsilon/2) , \end{aligned}$$

where the angles $\varepsilon = (\varepsilon_1, \varepsilon_2, \varepsilon_3) = \varepsilon\mathbf{n}$ characterise rotations about the three axes of the abstract isospin space. Here we can interpret this as a rotation through ε about an axis in the $\mathbf{n}$ direction.

The three components $T_i = \sigma_i/2$ of the vector $\mathbf{T}$ are matrices, or operators, in the language of quantum mechanics. They should be wearing a hat, but we shall not be dwelling on them for long. They span the Lie algebra of the symmetry group, and have commutation relations

$$[T_i, T_j] = T_i T_j - T_j T_i = i\varepsilon_{ijk}T_k ,$$

where ε_{ijk} is the totally antisymmetric tensor with $\varepsilon_{123} = +1$. The eigenstates of

$$T_3 = \begin{pmatrix} 1/2 & 0 \\ 0 & -1/2 \end{pmatrix}$$

are the isospin states corresponding to p and n, with eigenvalues $+1/2$ and $-1/2$, respectively. Another way to write these isospin states is

$$|\mathrm{p}\rangle = |T = 1/2, T_3 = 1/2\rangle \,, \qquad |\mathrm{n}\rangle = |T = 1/2, T_3 = -1/2\rangle \,,$$

where T is found from the eigenvalue $T(T+1)$ of the operator $\mathbf{T}^2$. The charge operator Q is related to T_3 by

$$Q := e \begin{pmatrix} 1 & 0 \\ 0 & 0 \end{pmatrix} = e(T_3 + 1/2) = \frac{1}{2}e(\sigma_3 + 1) \,.$$

The Lie algebra of a symmetry group is important in practical terms because it can be used to construct representations of the group itself. This is what is made explicit in [32].

But why are representations of the group important? Well, we just said that the proton and neutron states carried such a representation. Their importance is therefore just that a vector basis for a group representation can be put in correspondence with a set of particles of similar mass (and other similar properties), so this is what we need to understand. But first, let us see another classic example.

The pions π^-, π^0, and π^+ have masses 139.59, 135, and 139.59 MeV/c^2, respectively, putting them very close. They are similar in other ways. For example, they have spin 0. So the 3D vector space constructed by taking linear combinations of their states is considered to carry a representation of SU(2). This is a hypothesis, of course, and we have yet to see its utility. In detail, the hypothesis is that

$$|\pi^+\rangle = -|T = 1, T_3 = 1\rangle \,, \quad |\pi^0\rangle = |T = 1, T_3 = 0\rangle \,, \quad |\pi^-\rangle = |T = 1, T_3 = -1\rangle \,,$$

where the minus sign in the first relation is just a choice of phase made for convenience. (One needs to see the details to see why that is convenient.) The three vectors $|T = 1, T_3 = \pm, 0\rangle$ are precisely the same as the three vectors one constructs when studying angular momentum, with $L = 1$ and three values for L_3. They are all eigenvectors of the operator $\mathbf{T}^2$ with eigenvalue $T(T+1) = 2$, and eigenvectors of T_3 with values ± 1, 0. We have a carbon copy of the theory of angular momentum, and the 3D representation of the rotation group.

Now the crux of all this is that the isospin group is hypothesised to be a symmetry group of the strong interaction. This means that, if we have a Hamiltonian H_{strong} for the strong interaction, then we have

$$[H_{\text{strong}}, T_i] = 0 \,, \qquad i = 1, 2, 3 \,, \tag{13.20}$$

i.e., this Hamiltonian commutes with all the generators T_i of the isospin group. This means that it commutes with all members of the group, viz.,

$$[H_{\text{strong}}, U_{\text{iso}}(\varepsilon)] = 0 \,, \qquad \forall \varepsilon \,.$$

Then in turn the matrix operator $S_{strong} := \exp(iH_{strong}t/\hbar)$ determining the time evolution of states under the strong interaction also commutes with all the T_i and with all members of the isospin SU(2) group:

$$[S_{strong}, T_i] = 0, \quad i = 1, 2, 3, \qquad [S_{strong}, U_{iso}(\varepsilon)] = 0, \quad \forall \varepsilon. \tag{13.21}$$

The argument here works backwards, so that the latter imply (13.20), i.e., (13.20) and (13.21) are equivalent.

Qualitatively, this means that the strong interactions make no distinction between the states of a multiplet. Other interactions do, of course, and this is a key point in understanding why masses should vary over a multiplet. For example, the members of the two isospin multiplets discussed here have different charges, so the electromagnetic interaction will certainly make a distinction between them. And now the time has come to be absolutely explicit about the connection between this idea of symmetry or otherwise for different types of interaction and the way mass is treated in particle physics. Because this book is about inertial mass.

The point is that one considers the mass of a particle described by the state $|\psi\rangle$ to be just

$$M_\psi = \langle \psi | H | \psi \rangle,$$

up to a factor of c^2, i.e., it is basically just the expected value of the energy. So this involves the rule of thumb described in Sect. 13.1, and it requires us to consider the sources of energy within the system we consider here to be a particle, as modelled by some Hamiltonian H. Now let us see how this idea combined with SU(2) symmetry gets constancy of mass within an SU(2) multiplet.

Consider the proton $|1/2, 1/2\rangle$ when it is transformed in a purely mathematical way by $A := e^{-i\pi T_2} \in$ SU(2):

$$\begin{aligned} A|1/2, 1/2\rangle &= e^{-i\pi\sigma_2/2}|1/2, 1/2\rangle \\ &= \left(\cos\frac{\pi}{2} - i\sigma_2\sin\frac{\pi}{2}\right)|1/2, 1/2\rangle \\ &= -i\begin{pmatrix} 0 & -i \\ i & 0 \end{pmatrix}\begin{pmatrix} 1 \\ 0 \end{pmatrix} = \begin{pmatrix} 0 \\ 1 \end{pmatrix} = |1/2, -1/2\rangle. \end{aligned}$$

This is of course the neutron state. So we have shown that

$$A|p\rangle = |n\rangle.$$

We now have the simple argument

$$\begin{aligned} m_p^{strong} &= \langle p | H_{strong} | p \rangle \\ &= \langle n | A H_{strong} A^\dagger | n \rangle \\ &= \langle n | H_{strong} | n \rangle \\ &= m_n^{strong}, \end{aligned}$$

because A commutes with H_{strong}, and $A^{\dagger}A = \mathbb{I}$, the identity 2×2 complex matrix.

The notation $m_{\text{n}}^{\text{strong}}$ and $m_{\text{p}}^{\text{strong}}$ suggests that there may be other contributions to the masses of these particles, e.g., due to the electromagnetic interaction, which has its own Hamiltonian H_{EM}. If most of the energy in the system stems from strong interactions, then m^{strong} will be the dominant contribution to the mass, and other interactions may be treated as perturbations.

This is what happens with the EM interaction within isospin multiplets like $\{\text{n,p}\}$ and $\{\pi^{-}, \pi^{0}, \pi^{+}\}$. Indeed, the EM interaction is used to explain the mass splitting within these multiplets, i.e., the relative mass difference, which is of the order of $\Delta M / M \sim 1/1000$ in the nucleon multiplet and $\Delta M / M \sim 5/100$ in the pion multiplet. Explicitly, if the Hamiltonian has the form $H = H_{\text{strong}} + H_{\text{EM}}$, and if these interactions really are the only ones we need to take into account, then

$$m_{\text{n}} = \langle \text{n}|H|\text{n} \rangle = m_{\text{n}}^{\text{strong}} + m_{\text{n}}^{\text{EM}} \,,$$

where

$$m_{\text{n}}^{\text{EM}} := \langle \text{n}|H_{\text{EM}}|\text{n} \rangle \,,$$

and likewise,

$$m_{\text{p}} = \langle \text{n}|H|\text{p} \rangle = m_{\text{p}}^{\text{strong}} + m_{\text{p}}^{\text{EM}} \,,$$

where

$$m_{\text{p}}^{\text{EM}} := \langle \text{p}|H_{\text{EM}}|\text{p} \rangle \,.$$

This is precisely the quantum version of the self-force idea.

But let us see what Griffiths has to say about the classical idea [10, Sect. 4.5]. In his introduction to the idea of isospin, he mentions the original hope that the small difference between the proton and neutron masses might be attributed to the fact that the proton is charged, because the energy stored in its electric field should contribute to its inertia according to the rule of thumb in Sect. 13.1. But then he points out that the proton ought to be the more massive of the two particles, according to this, whereas it is in fact the neutron that is more massive. He adds in a footnote that, because the neutron–proton mass splitting is in the wrong direction to be purely electromagnetic, SU(2) is now taken to be only an approximate symmetry of the strong interactions.

It is not difficult to counter these statements. If the neutron and proton really are bound states of three quarks, viz., uud for the proton and udd for the neutron, and if each quark is charged, it is not obvious which particle should gain more mass from its internal EM interactions. It is not a simple calculation at all. For one thing, in either p or n, the EM binding energy may be negative or positive, hence contributing negatively or positively to the inertia. Of course, the suggestion here is not necessarily to try to do such a hard calculation, but just to realise that EM self-forces within a particle will contribute to its inertia, and in ways that are not always obvious.

Regarding the statement that SU(2) is now taken to be only an approximate symmetry of the strong interactions, this is undoubtedly the currently held view, but it is a *non sequitur* in the above. Indeed, we now consider that the components u and d themselves contribute different amounts of inertia to the system just by having different inertial masses. So there are several ingredients available for explaining the mass splitting in the neutron–proton system. Furthermore, it is not obvious what either of these ingredients actually contributes, because as we shall see, there is debate over the inertial masses of the u and d.

In discussing the isospin model of the pion multiplet, Greiner and Müller make the following remark [32, Sect. 5.4]. They say that the near equality in the pion masses may be interpreted as meaning that the strong interaction, which determines the dominant part of the mass, is invariant in isospin space, and that the small mass differences are caused by the EM or other interactions. They then note that the Coulomb energy of a homogeneously charged sphere with radius

$$ r_{\text{Compton}} := \frac{\hbar}{m_\pi c} , $$

the Compton wavelength of the pion, is found to be about 0.6 MeV. This is to be compared with the mass splitting of 4.59 MeV. There is a factor of ten difference, but then why take the Compton wavelength? At least in this case, the EM energy appears to be taking the mass in the right direction. If one considers that one should just add the Coulomb energy of the pion whenever it is charged, then the π^- and the π^+ should be more massive than the π^0, and why not by the same amount into the bargain?

But now go back to the idea that these pions are actually quark–antiquark bound states, i.e., π^- is d$\bar{\text{u}}$, π^0 is a mixture of d$\bar{\text{d}}$ and u$\bar{\text{u}}$, and π^+ is u$\bar{\text{d}}$. Perhaps our charge dumbbell model for self-forces would tell us something about this? We discussed just this issue in Sect. 5.4. In the dumbbell model, we would expect a certain electromagnetic mass from whatever tiny spatial structure the individual quarks might have, and a much smaller addition to this value due to the presence of two charge centers with like charge values. The latter goes as $1/d$, where d is the separation of the two charge centers. So for the charged pions, each composed of two like charges, we expect the $1/d$ contribution to be positive. Contrast with the neutral pion π^0 which is supposed to be a superposition of states like u$\bar{\text{u}}$, d$\bar{\text{d}}$ and so on. In this case, the charges have opposite signs, so the $1/d$ contribution will be a reduction.

In Sect. 5.4, we used this extremely crude classical model to estimate the separation between the quarks and found $d \sim 10^{-14}$ cm, something Feynman claims to accord with the diameters pions appear to have from cross-section measurements in high energy collisions [2, Chap. 28]. The point here is not to claim that we have made a great discovery with the dumbbell model. In fact there are two points:

- Mass splittings in multiplets of bound state particles like the isospin doublet and triplet discussed above are today explained by a quantum theoretical version of the classical self-force idea.
- The classical idea can be more sophisticated than just saying that the bound state has an overall charge and then working out the energy in its fields as though it were composed of just one point charge.

Another thing to reflect on given the first statement is the way energy involved in interactions like the EM interaction within bound states can cause the system to react to acceleration by always exactly opposing it. It is not obvious why that should work out so neatly, and this was the point of doing the four calculations in Chaps. 6–9. Although the system used for those calculations was very simple, and the situations were carefully chosen to be tractable, the calculations were not so simple, and yet still yield this simple opposition to acceleration.

Today we appeal to the rule of thumb $E = mc^2$ discussed in Sect. 13.1, so we do not need to think about what happens within the bound state when it is accelerated. But hopefully the reader will see that the standard arguments in favour of this rule do not necessarily imply that it will work. If it does always work, there is presumably something about the field theories for the interactions within the bound state, viz., QED, QCD, etc., that make this happen. Is it just their Lorentz symmetry? Could that be enough? Heuristically, Lorentz symmetry is only concerned with velocities, not accelerations.

One idea is that it may be the fact that the corresponding quantum field theories are renormalisable, which itself traces back to their being gauge theories. The point is just that the classical self-forces can be treated as contributing to the inertia of a system precisely because they have the right form to get absorbed on the $m\mathbf{a}$ side of Newton's second law, i.e., to renormalise the value of m.

That last idea is speculative. In the next couple of sections, we return to standard practice, and move on to bigger multiplets, viz., the meson nonets and baryon octet and decuplet. The aim is the same as with the isospin doublet and triplet: to classify some of the creatures in the particle zoo by finding something in common, i.e., approximately equal masses within the multiplet, and then to explain their differences.

13.3.4 Light Quark Mesons

We consider only ground state mesons with $l = 0$, composed of quarks u, d, and s. The spins of the two quarks can be antiparallel, in which case the system is in the spin singlet state with $s = 0$, or parallel, in which case it is in the spin triplet state with $s = 1$. The spin 0 light quark mesons constitute the pseudoscalar nonet, while the spin 1 light quark mesons constitute the vector nonet. These are shown in Fig. 13.2.

This is unfortunately rather jargonistic, but that is the way things are in any field. The words 'pseudoscalar' and 'vector' refer to the spins of the bound states, i.e., spin 0 states are scalar under space rotations, while spin 1 are vector. To see the

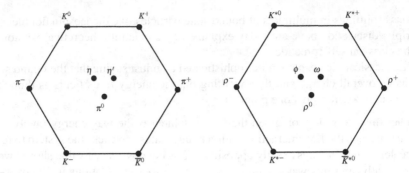

Fig. 13.2 Light quark mesons with $l = 0$. *Left*: Pseudoscalar nonet (spin 0). *Right*: Vector nonet (spin 1)

connection with the usual notions of scalar and vector, one needs to know a little about spinors. The prefix 'pseudo' is related to the fact that all these bound states happen to have negative parity, and while one would expect a vector to change sign under inversion through the origin, one would expect a scalar to remain unchanged in value.

We obtain nine mesons by combining one quark u, d, or s, with one antiquark $\bar{\text{u}}$, $\bar{\text{d}}$, or $\bar{\text{s}}$. Table 13.2 gives the quark flavour content of each meson. The word 'flavour' refers to upness, downness, and strangeness. The full meson wave functions are built up as tensor products of flavour wave functions, spin wave functions, space wave functions, and colour wave functions:

$$\psi = \psi(\text{flavour})\,\psi(\text{spin})\,\psi(\text{space})\,\psi(\text{colour}) . \qquad (13.22)$$

We will not be able to go into all the details here. Consider this as a taster. Full details of the construction of spin–flavour wave functions for mesons (and also baryons) can be found in the excellent theoretical reference [32]. This is known as the SU(6) model of the quarks and their bound states, because the three flavour states u, d, and s carry representations of SU(3), spin states carry representations of SU(2), and SU(3) × SU(2) is a subgroup of SU(6).

The up and down quarks form an isospin doublet

$$\text{u} = |1/2, 1/2\rangle , \qquad \text{d} = |1/2, -1/2\rangle ,$$

using the usual ket notation, where the first number 1/2 appearing in each indicates that the isospin has value $I = 1/2$, while the second gives the value of the component $I_3 = \pm 1/2$. The antiquarks also constitute an isospin doublet

$$\bar{\text{d}} = -|1/2, 1/2\rangle , \qquad \bar{\text{u}} = |1/2, -1/2\rangle ,$$

whence the $\bar{\text{d}}$ has $I_3 = +1/2$ and the $\bar{\text{u}}$ has $I_3 = -1/2$.

When a quark and an antiquark are combined, we obtain an isotriplet

Table 13.2 Quark flavour content of the mesons in the pseudoscalar and vector nonets. Concerning the flavour called strangeness, note that conventionally a meson containing one strange quark s has strangeness -1, while a meson containing one antiquark $\bar{s}$ has strangeness $+1$

Pseudoscalar meson	Quark flavour content	Vector meson	Quark flavour content
K^0	$d\bar{s}$	K^{*0}	$d\bar{s}$
K^+	$u\bar{s}$	K^{*+}	$u\bar{s}$
π^-	$d\bar{u}$	ρ^-	$d\bar{u}$
π^0	$(u\bar{u} - d\bar{d})/\sqrt{2}$	ρ^0	$(u\bar{u} - d\bar{d})/\sqrt{2}$
η	$(u\bar{u} + d\bar{d} - 2s\bar{s})/\sqrt{6}$	ω	$(u\bar{u} + d\bar{d})/\sqrt{2}$
η'	$(u\bar{u} + d\bar{d} + s\bar{s})/\sqrt{3}$	ϕ	$s\bar{s}$
π^+	$-u\bar{d}$	ρ^+	$-u\bar{d}$
K^-	$s\bar{u}$	K^{*-}	$s\bar{u}$
$\overline{K}^0$	$-s\bar{d}$	$\overline{K}^{*0}$	$-s\bar{d}$

$$\begin{cases} |1,1\rangle = -u\bar{d}\,, \\ |1,0\rangle = (u\bar{u} - d\bar{d})/\sqrt{2}\,, \\ |1,-1\rangle = d\bar{u}\,, \end{cases} \tag{13.23}$$

and an isosinglet

$$|0,0\rangle = (u\bar{u} + d\bar{d})/\sqrt{2}\,, \tag{13.24}$$

where the first number in each ket indicates the value of I and the second the value of I_3. Mathematically, combining means taking the tensor product $2 \otimes \bar{2}$ of the quark and antiquark isospin doublets, and the four kets just displayed are an alternative basis for that, which happen to be eigenstates of operators I and I_3 associated with the total isospin. This parallels the construction of spin states.

For the pseudoscalar mesons, the triplet is identified as the trio of pions, while for the vector mesons, it corresponds to the three ρ mesons. One makes the hypothesis that the flavour parts of the wave functions (13.22) for π^0 and the ρ^0 are neither $u\bar{u}$ nor $d\bar{d}$, but

$$\pi^0,\ \rho^0 = (u\bar{u} - d\bar{d})/\sqrt{2}\,,$$

which means physically that, if one could pull either π^0 or ρ^0 apart, one would find a u and a $\bar{u}$ half the time, and a d and a $\bar{d}$ the rest of the time. There is obviously a physical hypothesis hiding away in this kind of construction.

In the isospin model, the strange quark is assumed to have isospin 0. So among the quark–antiquark bound states involving u, d, and s, we have two isospin singlet ($I = 0$) states, viz., $s\bar{s}$ and the one in (13.24). We make the hypothesis that the physical states η and η' in the pseudoscalar nonet are some linear combinations of these isosinglet states, and likewise for the physical states ω and ϕ in the vector nonet. It turns out that

$$\eta = (u\bar{u} + d\bar{d} - 2s\bar{s})/\sqrt{6}\,, \qquad \eta' = (u\bar{u} + d\bar{d} + s\bar{s})/\sqrt{3}\,, \qquad (13.25)$$

while

$$\omega = (u\bar{u} + d\bar{d})/\sqrt{2}\,, \qquad \phi = s\bar{s}\,. \qquad (13.26)$$

The fact that the mixing is different for the pseudoscalar and vector nonets reminds us that these are physical hypotheses, checked by experiment.

The pseudoscalar mixing (13.25) makes η' an SU(3) singlet, i.e., it carries a 1D representation of this group, while η and the rest of the pseudoscalar mesons constitute an SU(3) octet, i.e., they carry an irreducible 8D representation of this group. This is the original pseudoscalar octet of the famous Eightfold Way. But according to (13.26), neither ω nor ϕ is an SU(3) singlet. They are maximally mixed in the sense that the strange–antistrange combination is kept out in the cold. One wonders to what extent this kind of feature can be explained by the Standard Model of particle physics.

As shown in Table 13.2, the strange mesons combine a strange quark with an up or a down. The three light quarks are said to carry the fundamental representation of the (approximate) SU(3) symmetry group, denoted by the symbol 3 [just as the isospin doublet representation of SU(2) was denoted by the symbol 2 when we took the tensor product $2 \otimes 2$ of the quark and antiquark isospin doublets]. The three antiquarks $\bar{u}$, $\bar{d}$, and $\bar{s}$ also carry a fundamental 3D representation of SU(3) denoted by $\bar{3}$. Combining the two, as we have effectively done to construct the nonets, we find that the tensor product of the 3 and $\bar{3}$ representations can be expressed as a direct sum of a 1D representation of SU(3) and an irreducible 8D representation:

$$3 \otimes \bar{3} = 8 \oplus 1\,.$$

But unfortunately, although we are guided by these considerations, SU(3) is only an approximate symmetry. What shows that it is not exact is the obvious fact that the particles in the given supermultiplets have different masses.

This is where we come to the thing that interests us here, namely the way inertial mass is treated in particle physics. One explanation for why the members of these nonets have a range of masses is simply that the quark components have different masses. The u and d masses are thought to be similar, one reason why isospin turns out to be a good SU(2) symmetry, but the s is much more massive, so there is no surprise that the K mesons are more massive than the π mesons, for example.

On the other hand, the ρ mesons are more massive than the π mesons, despite the fact that we assume them to have the same quark flavour content. Furthermore, we assume them to have the same spatial state ψ(space) in (13.22), corresponding to the $n = 1$, $l = 0$ state for whatever potential models the strong force. But, of course, their spin states ψ(spin) are different, because the π are spin 0, while the ρ are spin 1. The difference in mass of the π and ρ mesons has to be due to the different relative spin orientations of the quark and antiquark components, i.e., it has to be due to a spin–spin interaction.

Note that we are not talking about mass differences within the pion trio, or within the ρ meson trio, which are put down to differences between the u and d quark masses and electromagnetic effects. Here we are discussing the mass splitting that separates the two trios, i.e., the QCD equivalent of hyperfine splitting in the hydrogen ground state, which was basically a QED effect. Recall (13.12) on p. 321, viz.,

$$\Delta E_{\text{hf}} = \frac{8\pi\gamma_{\text{p}}e^2}{3mm_{\text{p}}c^2}(\mathbf{S}_{\text{p}}\cdot\mathbf{S}_{\text{e}})|\psi_{100}(0)|^2 \ .$$

If the spin–spin coupling in QCD has a similar structure, then it is reasonable to assume that it will be proportional to the scalar product of the spin vectors $\mathbf{S}_1$ and $\mathbf{S}_2$ of the quark and antiquark, and inversely proportional to each of their masses. So we try to fit the hypothesis

$$M(\text{meson}) = m_1 + m_2 + A\frac{\mathbf{S}_1\cdot\mathbf{S}_2}{m_1 m_2} \ , \tag{13.27}$$

where A is treated as a fitting parameter, assumed the same for all vector and pseudoscalar mesons.

This is a rather optimistic model (the following remarks are due to Griffiths). The idea is that the factor $|\psi_{100}(0)|^2$ is the same for all these mesons, because they are supposed to be in the same spatial quantum state. However, we found in (13.13) that it goes as $1/a^3$ in the QED case, where a is the Bohr radius, and we found in (13.19) that the Bohr radius goes as $a \sim m^{-1}$ for a Coulomb potential, where m was the reduced mass. This would give an m^3 mass dependence in the numerator. For a linear potential, it turns out that $a \sim m^{-1/3}$, so $|\psi_{100}(0)|^2 \sim m$ in that case. We keep the mass dependence in the denominator of (13.27), but we ignore the likely mass dependence in the numerator.

Another worry with our model (13.27) is that some of the mesons we are discussing contain combinations of quarks with their own antiquarks. In fact, any meson with $I_3 = 0$ is a superposition of such states, and there is likely to be an annihilation contribution to the mass, as in (13.16) on p. 325.

The model nevertheless gives quite good results. Since $\mathbf{S} = \mathbf{S}_1 + \mathbf{S}_2$,

$$\mathbf{S}_1\cdot\mathbf{S}_2 = \frac{1}{2}(\mathbf{S}^2 - \mathbf{S}_1^2 - \mathbf{S}_2^2) \ ,$$

it is easy to show that

$$\mathbf{S}_1\cdot\mathbf{S}_2 = \begin{cases} \dfrac{1}{4}\hbar^2 \ , & \text{for vector mesons } (s = 1), \\[2mm] -\dfrac{3}{4}\hbar^2 \ , & \text{for pseudoscalar mesons } (s = 0). \end{cases}$$

If we now insert the values $m_{\text{u}} = m_{\text{d}} = 310$ MeV/c^2 and $m_{\text{s}} = 483$ MeV/c^2, a good fit is obtained by

Table 13.3 Comparison between experimental pseudoscalar and vector meson masses and mass estimates using the spin–spin interaction model (13.27). Adapted from [10]

Meson	Theoretical mass [MeV/c^2]	Experimental mass [MeV/c^2]
π	139	138
K	487	496
η	561	548
ρ	775	776
ω	775	783
K^*	892	894
ϕ	1031	1020

$$A = 160 \left(\frac{2m_{\mathrm{u}}}{\hbar} \right)^2 \ \mathrm{MeV}/c^2 \,.$$

It leads to the values in Table 13.3. The estimate for η' is not included because it is so poor.

What do we conclude for the purposes of the present book? The thesis here is that it is worth thinking of inertia as resulting from internal forces within particles. The notion of force is replaced by the notion of interaction in particle physics, something that quantum theories are quite capable of handling. Whether or not the above model is a good one from a quantitative point of view, it is clearly relevant qualitatively. Indeed, it is standard practice in particle physics to estimate particle masses, and in particular, explain differences in the masses of otherwise similar particles, on the basis of internal interactions. This only works, of course, when those particles are considered to be bound states of other particles, so that there is some internal structure to play around with.

However, it is still not so obvious why this works. In particle physics, one applies a rule of thumb: the 'equivalence' of mass and energy discussed critically in Sect. 13.1. On the other hand, the classical (pre-quantum) physics of electromagnetism gives us a good reason, dare one say, a mechanism, for these inertial effects through the idea of self-force. It appears that this mechanism extends to strong forces. An interesting line of research would be to provide a quantum picture of this mechanism, since it is so widely applied.

It is also worth remembering that it is not obvious why EM self-forces always act so as to oppose accelerations. This was the point in exhibiting the calculations of Chaps. 6–9. Since self-forces appear to operate in this way for colour-sourced forces, i.e., the strong force (at least that is what is always assumed), it would be interesting to know what it is about these theories QED and QCD that leads to such a qualitatively simple phenomenological result, i.e., a self-force that always exactly opposes accelerations. It would seem that it must be either something very trivial, or something very deep. One possible idea was given at the end of the last section.

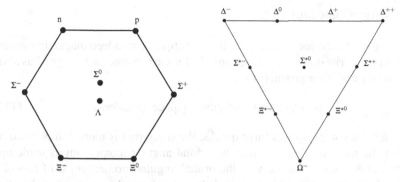

Fig. 13.3 *Left*: Baryon octet. *Right*: Baryon decuplet. In each figure, rows correspond to constant strangeness, with the top row being strangeness 0 (no strange quark in the baryon), the second row strangeness −1 (one strange quark in the baryon), and so on

Table 13.4 Quark flavour content of the baryons in the lowest energy octet and decuplet. In contrast to Table 13.2, no attempt is made here to give the quark flavour wave functions, which are more complicated than for the mesons. Octet baryons have spin 1/2, while decuplet baryons have spin 3/2. Note that conventionally a baryon containing one strange quark s has strangeness −1, while a baryon containing one antiquark s̄ has strangeness +1

Octet baryon	Quark flavour content	Decuplet baryon	Quark flavour content
		Δ^-	ddd
n	udd	Δ^0	udd
p	uud	Δ^+	uud
		Δ^{++}	uuu
Σ^-	sdd	Σ^{*-}	sdd
Σ^0, Λ	sud	Σ^{*0}	sud
Σ^+	suu	Σ^{*+}	suu
Ξ^-	ssd	Ξ^{*-}	ssd
Ξ^0	ssu	Ξ^{*0}	ssu
		Ω^-	sss

13.3.5 Baryons

For completeness, let us see how baryon masses are handled. Once again the treatment here is largely qualitative and based on [10] unless otherwise stated. We shall consider the baryon octet containing the neutron and the proton, and the baryon decuplet containing the Δ resonances. These are shown in Fig. 13.3, and their quark flavour contents are shown in Table 13.4. The octet and decuplet baryons differ in their spin, which is 1/2 for the octet and 3/2 for the decuplet. The decuplet contains three quark flavour combinations that do not fit into the octet, viz., uuu, ddd, and sss.

Baryon Wave Functions

In order to model the baryon masses within multiplets, or indeed magnetic moments, one must first write down wave functions. As for the mesons, each baryon has a wave function of the (tensor product) form

$$\psi = \psi(\text{flavour})\,\psi(\text{spin})\,\psi(\text{space})\,\psi(\text{colour})\,. \tag{13.28}$$

Since each baryon comprises three quarks, the space part is more complicated here than for the mesons. One can treat the orbital angular momentum as made up of two orbital angular momenta, viz., the orbital angular momentum $\mathbf{L}$ of two of the quarks about their center of mass, and the orbital angular momentum $\mathbf{L}'$ of this combination and the third quark about the center of mass of all three. Here we only consider $l = 0 = l'$.

Under this hypothesis, the total angular momentum comes entirely from the spins of the quarks. Each quark has spin 1/2, so their are eight possible spin states for the baryon:

$$|\uparrow\uparrow\uparrow\rangle,\ |\uparrow\uparrow\downarrow\rangle,\ |\uparrow\downarrow\uparrow\rangle,\ |\downarrow\uparrow\uparrow\rangle,\ |\uparrow\downarrow\downarrow\rangle,\ |\downarrow\uparrow\downarrow\rangle,\ |\downarrow\downarrow\uparrow\rangle,\ |\downarrow\downarrow\downarrow\rangle\,.$$

These are not eigenstates of the total angular momentum. The quark spins can combine to give a total of 1/2 or 3/2. With the usual notation:

$$\left.\begin{aligned}
|3/2,3/2\rangle_s &= |\uparrow\uparrow\uparrow\rangle\\
|3/2,1/2\rangle_s &= (|\uparrow\uparrow\downarrow\rangle + |\uparrow\downarrow\uparrow\rangle + |\downarrow\uparrow\uparrow\rangle)/\sqrt{3}\\
|3/2,-1/2\rangle_s &= (|\uparrow\downarrow\downarrow\rangle + |\downarrow\uparrow\downarrow\rangle + |\downarrow\downarrow\uparrow\rangle)/\sqrt{3}\\
|3/2,-3/2\rangle_s &= |\downarrow\downarrow\downarrow\rangle
\end{aligned}\right\} \quad \psi_s\ (\text{spin } 3/2)\,, \tag{13.29}$$

$$\left.\begin{aligned}
|1/2,1/2\rangle_{12} &= (|\uparrow\downarrow\uparrow\rangle - |\downarrow\uparrow\uparrow\rangle)/\sqrt{2}\\
|1/2,-1/2\rangle_{12} &= (|\uparrow\downarrow\downarrow\rangle - |\downarrow\uparrow\downarrow\rangle)/\sqrt{2}
\end{aligned}\right\} \quad \psi_{12}\ (\text{spin } 1/2)\,, \tag{13.30}$$

$$\left.\begin{aligned}
|1/2,1/2\rangle_{23} &= (|\uparrow\uparrow\downarrow\rangle - |\uparrow\downarrow\uparrow\rangle)/\sqrt{2}\\
|1/2,-1/2\rangle_{23} &= (|\downarrow\uparrow\downarrow\rangle - |\downarrow\downarrow\uparrow\rangle)/\sqrt{2}
\end{aligned}\right\} \quad \psi_{23}\ (\text{spin } 1/2)\,. \tag{13.31}$$

The states ψ_s are fully symmetric in the three quarks, while ψ_{12} are antisymmetric in the first two and ψ_{23} are antisymmetric in the last two. The states

$$\left.\begin{aligned}
|1/2,1/2\rangle_{31} &= (|\downarrow\uparrow\uparrow\rangle - |\uparrow\uparrow\downarrow\rangle)/\sqrt{2}\\
|1/2,-1/2\rangle_{31} &= (|\downarrow\downarrow\uparrow\rangle - |\uparrow\downarrow\downarrow\rangle)/\sqrt{2}
\end{aligned}\right\} \quad \psi_{31}\ (\text{spin } 1/2)\,, \tag{13.32}$$

are not needed to span the space of possible spin states because the states ψ_s, ψ_{12} and ψ_{23} already span the whole space. In fact,

$$2 \otimes 2 \otimes 2 = 4 \oplus 2 \oplus 2 \,,$$

where numbers on the right give the dimensions of the representations ψ_s, ψ_{12} and ψ_{23}, respectively. In fact, it is easy to see that the states ψ_{31} are not independent in the vector space sense. For example,

$$|1/2, 1/2\rangle_{12} + |1/2, 1/2\rangle_{23} + |1/2, 1/2\rangle_{31} = 0 \,.$$

Naturally, one could choose other breakdowns of the full spin space $2 \otimes 2 \otimes 2$.

The symmetries of these states are important because of the Pauli exclusion principle, which follows from the full quantum field theory for spin 1/2 particles like quarks. However, we shall make another bold hypothesis here, namely that the quarks in the baryon are actually identical fermions, despite the fact that they may have different colours and flavours. Griffiths describes this as a subtle extension of the notion of identical particle, because all quarks are treated as different states of a single particle, regardless of their colour or flavour. However, it is only subtle if we do it implicitly, as is often the case. Many accounts of physics become subtle in this way. But physics works by hypothesis, and this is just another hypothesis, with its weak and its strong points. Its weak point is presumably that flavour is a heavily broken symmetry.

The Pauli exclusion principle stipulates that our 3-quark states be totally antisymmetric. What we are saying then is that the full wave function of the baryon given schematically in (13.28) has to be completely antisymmetric in the three quark positions, just like the spin states ψ_s given in (13.29). (By quark positions, here and in the following, we refer to their mathematical positions in the tensor product, i.e., position 1, 2, or 3, not position in space.) We do not know the functional form of the spatial component $\psi(\text{space})$ because we do not know the interquark potential, or what stands in for it in the full theory of quantum chromodynamics. However, we guess that, for the ground state with $l = 0 = l'$, this part of the baryon wave function is likely to be symmetric in the three quark positions.

We have just seen that the spin states can be completely symmetric ($j = 3/2$), or partially antisymmetric ($j = 1/2$), where j is the total angular momentum in this case. Regarding the flavour component, there are 27 possible combinations uuu, uud, udu, duu, udd, ..., sss, which can be combined linearly into bases for vector spaces carrying irreducible representations of SU(3), just as the spin combinations given above carry irreducible representations of SU(2). The bases for the various possible irreducible representations are displayed in the familiar eightfold-way patterns in Figs. 13.4–13.6.

One more state is required to span the whole space $3 \otimes 3 \otimes 3$. We may take the completely antisymmetric state

$$\psi_a = (\text{uds} - \text{usd} + \text{dsu} - \text{dus} + \text{sud} - \text{sdu})/\sqrt{6} \,. \tag{13.33}$$

What we have then is a decomposition of the form

$$3 \otimes 3 \otimes 3 = 10 \oplus 8 \oplus 8 \oplus 1 \,.$$

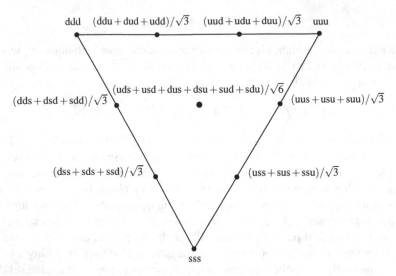

Fig. 13.4 Fully symmetric baryon flavour states ψ_s

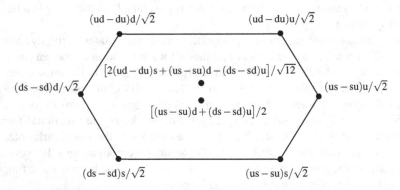

Fig. 13.5 Partially antisymmetric flavour states ψ_{12}

Another partially antisymmetric octet can be constructed, which would be denoted ψ_{31}, antisymmetric in the first and third quark positions, but the basis elements for this irreducible representation of flavour SU(3) are of course expressible as linear combinations of the basis elements we have already found, since the latter spanned the whole 27 dimensional space. In fact, $\psi_{12} + \psi_{23} + \psi_{31} = 0$. The following constructions could choose any pair from these partially antisymmetric SU(3) representations.

Referring to (13.28), we still have to talk about colour. There are more states now, colour states, and they carry another representation of SU(3), the colour symmetry group of quantum chromodynamics. This is considered to be an exact symmetry, in the sense that quarks of different colours but the same flavour have the same mass, while the flavour SU(3) symmetry is an approximation, because quarks of

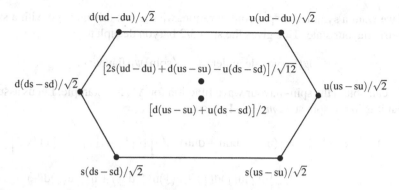

$$d(ud - du)/\sqrt{2} \qquad\qquad u(ud - du)/\sqrt{2}$$

$$[2s(ud - du) + d(us - su) - u(ds - sd)]/\sqrt{12}$$

$$d(ds - sd)/\sqrt{2} \qquad\qquad u(us - su)/\sqrt{2}$$

$$[d(us - su) + u(ds - sd)]/2$$

$$s(ds - sd)/\sqrt{2} \qquad\qquad s(us - su)/\sqrt{2}$$

Fig. 13.6 Partially antisymmetric flavour states ψ_{23}

different flavours have different masses. Mass is all important here. When masses are different within a multiplet, i.e., within a set of states carrying an irreducible representation of the putative symmetry group, this means that the operators generating the latter are not commuting properly with a mass (energy) operator. It is in precisely this sense that it is not a good symmetry.

However, the current hypothesis is that mass is constant over colour multiplets. So we can construct a colour decuplet, colour octets, and so on. Looking at the flavour multiplets, we only have to systematically replace u → red, s → green, and d → blue, for example, and we have a suitable set of irreducible representations of the colour SU(3) symmetry group. But now we make a further hypothesis (see p. 327): all naturally occurring particles are colourless (or strictly, in some cases, white). A meson can contain a red and an antired quark, and a baryon must contain one of each colour, whence we ought to say that it is white.

Well, the colour analogy is not perfect, even if the symmetry may be! It is better to stick to the mathematical formulation and state the hypothesis like this: every naturally occurring particle is in a colour singlet state. This state is the analogue of (13.33), viz.,

$$\psi(\text{colour}) = (rgb - rbg + gbr - grb + brg - bgr)/\sqrt{6}. \qquad (13.34)$$

Since each baryon state of the form (13.28) is going to contain this same state as a (tensorial) factor, it is usually just dropped from the proceedings, and we retain only the absolutely crucial fact that it is totally antisymmetric in the three quark positions.

This is crucial because it means that, in combination with the Pauli principle, which follows from quantum field theory, the rest of the baryon state, i.e., the tensor product of all the other factors, must be totally symmetric in the three quark positions. Since we guess that $\psi(\text{space})$ is symmetric for states with $l = 0 = l'$, the product of $\psi(\text{spin})$ and $\psi(\text{flavour})$ must be symmetric. Following Griffiths [10], we give examples of a decuplet state and an octet state:

- If we want a symmetric spin configuration, we see that this must go with a symmetric flavour state. This gives the spin 3/2 baryon decuplet

$$\psi(\text{baryon decuplet}) = \psi_s(\text{spin})\,\psi_s(\text{flavour})\,.$$

We can obtain the spin–flavour wave function for Δ^+, for example. Let us assume that it is in the spin state $m_j = -1/2$, so

$$|\Delta^+;3/2,-1/2\rangle = \Big[(uud+udu+duu)/\sqrt{3}\Big]\Big[(\downarrow\downarrow\uparrow + \downarrow\uparrow\downarrow + \uparrow\downarrow\downarrow)/\sqrt{3}\Big]$$

$$= \frac{1}{3}\Big[u(\downarrow)u(\downarrow)d(\uparrow)+u(\downarrow)u(\uparrow)d(\downarrow)+u(\uparrow)u(\downarrow)d(\downarrow)$$

$$+u(\downarrow)d(\downarrow)u(\uparrow)+u(\downarrow)d(\uparrow)u(\downarrow)+u(\uparrow)d(\downarrow)u(\downarrow)$$

$$+d(\downarrow)u(\downarrow)u(\uparrow)+d(\downarrow)u(\uparrow)u(\downarrow)+d(\uparrow)u(\downarrow)u(\downarrow)\Big]\,.$$

In practical terms, this means that if we could sample the quarks in such a state, we would expect to find that in 1/9 of cases the first quark would be a d with spin up, while in 4/9 of cases it would be a u with spin down.

- The baryon octet states are more complicated than this. A fully symmetric combination is obtained by putting together states of mixed symmetry, using the fact that the product of two antisymmetric wave functions is symmetric. We thus observe that $\psi_{12}(\text{spin})\,\psi_{12}(\text{flavour})$ is symmetric in the first two quark positions, while $\psi_{23}(\text{spin})\,\psi_{23}(\text{flavour})$ is symmetric in the second two, and $\psi_{31}(\text{spin})\,\psi_{31}(\text{flavour})$ in the third. It turns out that, if we add these three combinations, the result is fully symmetric. So with a suitable normalisation coefficient,

$$\psi(\text{baryon octet}) = \frac{\sqrt{2}}{3}\Big[\psi_{12}(\text{spin})\,\psi_{12}(\text{flavour}) + \psi_{23}(\text{spin})\,\psi_{23}(\text{flavour})$$

$$+ \psi_{31}(\text{spin})\,\psi_{31}(\text{flavour})\Big]\,.$$

As an example, consider the spin–flavour wave function for a spin up proton:

$$|p;1/2,1/2\rangle = \frac{\sqrt{2}}{3}\Big[\frac{1}{2}(\uparrow\downarrow\uparrow - \downarrow\uparrow\uparrow)(udu-duu) + \frac{1}{2}(\uparrow\uparrow\downarrow - \uparrow\downarrow\uparrow)(uud-udu)$$

$$+\frac{1}{2}(\uparrow\uparrow\downarrow - \downarrow\uparrow\uparrow)(uud-duu)\Big]$$

$$= \frac{2}{3\sqrt{2}}u(\uparrow)u(\uparrow)d(\downarrow) - \frac{1}{3\sqrt{2}}u(\uparrow)u(\downarrow)d(\uparrow)$$

$$-\frac{1}{3\sqrt{2}}u(\downarrow)u(\uparrow)d(\uparrow) + \text{permutations}\,. \tag{13.35}$$

The task of constructing these bound states is not really trivial. However, as presented here, it is just a mathematical game. The reader should be asking physical questions about it. There is a tendency in textbooks to get on with the theory and apply it, and forget the intellectual struggle of those who originally built it up, or even the original problem they were trying to solve. Not only does this give a misleading aura of finality to physical theory, but it encourages students to become mathematicians rather than physicists. Even though the above manipulations are quite sophisticated, the mathematics of physics is much easier than physics.

The book by Griffiths stands out from the crowd because it does mention the physics from time to time. Here is one of the insights. The corners of the baryon decuplet contain three quarks of the same flavour, viz., uuu, ddd, and sss. They are necessarily symmetric flavour states. This means that they must go with symmetric spin states, so they must have $j = 3/2$. When there are two identical flavours, e.g., uud, there are three arrangements, viz., uud, udu, and duu, so we can build a symmetric linear combination that belongs to the decuplet and two linear combinations of mixed symmetry that belong to the octets. Finally, when all three of the quarks have different flavours, e.g., uds, there are six possibilities: the completely symmetric linear combination completes the decuplet, the completely antisymmetric linear combination is an SU(3) singlet, and the other four go into building the two octets.

. But colour plays a key role here, even if the colour component of the wave function has not been included explicitly. If colour were not part of this, we would be seeking antisymmetric spin–flavour wave functions. The spin 3/2 states would have to go with the flavour singlet (13.33), which is antisymmetric. A spin 1/2 octet can be constructed, but instead of the decuplet, we would have only one spin 3/2 baryon. This was why colour was originally introduced, because it was felt that the Pauli principle had to be maintained. This was the first problem that colour solved.

Another good reference for concrete applications of symmetry groups in physics is [32]. The baryon states are constructed in much more detail, with more explanation of the mathematical features, although the physics is often taken as understood.

Baryon Magnetic Moments

The magnetic dipole moment of a spin 1/2 particle of charge q and mass m is

$$\mu = \frac{q}{mc}\mathbf{S},\qquad(13.36)$$

where $\mathbf{S}$ is its spin vector. It is because this depends on the mass that it is interesting to discuss it in the present context. We have been discussing baryons in their ground state, with $l = 0 = l'$, so the net magnetic moment will be the vector sum of the moments of the three quarks taken separately, viz.,

$$\mu = \mu_1 + \mu_2 + \mu_3 .$$

According to (13.36), the different quark flavours correspond to different magnetic dipole moments, because μ depends on the charge and the mass. It has magnitude

$$\mu = \frac{q\hbar}{2mc} , \tag{13.37}$$

so μ_z has this value in the spin up state, for which $S_z = \hbar/2$, and minus this value in the spin down state. If the charge is negative, the magnetic moment points in the opposite direction to the spin. For the three quark flavours, we have

$$\mu_u = \frac{2}{3}\frac{e\hbar}{2m_u c} , \qquad \mu_d = -\frac{1}{3}\frac{e\hbar}{2m_d c} , \qquad \mu_s = -\frac{1}{3}\frac{e\hbar}{2m_s c} . \tag{13.38}$$

The magnetic moment of an octet baryon B in its spin up state (recall that the octet baryons have spin 1/2) is

$$\mu_B = \langle B\uparrow|(\mu_1 + \mu_2 + \mu_3)_z|B\uparrow\rangle = \frac{2}{\hbar}\sum_{i=1}^{3}\langle B\uparrow|\mu_i S_{iz}|B\uparrow\rangle , \tag{13.39}$$

where the sum is over the three quarks in the baryon. We could also do this for decuplet baryons, but we have implicitly considered spin 1/2 here rather than spin 3/2.

Using the wave function (13.35), we can calculate the magnetic moment of the proton as follows. The first term in (13.35) is

$$\frac{2}{3\sqrt{2}}u(\uparrow)u(\uparrow)d(\downarrow) .$$

We thus evaluate

$$(\mu_1 S_{1z} + \mu_2 S_{2z} + \mu_3 S_{3z})|u(\uparrow)u(\uparrow)d(\downarrow)\rangle = \left(\mu_u\frac{\hbar}{2} + \mu_u\frac{\hbar}{2} - \mu_d\frac{\hbar}{2}\right)|u(\uparrow)u(\uparrow)d(\downarrow)\rangle .$$

According to (13.39), the first term in the proton wave function thus contributes

$$\left(\frac{2}{3\sqrt{2}}\right)^2 \frac{2}{\hbar}\sum_{i=1}^{3}\langle u(\uparrow)u(\uparrow)d(\downarrow)|\mu_i S_{iz}|u(\uparrow)u(\uparrow)d(\downarrow)\rangle = \frac{2}{9}(2\mu_u - \mu_d) .$$

Needless to say, the quark wave functions are normalised so that, for example,

$$\langle u(\uparrow)|u(\uparrow)\rangle = 1 , \qquad \langle u(\uparrow)|u(\downarrow)\rangle = 0 .$$

The second and third terms in (13.35) are each found to give $\mu_d/18$. The other six terms in (13.35) are found by permutations of the first three, hence give the same results, and the grand total for the magnetic moment of the proton is

$$\mu_p = 3\left[\frac{2}{9}(2\mu_u - \mu_d) + \frac{1}{18}\mu_d + \frac{1}{18}\mu_d\right] = \frac{1}{3}(4\mu_u - \mu_d) . \tag{13.40}$$

Table 13.5 Baryon octet magnetic moments as calculated from the octet wave functions and expressed in terms of the quark magnetic moments μ_u, μ_d, and μ_s

Baryon	Calculated magnetic moment
p	$\frac{1}{3}(4\mu_u - \mu_d)$
n	$\frac{1}{3}(4\mu_d - \mu_u)$
Λ	μ_s
Σ^+	$\frac{1}{3}(4\mu_u - \mu_s)$
Σ^0	$\frac{1}{3}(2\mu_u + 2\mu_d - \mu_s)$
Σ^-	$\frac{1}{3}(4\mu_d - \mu_s)$
Ξ^0	$\frac{1}{3}(4\mu_s - \mu_u)$
Ξ^-	$\frac{1}{3}(4\mu_s - \mu_d)$

The other octet magnetic moments can be calculated and expressed in terms of the quark magnetic moments μ_u, μ_d, and μ_s, which we do not know, of course, because they depend on the quark masses. Table 13.5 gives the results.

One approach is to take the masses of the quarks when they are components of baryons as given in Sect. 13.3.1, feed them into (13.38), and compare the calculated values with measured values. For example, with $m_u = m_d = 363$ MeV$/c^2$, (13.40) gives

$$\mu_p = 2.79\mu_{\text{nuclear}} ,$$

where μ_{nuclear} is the nuclear magneton, i.e., the magnetic diple moment of the proton according to (13.37) if one does not take into account its being a composite particle, given by

$$\mu_{\text{nuclear}} := \frac{e\hbar}{2m_p c} = 3.152 \times 10^{-18} \text{ MeV/gauss} .$$

The measured value for μ_p is $2.793\mu_{\text{nuclear}}$. Doing this for the other octet baryons gives reasonable results, which is presumably a confirmation of sorts for the values of the quark masses given in Sect. 13.3.1. But where did those mass values come from? Griffiths does not say, but we can be sure of one thing: they are already model dependent. So this would in fact be a confirmation for some model of the bound state.

Another approach here is simply to assume that the up and down quarks have equal mass, i.e., $m_u = m_d$, whence (13.38) implies that

$$\mu_u = -2\mu_d ,$$

then note that, from the two formulas in Table 13.5,

$$\frac{\mu_n}{\mu_p} = -\frac{2}{3} .$$

The experimental value, which is very accurately known, is

$$\left.\frac{\mu_n}{\mu_p}\right|_{\text{exp}} = 0.68497945 \pm 0.00000058 ,$$

which compares reasonably well. This presumably confirms the idea that the up and down quarks have similar masses.

However, it does raise the question of what masses should go in the formulas (13.38). Should it be the mysterious values of the quark masses when they are bound in baryons? Or should it perhaps be their masses when free, if indeed they can be free? The formulas (13.38) assume that the individual quarks are truly elementary spin 1/2 particles, i.e., not themselves composite, and the mass one usually feeds into this formula is normally the rest mass of the particle. Of course, when the quarks are bound in a baryon, one would expect them to have kinetic energies on a par with the binding potential (virial theorem). Is it due to their motions within the system that their masses differ from their free masses, by the usual γ factor of special relativistic dynamics? And if that is so, should it not be their free masses that we are proposing to be equal by the above result?

Another discussion of these same issues can be found in [32, Sect. 8.13]. A further possibility here is to express μ_u, μ_d, and μ_s in terms of the known value μ_{nuclear} of the nuclear magneton by first solving the linear equations

$$\begin{cases} \mu_p = \dfrac{1}{3}(4\mu_u - \mu_d) , \\ \mu_n = \dfrac{1}{3}(4\mu_d - \mu_u) , \\ \mu_\Lambda = \mu_s , \end{cases}$$

to obtain

$$\begin{cases} \mu_u = \dfrac{1}{5}(\mu_n + 4\mu_p) , \\ \mu_d = \dfrac{1}{5}(\mu_p + 4\mu_n) , \\ \mu_s = \mu_\Lambda , \end{cases}$$

then using the measured values

$$\mu_p = 2.793\mu_{\text{nuclear}} , \qquad \mu_n = -1.913\mu_{\text{nuclear}} , \qquad \mu_\Lambda = -0.613\mu_{\text{nuclear}} .$$

This leads to

Table 13.6 Comparison between measured baryon octet magnetic moments and predictions from the octet wave functions in conjunction with the results of (13.41) [32]

Baryon	Calculated magnetic moment	Predicted [μ_{nuclear}]	Measured [μ_{nuclear}]
Σ^+	$\frac{1}{3}(4\mu_u - \mu_s)$	2.67	2.379 ± 0.020
Σ^0	$\frac{1}{3}(2\mu_u + 2\mu_d - \mu_s)$	0.76	–
Σ^-	$\frac{1}{3}(4\mu_d - \mu_s)$	−1.09	-1.14 ± 0.05
Ξ^0	$\frac{1}{3}(4\mu_s - \mu_u)$	−1.44	-1.250 ± 0.014
Ξ^-	$\frac{1}{3}(4\mu_s - \mu_d)$	−0.49	-0.69 ± 0.04

$$\mu_u = 1.852\mu_{\text{nuclear}}, \qquad \mu_d = -0.972\mu_{\text{nuclear}}, \qquad \mu_s = -0.613\mu_{\text{nuclear}}. \quad (13.41)$$

With these values, we can then go back to Table 13.5 and make predictions for the magnetic moments of the remaining octet baryons. We obtain Table 13.6 [32].

The agreement in the table is reasonable. But there is another thing we can do with the results (13.41). If we assume that the quarks are elementary fermions with spin 1/2, then we have the relations (13.38) for their magnetic moments, so we can actually directly estimate m_u, m_d, and m_s, with the results [32]:

$$m_u = 338 \text{ MeV}/c^2, \qquad m_d = 322 \text{ MeV}/c^2, \qquad m_s = 510 \text{ Mev}/c^2. \quad (13.42)$$

These look similar in order of magnitude to the quark masses given by Griffiths (see Sect. 13.3.1) when the quarks participate in baryons. But are these masses supposed to be rest masses?

It is interesting to read the comment by Greiner and Müller in [32]. They find these masses rather small. The point is that no particles with masses below 10 GeV/c^2 have been found in particle accelerators that could be interpreted as quarks. They should be relatively easy to spot, because of their fractional charge. This is the problem of quark confinement. They suggest that free quarks might have masses of 100 GeV/c^2 or more, and suggest that this does not contradict the statement that $m_{\text{quark}} \sim 330$ MeV, because quarks inside hadrons are bound particles and therefore their mass is reduced by the binding energy. However, what is reduced in a bound state is only the mass of the bound state as compared with the sum of the masses of its components. Surely, the mass of a point particle component could not be changed by its being bound?

Greiner and Müller then shelter behind a proviso: one should really be talking about m_{quark} as an energy eigenvalue, i.e., one should make the switch from mass to energy here. However, there remains our question of Sect. 13.1: why does energy have inertia? There is an avoidance of questions raised by our classical understanding, and also the perduring physical question as to why some objects resist acceleration. And coming back to the specific route to the results (13.42), we have to ask what physical quantity m_{quark} should go into a relation like

$$\mu_{\text{quark}} = \frac{Q_{\text{quark}} e\hbar}{2 m_{\text{quark}} c} \, ,$$

where Q_{quark} takes the value 2/3 for u and $-1/3$ for d and s.

Baryon Masses

Even if flavour were a perfect SU(3) symmetry, all the octet baryons would still not have the same mass, and neither would all the decuplet baryons have the same mass. The three quark flavours cause the quarks in which they reside to have different masses, and this already messes up the flavour SU(3) symmetry. But Table 13.4 shows that Λ and Σ^0 in the baryon octet have the same content with regard to flavour, and they are found experimentally to have different masses, viz., $m_\Lambda = 1114\,\text{MeV}/c^2$ and $m_{\Sigma^0} = 1193\,\text{MeV}/c^2$. And perhaps even less surprisingly, a particle (resonance) like Δ^+ has a different mass to the proton, despite being made up of two u quarks and one d quark.

We know what is going on here from the discussion of the hyperfine splitting of energy levels in the hydrogen atom (see Sect. 13.2.2). There is a spin–spin interaction within the bound system. We shall adopt the same rather cavalier approach to modelling this as we did in Sect. 13.3.4 [see in particular (13.27) on p. 343], making the hypothesis that

$$m(\text{baryon}) = m_1 + m_2 + m_3 + A\left(\frac{\mathbf{S}_1 \cdot \mathbf{S}_2}{m_1 m_2} + \frac{\mathbf{S}_2 \cdot \mathbf{S}_3}{m_2 m_3} + \frac{\mathbf{S}_3 \cdot \mathbf{S}_1}{m_3 m_1}\right) , \qquad (13.43)$$

where m_i and $\mathbf{S}_i$ are the mass and spin of the ith quark in the baryon, and A is a constant to be determined by looking for a best fit (not the same as the one we had for the meson masses). All the comments made with regard to (13.27) are relevant here too.

Since the total angular momentum $\mathbf{J}$ is just $\mathbf{J} = \mathbf{S}_1 + \mathbf{S}_2 + \mathbf{S}_3$ for the case we are considering in this chapter where there is absolutely no orbital angular momentum, we have a useful result

$$J^2 = (\mathbf{S}_1 + \mathbf{S}_2 + \mathbf{S}_3)^2 = S_1^2 + S_2^2 + S_3^2 + 2(\mathbf{S}_1 \cdot \mathbf{S}_2 + \mathbf{S}_2 \cdot \mathbf{S}_3 + \mathbf{S}_3 \cdot \mathbf{S}_1) ,$$

which leads to

$$\mathbf{S}_1 \cdot \mathbf{S}_2 + \mathbf{S}_2 \cdot \mathbf{S}_3 + \mathbf{S}_3 \cdot \mathbf{S}_1 = \frac{1}{2}\hbar^2\left[j(j+1) - \frac{9}{4}\right] ,$$

and therefore,

$$S_1 \cdot S_2 + S_2 \cdot S_3 + S_3 \cdot S_1 = \begin{cases} \dfrac{3}{4}\hbar^2 , & \text{for } j = 3/2 \text{ (decuplet)}, \\[2mm] -\dfrac{3}{4}\hbar^2 , & \text{for } j = 1/2 \text{ (octet)}. \end{cases}$$

This is useful in analysing (13.43) when $m_1 = m_2 = m_3$, a rather special case admittedly. However, we shall take $m_u = m_d$ as a reasonable approximation. This gives the neutron and proton masses as

$$m_p = m_n = 3m_u - \frac{3}{4}\frac{\hbar^2}{m_u^2}A ,$$

while all four Δ resonances in the decuplet get the mass

$$m_\Delta = 3m_u + \frac{3}{4}\frac{\hbar^2}{m_u^2}A ,$$

and Ω gets the mass

$$m_\Omega = 3m_s + \frac{3}{4}\frac{\hbar^2}{m_s^2}A .$$

In the decuplet, the spins of any pair of quarks are parallel, since each baryon has spin 3/2, so

$$(S_1 + S_2)^2 = (S_2 + S_3)^2 = (S_3 + S_1)^2 = 2\hbar^2 .$$

Hence, from results like

$$(S_1 + S_2)^2 = S_1^2 + S_2^2 + 2S_1 \cdot S_2 ,$$

we have

$$S_1 \cdot S_2 = S_2 \cdot S_3 = S_3 \cdot S_1 = \hbar^2/4 .$$

We thus obtain

$$m_{\Sigma^*} = 2m_u + m_s + \frac{\hbar^2}{4}\left(\frac{1}{m_u^2} + \frac{2}{m_u m_s}\right)A$$

and

$$m_{\Xi^*} = m_u + 2m_s + \frac{\hbar^2}{4}\left(\frac{1}{m_s^2} + \frac{2}{m_u m_s}\right)A .$$

By similar, slightly more complicated reasoning, we obtain for the remaining octet masses [10]:

Table 13.7 Estimating baryon octet and decuplet masses with the assumption $m_u = m_d$ and using the spin–spin interaction model (13.43) with $A = 50(2m_u/\hbar)^2$ MeV/c^2

Baryon	Predicted mass [MeV/c^2]	Measured mass [MeV/c^2]
n, p	939	939
Λ	1116	1114
Σ	1179	1193
Ξ	1327	1318
Δ	1239	1232
Σ^*	1381	1384
Ξ^*	1529	1533
Ω	1682	1672

$$m_\Sigma = 2m_u + m_s + \frac{\hbar^2}{4}\left(\frac{1}{m_u^2} - \frac{4}{m_u m_s}\right)A\,,$$

$$m_\Lambda = 2m_u + m_s - \frac{3}{4}\frac{\hbar^2}{m_u^2}A\,,$$

and

$$m_\Xi = 2m_s + m_u + \frac{\hbar^2}{4}\left(\frac{1}{m_s^2} - \frac{4}{m_u m_s}\right)A\,.$$

If we take the quark masses as given in Sect. 13.3.1, then we have only one parameter to play around with, viz., A. It turns out that the best value is $A = 50(2m_u/\hbar)^2$ MeV/c^2, which gives Table 13.7.

Since we do not know where Griffiths obtained the values for m_u, m_d, and m_s when they belong to a baryon, a better approach here might be to adjust these too. Perhaps that is where they come from. In any case, the fit is good, despite the crudity of the model (see the comments due to Griffiths in the meson case). So, not unexpectedly, spin–spin interactions contribute to the inertia of the system as a whole, either positively or negatively. Here we have another example of the classical self-force mechanism in a quantum guise.

But once again, looked at from this angle, and having seen the self-force calculations in Chaps. 6–9, there remains something of a mystery about why these spin–spin interactions should always make just the right self-force contribution, whatever motion the baryon has, to contribute to its inertia in this way. The explanation, if not trivial, is presumably deep.

13.3.6 Coleman–Glashow Relation

The Coleman–Glashow relation relates mass splittings between several pairs of particles within a multiplet like the baryon octet [32]:

$$m_n - m_p + m_{\Xi^-} - m_{\Xi^0} = m_{\Sigma^-} - m_{\Sigma^+} . \tag{13.44}$$

The aim in this section is just to see that this also derives by considering the possible effects of electromagnetic interactions within the multiplet. The discussion will be somewhat qualitative. Full details can be found in [32].

There are two isospin doublets and one isospin triplet in the baryon octet:

$$
\begin{array}{llll}
T = 1/2 & n,\, p & \Delta M \sim 2\,\mathrm{MeV}/c^2 & Y = 1 \\
T = 1 & \Sigma^-,\, \Sigma^0,\, \Sigma^+ & \Delta M \sim 8\,\mathrm{MeV}/c^2 & Y = 0 \\
T = 1/2 & \Xi^-,\, \Xi^0 & \Delta M \sim 7\,\mathrm{MeV}/c^2 & Y = -1
\end{array}
$$

These constitute the rows of the display in Fig. 13.3 left, which uses the value of the isospin component T_3 as the horizontal axis and the value of the hypercharge Y as the vertical axis. Note that Λ is an isospin singlet.

The mass splittings ΔM within each isospin multiplet can be at least partly explained by electromagnetic interactions, precisely because the charge Q is not constant over any given isospin multiplet. They are also partly explained by the different masses of the constituent quarks. These mass splittings correspond to about 1% of the total particle mass in each case, and the isospin symmetry is said to be weakly broken.

States of a given isospin multiplet are transformed into one another by applying the operators

$$T_\pm := \frac{1}{2}(T_1 \pm iT_2) ,$$

where T_1 and T_2 were introduced in Sect. 13.3.3. For example, with $\tau_i := 2T_i$, $i = 1,2,3$, which are just the Pauli matrices, and with $\tau_\pm := (\tau_1 \pm i\tau_2)/2$,

$$\tau_+|n\rangle = |p\rangle , \qquad \tau_-|p\rangle = |n\rangle .$$

The T_i, $i = 1,2,3$, form a basis for the Lie algebra of SU(2).

But the flavour symmetry group we implement here, at least approximately, is SU(3), which contains SU(2) as a subgroup in various ways. The Lie algebra of SU(3) contains the Lie algebra of SU(2) in various ways too. Another version of the Lie algebra of SU(2) is spanned and generated by the three U-spin operators U_i, $i = 1,2,3$, which can be used to analyse the octet representation of SU(3) into U-spin multiplets in an exactly analogous way to the the isospin operators, but with the charge operator Q in the place of the hypercharge operator Y. The octet baryons then correspond to the following three U-spin multiplets:

$$U = 1/2 \qquad \Sigma^-, \Xi^- \qquad \Delta M \sim 124 \text{ MeV}/c^2 \qquad Q = -1$$
$$U = 1 \qquad \text{n}, \Sigma^0, \Xi^0 \qquad \Delta M \sim 374 \text{ MeV}/c^2 \qquad Q = 0$$
$$U = 1/2 \qquad \text{p}, \Sigma^+ \qquad \Delta M \sim 251 \text{ MeV}/c^2 \qquad Q = 1$$

The states of any given U-spin multiplet have the same charge. One can draw a new picture of the baryon octet in the (U, Q) plane, analogous to the one in the (T, Y) plane, only now, the rows are U-spin multiplets, over each of which the charge is constant. Strictly speaking, Σ^0 is not an eigenstate of the U-spin operator $\mathbf{U}^2$, and Λ is not a U-spin singlet. Greiner and Müller show that the $U_3 = 0$ eigenstate of the U-spin triplet is in fact [32]

$$\chi := \frac{1}{2}\left(\Sigma^0 + \sqrt{3}\Lambda\right) ,$$

while the U-spin singlet is

$$\phi := \frac{1}{2}\left(\sqrt{3}\Sigma^0 - \Lambda\right) .$$

So far this is just mathematics. What is the use of it? The reason that charge is constant over each U-spin multiplet is of course that $[Q, U_3] = 0$. The electromagnetic interaction should not break U-spin symmetry, but the strong interaction may do so. Indeed, the mass splittings for the U-spin multiplets given above are of order 100 MeV, or roughly 10% of the mass of each particle. How can we deduce anything here?

Well, let us assume that the mass of a baryon results partly from the strong interaction, which conserves isospin, and partly from the EM interaction, which conserves U-spin. By the latter, it seems reasonable to conclude that EM contributions to the masses are equal within any given U-spin multiplet:

$$\delta m_{\text{p}} = \delta m_{\Sigma^+} , \qquad \delta m_{\text{n}} = \delta m_{\Xi^0} , \qquad \delta m_{\Sigma^-} = \delta m_{\Xi^-} , \qquad (13.45)$$

where δm refers to the EM contribution to the baryon mass in each case. But if there is no other interaction to interfere with the degeneracy of the multiplet, i.e., the common mass across the multiplet, then we expect, for example,

$$m_{\text{n}} - \delta m_{\text{n}} = m_{\text{p}} - \delta m_{\text{p}} , \qquad (13.46)$$

since the left-hand side is just the strong contribution to the neutron mass and the right-hand side is just the strong contribution to the proton mass, and n and p belong to the same isospin multiplet, so their masses are expected to include the same contributions from the strong interaction. Likewise,

$$m_{\Xi^-} - \delta m_{\Xi^-} = m_{\Xi^0} - \delta m_{\Xi^0} , \qquad m_{\Sigma^-} - \delta m_{\Sigma^-} = m_{\Sigma^+} - \delta m_{\Sigma^+} . \qquad (13.47)$$

Combining (13.45), (13.46), and (13.47), we deduce the Coleman–Glashow relation (13.44) given at the beginning of this section. What we find experimentally is

$$m_n - m_p = 1.3 \, \text{MeV}/c^2, \qquad m_{\Xi^-} - m_{\Xi^0} = 6.5 \pm 0.7 \, \text{MeV}/c^2,$$

$$m_{\Sigma^-} - m_{\Sigma^+} = 8.0 \, \text{MeV}/c^2,$$

which accords reasonably with (13.44).

There is thus some evidence for the above assumptions. Let us just make two points:

- When Greiner and Müller assume that the EM contributions to the baryon masses are equal within any given U-spin multiplet, they say explicitly that this implies that the radii of the particles have to be the same. They are clearly thinking of the Coulomb energy in the EM fields around charged spheres. This is fair enough in the context, because their discussion at this point is intended to lie outside the framework of the quark model.

- Once we consider the baryons in the octet to be made up of u, d, and s quarks, the way we view the EM contributions to mass changes somewhat, as stressed on several occasions in this book. The point is that the baryons are now considered to be made of three point particles with different charges. However, within a given U-spin multiplet, two baryons, e.g., n and Ξ^0, are each made up of a trio of quarks with the same set of charges, e.g., n is udd and Ξ^0 is uss, so each contains a particle of charge $+2/3$ and two particles of charge $-1/3$. In this respect, we can understand that the EM contribution to mass should be similar for each baryon in a given U-spin multiplet.

Concerning the second point there, recall the situation with n and p. One could not suppose that the EM contribution to m_n would be zero just because the neutron is electrically neutral.

So what messes up the U-spin symmetry? The operator U may or may not commute with the Hamiltonian H_{strong} of the strong interaction, but in any case, the u, d, and s have different intrinsic masses. So we come back to a problem that awaits us even if we deal with all issues of symmetry breaking in the particle multiplets: why do the quark flavours have different masses? Note on the other hand that, in (13.46) for example, the sum of the quark masses is considered to be roughly equal in both n and p if we assume that $m_u \approx m_d$. Likewise for each relation in (13.47), the sums of the quark masses are roughly equal in the particles on either side of each relation.

13.3.7 Gell-Mann–Okubo Mass Formula

If the flavour SU(3) symmetry were an exact symmetry of the strong interactions, then all states belonging to one SU(3) multiplet would be energetically degenerate, i.e., they would have the same mass. In the baryon octet, the mass splitting is of the order of 10%, so the SU(3) symmetry is more strongly broken than the isospin symmetry by the EM interaction, or whatever else breaks it. Mass splittings in isospin multiplets are of the order of 1%.

Greiner and Müller imagine the Hamiltonian of the strong interaction as splitting into two parts [32, Sect. 8.9]:

$$H_{strong} = H_{ss} + H_{ms} \, ,$$

where H_{ss} is superstrong, meaning that it is SU(3) invariant, and H_{ms} is medium strong, meaning that it breaks SU(3) invariance. The mass of a given baryon is

$$M = \langle H_{strong} \rangle = \langle H_{ss} \rangle + \langle H_{ms} \rangle \, . \tag{13.48}$$

In this picture, the first term on the right-hand side is constant over any multiplet, i.e., H_{ss} sees the multiplet as degenerate, while the second term involving H_{ms} breaks that degeneracy.

This is most easily arranged by having H_{ss} commute with all members of the flavour SU(3) group. Then if any member $|\psi\rangle$ of a given multiplet is an eigenvector of H_{ss} with a given eigenvalue E, since a multiplet carries an irreducible representation of SU(3), so that any member of the multiplet can be obtained as $A|\psi\rangle$ for some $A \in SU(3)$, we find that all members of the multiplet are eigenvalues of H_{ss} with the same eigenvalue E. Even if the member $|\psi\rangle$ of the multiplet is not an eigenvector of H_{ss}, it is easy to see that

$$\langle \varphi | H_{ss} | \varphi \rangle = \langle \psi | H_{ss} | \psi \rangle \, ,$$

for all other φ in the multiplet if H_{ss} commutes with all members of the flavour SU(3) group, precisely because a multiplet carries an irreducible representation of the group. And one way to get H_{ss} to commute with all members of SU(3) is to build it from the Casimir operators, which are matrices commuting with all the group generators, hence with all elements of the group. See [32] for the details of all this theory.

We thus consider H_{ms} as being constructed from generators of the flavour SU(3) symmetry group, rather than just Casimir operators. As a generator is an arbitrary matrix in the Lie algebra of the group, it is clear that no generator can commute with all elements of the group. For the present purposes, we neglect EM mass splitting, so we consider members of an isospin multiplet to have the same mass. We arrange for this by having the symmetry breaking part H_{ms} of the strong interaction Hamiltonian $H_{strong} = H_{ss} + H_{ms}$ commute with T_i, $i = 1, 2, 3$, i.e.,

$$[H_{ms}, T_i] = 0 \, , \qquad i = 1, 2, 3 \, .$$

Now it transpires that only one generator of SU(3) commutes with all the T_i, and this is the hypercharge operator Y, which is just

$$Y = 2(Q - T_3) \, ,$$

where Q is the charge operator. (Note that, for the baryons, the hypercharge is related to the strangeness S by $Y = S + 1$, with the strange convention that the strange quark has strangeness -1.) We thus propose

$$H_{\text{ms}} = bY ,$$

for some constant b.

Multiplets like the baryon octet or decuplet are sets of states $|T T_3 Y\rangle$ that are simultaneously eigenstates of $\mathbf{T}^2$, T_3, and Y. We consider H_{ms} as a perturbation, i.e., we assume $\langle H_{\text{ms}}\rangle \ll \langle H_{\text{ss}}\rangle$, whence the perturbation to the main mass contribution $a := \langle T T_3 Y | H_{\text{ss}} | T T_3 Y\rangle$, constant over the whole multiplet, will be

$$\langle T T_3 Y | H_{\text{ms}} | T T_3 Y\rangle = bY ,$$

according to first order perturbation theory. (We abuse notation here by using the same symbol Y for the hypercharge operator and its eigenvalue.)

So the outcome of this reasoning is a mass formula $M = a + bY$, where a and b are constant over any multiplet, but Y varies. In fact, a is the constant expectation value of H_{ss} in the given multiplet. Now consider the baryon decuplet. We are saying that

$$m_{\Omega^-} = a - 2b , \quad m_{\Xi^*} = a - b , \quad m_{\Sigma^*} = a , \quad m_\Delta = a + b ,$$

whence we would hope to find that

$$m_{\Omega^-} - m_{\Xi^*} = m_{\Xi^*} - m_{\Sigma^*} = m_{\Sigma^*} - m_\Delta . \tag{13.49}$$

Note that this was originally used by Gell-Mann to predict the mass of the as yet undiscovered Ω^-. Earlier, we ignored mass splitting due to EM interactions, but at this point we can take it into account by comparing only decuplet baryons with the same charge, i.e., in the same U-spin multiplet. The result is:

$$m_{\Omega^-} - m_{\Xi^{*-}} = 137 \pm 1 \text{ MeV}/c^2 ,$$
$$m_{\Xi^{*-}} - m_{\Sigma^{*-}} = 148 \pm 1 \text{ MeV}/c^2 ,$$
$$m_{\Sigma^{*-}} - m_{\Delta^-} = 148 \pm 5 \text{ MeV}/c^2 .$$

This is a good agreement, but the idea fails for the baryon octet. The problem is that $Y_{\Sigma^0} = Y_\Lambda$, so it would predict $m_{\Sigma^0} = m_\Lambda$, whereas in fact $m_{\Sigma^0} - m_\Lambda = 77$ MeV.

To obtain a better model, we observe that both $\mathbf{T}^2$ and Y^2 commute with T_i, $i = 1, 2, 3$, so we make the hypothesis that

$$H_{\text{ms}} = bY + c\mathbf{T}^2 + dY^2 ,$$

for some constants b, c, and d. In states $|T T_3 Y\rangle$, this has expectation value

$$M = a + bY + cT(T+1) + dY^2 , \tag{13.50}$$

with four parameters to fit now. The argument is much less convincing now. The more parameters we allow ourselves, the easier it is to get a good fit.

However, (13.50) also encounters a problem. In fact, it destroys the relations (13.49) that worked so well for the decuplet, for most choices of c and d. What one does is to demand that, in the decuplet, $cT(T+1)+dY^2$ have the form $x+yY$, for some constants x and y. This can be done [32] by making the choice $d = -c/4$, so this constraint removes one of our parameters. The result is the well known Gell-Mann–Okubo mass formula

$$M = a + bY + c\left[T(T+1) - \frac{1}{4}Y^2\right], \qquad (13.51)$$

where the three parameters are supposed to be constant over any given multiplet. For the baryon octet, it implies

$$m_N = a + b + \frac{1}{2}c, \quad m_\Xi = a - b + \frac{1}{2}c, \quad m_\Sigma = a + 2c, \quad m_\Lambda = a,$$

where N is either n or p, and hence,

$$\frac{1}{2}(m_N + m_\Xi) = \frac{3}{4}m_\Lambda + \frac{1}{4}m_\Sigma.$$

Once again, to minimise EM effects, we apply this to the neutral particles and find

$$\frac{1}{2}(m_n + m_{\Xi^0}) = 1127.1 \pm 0.7 \,\text{MeV}/c^2, \quad \frac{3}{4}m_\Lambda + \frac{1}{4}m_{\Sigma^0} = 1134.8 \pm 0.2 \,\text{MeV}/c^2,$$

which is a good result, because the 7.7 MeV discrepancy is much lower than the 100 MeV mass splitting within the baryon octet.

A point in favour of (13.51) is that constants a, b, and c can be found to make it fit other SU(3) multiplets, and not just baryon multiplets, but also meson multiplets. What is interesting is that the formula is modified for mesons: it does not apply to the masses, but to the squares of the masses. Greiner and Müller justify this by the fact that baryons are fermions satisfying the Dirac equation, which contains linear terms in the energy, while mesons are bosons satisfying the Klein–Gordon equation, which contains squares of masses. Griffiths just says that the reasons remain something of a mystery [10, Exercise 1.5].

There are two points to make about the Gell-Mann–Okubo mass relation in the context of the present book:

- The analysis here makes no reference to quark components of the particles, or any other components. In this sense, the particles are not treated explicitly as bound states.
- The key relation is (13.48), because it shows how inertial mass is understood for these particles as the expectation value of interaction energies. Since an interaction requires several interacting parts, these particles *are* being treated implicitly as bound states. This is therefore a quantum theoretical version of the self-force idea.

This mass relation shows just how far one can go by heuristic considerations involving symmetries, particularly in the context of the quantum theory, without explicitly modelling any internal structure.

13.3.8 Gürsey–Radicati Mass Formula

This is the generalisation of the Gell-Mann–Okubo mass formula to the SU(6) flavour–spin quark model discussed earlier. Here the octet and decuplet baryons are considered explicitly as being made up of quarks, although the masses of the different quark flavours making up the bound states remain unexplained. Furthermore, no very explicit use of the quark structure is made. One simply adds a spin term to the previous result (13.51) to allow for the formula to work for both octet and decuplet baryons at the same time:

$$M = a + bY + c \left[T(T+1) - \frac{1}{4}Y^2 \right] + dS(S+1) . \tag{13.52}$$

We now have an extra parameter d, but the formula has to work with a, b, c, and d constant over both the baryon octet and the baryon decuplet.

The rationale here is that the octet (spin 1/2) and the decuplet (spin 3/2) constitute a 56-plet of the SU(6) group (containing 2×8 octet states and 4×10 decuplet states), and the symmetry breaking part H_{ms} of the Hamiltonian should have the form

$$H_{\mathrm{ms}} = a + bY + c \left(\mathbf{T}^2 - \frac{1}{4}Y^2 \right) + d\mathbf{S}^2 ,$$

so that the last term distinguishes between octet and decuplet baryons. (Recall that the octet baryons have $\mathbf{S}^2 = 3\hbar^2/4$, while the decuplet baryons have $\mathbf{S}^2 = 15\hbar^2/4$.)

By considering the experimental values for the masses of n, Λ, Σ^0, Σ^{*0}, and Ξ^{*0}, we obtain [32]

$$a = 1066.6 \, \mathrm{MeV}/c^2 , \qquad b = -196.1 \, \mathrm{MeV}/c^2 ,$$

$$c = 38.8 \, \mathrm{MeV}/c^2 , \qquad d = 65.3 \, \mathrm{MeV}/c^2 ,$$

which imply $m_{\Xi^0} = 1331 \, \mathrm{MeV}/c^2$ (experimental value 1318 MeV/c^2), $m_{\Delta^0} = 1251.2 \, \mathrm{MeV}/c^2$ (experimental value 1232 MeV/c^2), and $m_{\Omega^-} = 1664.9 \, \mathrm{MeV}/c^2$ (experimental value 1672.4 MeV/c^2).

Despite the specific reference to the SU(6) model, the comments concerning the Gell-Mann–Okubo mass relation apply here too. We only pay lip service to the quark components. The force of the argument comes from symmetry considerations. The same game can be played when charm is included, with an SU(8) mass formula [32, Sect 11.7]. For the baryons, one just adds a constant multiple of the charm operator to (13.52). As far as charmed baryons have been identified, it seems to work reasonably well. The corresponding formula for mesons, with the mysterious squared masses in it, has to be built with a square of the charm operator, and the results are mediocre.

13.4 Higgs Mechanism

What do we know about the masses of those particles that are considered to be truly elementary today? Why do different quark flavours have different masses? Why are there generations of quarks: (u,d), (s,c), and (b,t), getting more and more massive? Why are there generations of leptons: e, μ, and τ, getting more and more massive? Why do the intermediate vector bosons $W^\pm$ and Z have the masses they do? Some of the answers, or partial answers, are provided by the Higgs mechanism to be outlined here. Once again we closely follow the masterful account in [10].

The starting point for modern field theory is the Lagrangian density (usually just called the Lagrangian) and the reader will need to have some knowledge of this to feel comfortable with the potted version that follows. The simplest such item is perhaps the Klein–Gordon Lagrangian for a scalar (spin 0) field ϕ:

$$\mathcal{L} = \frac{1}{2}(\partial_\mu\phi\,\partial^\mu\phi) - \frac{1}{2}\left(\frac{mc}{\hbar}\right)^2\phi^2\,,$$

where m is the mass of the particles supposed to correspond to the field, c is the speed of light (often set to 1), and $\hbar$ is Planck's constant (also often set to 1). This is the appropriate Lagrangian precisely because the Euler–Lagrange equations

$$\partial_\mu\left[\frac{\partial\mathcal{L}}{\partial(\partial_\mu\phi)}\right] - \frac{\partial\mathcal{L}}{\partial\phi} = 0$$

deliver the well-known Klein–Gordon equation for the field, viz.,

$$\partial_\mu\partial^\mu\phi + \left(\frac{mc}{\hbar}\right)^2\phi = 0\,. \tag{13.53}$$

The Dirac Lagrangian for a spin 1/2 field ψ is

$$\mathcal{L} = i\hbar c\overline{\psi}\gamma^\mu\partial_\mu\psi - (mc^2)\overline{\psi}\psi\,, \tag{13.54}$$

where γ^μ are the Dirac matrices and ψ is a 4-component object with $\overline{\psi} := \psi^\dagger\gamma^0$. Here there is an Euler–Lagrange equation for each component of ψ or $\overline{\psi}$, and ψ and $\overline{\psi}$ are treated as independent. The equations

$$\partial_\mu\left[\frac{\partial\mathcal{L}}{\partial(\partial_\mu\overline{\psi})}\right] - \frac{\partial\mathcal{L}}{\partial\overline{\psi}} = 0$$

deliver the usual Dirac equation for ψ, viz.,

$$i\gamma^\mu\partial_\mu\psi - \frac{mc}{\hbar}\psi = 0\,, \tag{13.55}$$

while the Euler–Lagrange equations

$$\partial_\mu \left[\frac{\partial \mathscr{L}}{\partial (\partial_\mu \psi)} \right] - \frac{\partial \mathscr{L}}{\partial \psi} = 0$$

deliver the adjoint of the Dirac equation for $\overline{\psi}$, viz.,

$$i\partial_\mu \overline{\psi} \gamma^\mu + \frac{mc}{\hbar} \overline{\psi} = 0 .$$

The other kind of field commonly considered is the vector or spin 1 field, for which the starting point is the Proca Lagrangian

$$\mathscr{L} = -\frac{1}{16\pi} (\partial^\mu A^\nu - \partial^\nu A^\mu)(\partial_\mu A_\nu - \partial_\nu A_\mu) + \frac{1}{8\pi} \left(\frac{mc}{\hbar} \right)^2 A^\mu A_\mu .$$

This is designed to deliver what is considered to be the right equation for such a field, viz., the Proca equation

$$\partial_\mu (\partial^\mu A^\nu - \partial^\nu A^\mu) + \left(\frac{mc}{\hbar} \right)^2 A^\nu = 0 . \tag{13.56}$$

It is useful to introduce the antisymmetric second rank tensor

$$F^{\mu\nu} := \partial^\mu A^\nu - \partial^\nu A^\mu .$$

The Lagrangian and field equation are then

$$\mathscr{L} = -\frac{1}{16\pi} F^{\mu\nu} F_{\mu\nu} + \frac{1}{8\pi} \left(\frac{mc}{\hbar} \right)^2 A^\mu A_\mu , \qquad \partial_\mu F^{\mu\nu} + \left(\frac{mc}{\hbar} \right)^2 A^\nu = 0 . \tag{13.57}$$

Note that the electromagnetic field is a massless Proca field, obtained by putting $m = 0$ in the above. At least, that gives us an EM field in the absence of any sources. And in fact, all three of the above fields are called free fields, because there are no sources around, and no interactions. A massless vector field with a source, like the EM field in the presence of a source J^μ, has the Lagrangian

$$\mathscr{L} = -\frac{1}{16\pi} F^{\mu\nu} F_{\mu\nu} - \frac{1}{c} J^\mu A_\mu ,$$

precisely because the Euler–Lagrange equations lead to the expected field equation

$$\partial_\mu F^{\mu\nu} = \frac{4\pi}{c} J^\nu .$$

Obviously, this implies that $\partial_\nu J^\nu = 0$, so only sources with this property would be appropriate here.

Notice how we talk about fields, but these equations (13.53), (13.55), and (13.56) each refer to a mass m. Although ostensibly about fields, quantum field theory is used to do particle physics, and the (unrenormalised) particle masses are fed in by hand. Given a Lagrangian, it is not always immediately obvious what the mass of the corresponding particle is. Consider for example [10]

$$\mathscr{L} = \frac{1}{2}(\partial_\mu \phi)(\partial^\mu \phi) + e^{-(\alpha\phi)^2},$$

for some constant $\alpha \in \mathbb{R}$. We have to expand the exponential and read off the mass term as the one that goes as the square of ϕ. So the Lagrangian is in fact more usefully written in the form

$$\mathscr{L} = \frac{1}{2}(\partial_\mu \phi)(\partial^\mu \phi) + 1 - \alpha^2 \phi^2 + \frac{1}{2}\alpha^4 \phi^4 + \cdots.$$

The constant 1 can be dropped because it does not affect the Euler–Lagrange equations. The mass of the particle associated with this field is taken to be $m = \sqrt{2}\alpha\hbar/c$. The terms in higher powers of ϕ are self-interaction terms. A term in ϕ^n, $n \in \mathbb{N}$, would lead to a vertex with n lines coming out of it in the Feynman perturbative expansion.

Another illustration of this method for reading off the particle mass, and a classic for introducing the idea of symmetry breaking, as we shall see shortly, is [10]

$$\mathscr{L} = \frac{1}{2}(\partial_\mu \phi)(\partial^\mu \phi) + \frac{1}{2}\mu^2 \phi^2 - \frac{1}{4}\lambda^2 \phi^4, \qquad (13.58)$$

where μ and λ are positive real constants. The mass of the corresponding particle is not $\mu\hbar/c$, however, because the mass term has to go as $-\phi^2$, not $+\phi^2$. Actually, if there were no other term after that in the Lagrangian, we would be in trouble. The Lagrangian would not be taken to describe any possible field. In the above case, what one can do is to think of $\mathscr{L}$ as having the form $\mathscr{L} = \mathscr{T} - \mathscr{U}$, the difference between a kinetic term and a potential term, just as one finds in classical particle dynamics (although these are really Lagrangian densities here). In the present case,

$$\mathscr{U} = -\frac{1}{2}\mu^2 \phi^2 + \frac{1}{4}\lambda^2 \phi^4.$$

The idea now is to note that this quartic in ϕ has minima at $\phi = \pm\mu/\lambda$, introduce a new field variable $\eta := \phi \pm \mu/\lambda$, and rewrite the Lagrangian in terms of this new field:

$$\mathscr{L} = \frac{1}{2}(\partial_\mu \eta)(\partial^\mu \eta) - \mu^2 \eta^2 \pm \mu\lambda\eta^3 - \frac{1}{4}\lambda^2\eta^4 + \frac{1}{4}\left(\frac{\mu^2}{\lambda}\right)^2. \qquad (13.59)$$

We take η to be the field actually described by $\mathscr{L}$, with an associated particle of mass $m = \sqrt{2}\mu\hbar/c$, and cubic and quartic self-interactions, corresponding to vertices with three and four lines emerging from them in the Feynman perturbative expansion.

We think of the minima of $\mathscr{U}$ as being something like ground states, and the Feynman diagrams as terms in an expansion about one of these ground states. Put like this, it all sounds a bit vague. In order to understand it, one needs to plough through the heavy machinery of quantum field theory, plagued by its own ontological vagaries. But the Lagrangian method is a great short cut through all that, because one can read off Feynman rules directly (with a little care at times), as Griffiths explains rather elegantly [10]. So for a potted version like this, we shall consider that as satisfactory.

We have lost some symmetry in going from (13.58) to (13.59). Or rather, $\mathscr{L}$ is invariant under $\phi \rightarrow -\phi$, while it is not invariant under $\eta \rightarrow -\eta$. We are going to call this spontaneous symmetry breaking. The set of all ground states, viz., $\{-\mu/\lambda, +\mu/\lambda\}$, is still symmetric, but we have to choose one of them to be the actual ground state. What is spontaneous here is just the way the field is going to drop into one of these ground states for reasons that go beyond our predictive ability, just as many other physical systems will flip into one of several available ground states when the conditions are right. One thinks of a ferromagnet, in which all the spins line up when environmental conditions will allow it. This then picks out some direction in space, and 3D rotational symmetry is lost.

The symmetry under $\phi \rightarrow -\phi$ is a discrete symmetry. A Lagrangian with a continuous symmetry is

$$\mathscr{L} = \frac{1}{2}(\partial_\mu \phi_1)(\partial^\mu \phi_1) + \frac{1}{2}(\partial_\mu \phi_2)(\partial^\mu \phi_2) + \frac{1}{2}\mu^2(\phi_1^2 + \phi_2^2) - \frac{1}{4}\lambda^2(\phi_1^2 + \phi_2^2)^2 .$$

What we mean by this is that the symmetry group for this Lagrangian, the group of rotations in (ϕ_1, ϕ_2) space, contains a continuum of elements, while the one we had before contains only two. The potential energy function here is

$$\mathscr{U} = -\frac{1}{2}\mu^2(\phi_1^2 + \phi_2^2) + \frac{1}{4}\lambda^2(\phi_1^2 + \phi_2^2)^2 . \tag{13.60}$$

Now any pair $(\phi_{1\,\text{min}}, \phi_{2\,\text{min}})$ minimising $\mathscr{U}$ satisfies

$$\phi_{1\,\text{min}}^2 + \phi_{2\,\text{min}}^2 = \frac{\mu^2}{\lambda^2} ,$$

so there are infinitely many possible ground states for our field to flip into. Indeed, they form a circle in (ϕ_1, ϕ_2) space. We note that the set of all possible ground states has the rotational symmetry of the Lagrangian itself.

The symmetry is spontaneously broken by the system when it selects one minimum to be the actual ground state or vacuum. In our notation, this ground state will be

$$\phi_{1\,\text{min}} = \frac{\mu}{\lambda} , \qquad \phi_{2\,\text{min}} = 0 .$$

We rewrite the Lagrangian in terms of new fields

$$\eta := \phi_1 - \frac{\mu}{\lambda} , \qquad \xi := \phi_2 ,$$

which gives

$$\mathscr{L} = \left[\frac{1}{2}(\partial_\mu \eta)(\partial^\mu \eta) - \mu^2 \eta^2\right] + \left[\frac{1}{2}(\partial_\mu \xi)(\partial^\mu \xi)\right]$$
$$- \left[\mu\lambda(\eta^3 + \eta\xi^2) + \frac{\lambda^2}{4}(\eta^4 + \xi^4 + 2\eta^2\xi^2)\right] + \frac{\mu^4}{4\lambda^2}.$$

Like the fields ϕ_1 and ϕ_2, the new fields are coupled by the terms in the second line, and there are also interactions and self-interactions. For example, the term in $\eta\xi^2$ would lead to vertices in the Feynman diagrams with one η and two ξ lines coming out of the them, while the term in η^3 would give vertices with three η lines coming out of them. But from the first line, we read off the masses of the corresponding particles. The particles for the η field have mass $m_\eta = \sqrt{2}\mu\hbar/c$, while the particles for the ξ field are massless.

Note that $\mathscr{L}$ no longer looks very symmetrical, but it is. We have just hidden the symmetry. What is not symmetrical is the physical situation, in which the actual ground state of the system lies at some point on the circle of possibilities. Note also that this exemplifies Goldstone's theorem: when a continuous global symmetry is spontaneously broken, we always obtain some massless scalar particles, known as Goldstone bosons. This looks bad for the idea of spontaneous symmetry breaking, because there are no massless bosons. At least, it seems unlikely, because such a thing ought to show up in the form of missing energy, in contrast to a massive boson, which might just be too massive to have been produced yet.

Now we come at last to the Higgs mechanism, which enters the scene when we apply the idea of spontaneous symmetry breaking to a local symmetry, or gauge symmetry. For this, we need a Lagrangian that is invariant under a spacetime dependent symmetry group. We can use the last example to show how this works, following Griffiths [10], although this presentation is rather standard. First we combine the two real fields ϕ_1 and ϕ_2 into one complex field

$$\phi := \phi_1 + i\phi_2,$$

whence

$$\phi^*\phi = \phi_1^2 + \phi_2^2.$$

With this change of notation, the Lagrangian takes the neat form

$$\mathscr{L} = \frac{1}{2}(\partial_\mu\phi)^*(\partial^\mu\phi) + \frac{1}{2}\mu^2\phi^*\phi - \frac{1}{4}\lambda^2(\phi^*\phi)^2.$$

The SO(2) symmetry we had previously becomes a U(1) symmetry, i.e., for any $e^{i\theta} \in U(1)$, the Lagrangian is invariant under the phase transformation

$$\phi \longrightarrow e^{i\theta}\phi.$$

We shall now tamper with the Lagrangian to make it invariant under local U(1) transformations, i.e., under transformations of the type

$$\phi \longrightarrow e^{i\theta(x)}\phi ,$$

where $\theta(x)$ is now allowed to be any reasonable function of the spacetime coordinates x.

We have not motivated this move, through lack of space. Basically, one does this by analogy with the fact that electromagnetic fields can be built into gauge invariant theories of charged particles. This in turn comes back to the gauge invariance of Maxwell's theory mentioned in Sect. 2.1. It transpires that the whole theory of electromagnetism merely constitutes what is known as a gauge field, of the kind we are about to introduce for the above example. Fortunately, there are many good books to describe this in more detail, and [10] is an obvious choice.

In the present case, we first observe why the above Lagrangian is *not* invariant under the spacetime dependent transformation. The problem is the partial derivatives in $\mathscr{L}$. The reader can check that $\mathscr{L}$ changes when $\phi \to e^{i\theta(x)}\phi$, except in special cases, e.g., θ a constant function. But there is a very similar Lagrangian that does not change. We obtain this by making the replacement

$$\partial_\mu \longrightarrow D_\mu := \partial_\mu + i\frac{q}{\hbar c}A_\mu , \tag{13.61}$$

where A^μ is a massless vector field and q is a constant. The new Lagrangian is

$$\mathscr{L} = \frac{1}{2}(D_\mu\phi)^*(D^\mu\phi) + \frac{1}{2}\mu^2\phi^*\phi - \frac{1}{4}\lambda^2(\phi^*\phi)^2 - \frac{1}{16\pi}F^{\mu\nu}F_{\mu\nu} , \tag{13.62}$$

including the usual term at the end for a massless vector field, where $F_{\mu\nu} := \partial_\mu A_\nu - \partial_\nu A_\mu$. We must of course mention how A^μ changes under our spacetime dependent transformations, and the reader will already have guessed that it must change according to

$$A^\mu \longrightarrow A^\mu - \frac{\hbar c}{q}\partial^\mu\theta . \tag{13.63}$$

Compare with the gauge transformations of the EM four-potential in Sect. 2.1. It is an easy, although somewhat lengthy exercise to check that the Lagrangian in (13.62) is indeed gauge invariant, i.e., does not change when we make the above changes. This is called local U(1) symmetry.

We now proceed to hide this hard-earned symmetry by exactly the same process discussed above. So we introduce new fields

$$\eta := \phi_1 - \frac{\mu}{\lambda} , \qquad \xi := \phi_2 ,$$

and obtain the new version

$$\mathscr{L} - \left[\frac{1}{2}(\partial_\mu \eta)(\partial^\mu \eta) - \mu^2 \eta^2 \right] + \left[\frac{1}{2}(\partial_\mu \xi)(\partial^\mu \xi) \right]$$

$$+ \left[-\frac{1}{16\pi} F^{\mu\nu} F_{\mu\nu} + \frac{1}{2} \left(\frac{q}{\hbar c} \frac{\mu}{\lambda} \right)^2 A_\mu A^\mu \right]$$

$$+ \left\{ \frac{q}{\hbar c} \left[\eta(\partial_\mu \xi) - \xi(\partial_\mu \eta) \right] A^\mu + \frac{\mu}{\lambda} \left(\frac{q}{\hbar c} \right)^2 \eta(A_\mu A^\mu) \right.$$

$$+ \frac{1}{2} \left(\frac{q}{\hbar c} \right)^2 (\xi^2 + \eta^2)(A_\mu A^\mu) - \lambda\mu(\eta^3 + \eta\xi^2) - \frac{1}{4}\lambda^2(\eta^4 + 2\eta^2\xi^2 + \xi^4) \right\}$$

$$+ \frac{\mu}{\lambda} \frac{q}{\hbar c} (\partial_\mu \xi) A^\mu + \left(\frac{\mu^2}{2\lambda} \right)^2 . \tag{13.64}$$

So we still have a scalar particle η of mass $m = \sqrt{2}\mu\hbar/c$ and the massless Goldstone boson ξ, but now the previously massless gauge field appears suddenly to have acquired a mass! Indeed, comparing the second line of (13.64) with (13.57),

$$m_A = 2\sqrt{\pi} \frac{q\mu}{\lambda c^2} .$$

This is quite an achievement, because gauge fields like A^μ, or their counterparts in Yang–Mills field theories which deal in more sophisticated symmetry groups than U(1), are *always* massless. This was one of the problems in devising the electroweak theory, also known as the Glashow–Weinberg–Salaam (GWS) theory of electromagnetic and weak interactions. It was expected to be a gauge theory, but the vector bosons mediating the weak interactions had to be massive, partly to explain the weakness of these interactions, and partly to explain why, unlike the photon, they had never been observed.

Where did the gauge field mass term come from? In the original form of the Lagrangian, there was a term $\phi^* \phi A_\mu A^\mu$ which would have corresponded to a Feynman vertex with two A lines and two ϕ lines emerging from it. But when the ground state is fixed at a minimum of the potential, the ϕ_1 field shifts, and this is what yields the mass term for the gauge field. So ultimately, the masses of the gauge fields come from an interaction with the Higgs field ϕ which only reveals itself when nature has selected some ground state for the Higgs field to drop into.

Returning to the above expression for the Lagrangian again, the last three lines describe interactions between the three fields A^μ, η, and ξ, or self-interactions. We still have the massless Goldstone boson ξ that has never been observed and there is the dubious term on the last line of the Lagrangian proportional to $(\partial_\mu \xi) A^\mu$, which looks like a Feynman vertex in which a ξ turns into an A, or vice versa. The latter suggests an incorrect identification of the fundamental particles.

So there is one last piece of trickery in the Higgs mechanism: we use the gauge invariance of $\mathscr{L}$ to transform the field ξ away completely. This is called gauge fixing. But in contrast to the gauge fixing of Sect. 2.1, this one has presumably been done for us by nature, since we do not see the ξ particle (or its counterpart in some more realistic theory). Explicitly, we make the gauge transformation

$$\phi \longrightarrow \phi' = (\cos\theta + i\sin\theta)(\phi_1 + i\phi_2)$$
$$= (\phi_1\cos\theta - \phi_2\sin\theta) + i(\phi_1\sin\theta + \phi_2\cos\theta),$$

but then make the choice

$$\theta = -\arctan\frac{\phi_2}{\phi_1}, \tag{13.65}$$

which makes ϕ' real. Hence, $\xi' := \phi_2' = 0$, as required. The gauge field A^μ transforms to some new field according to (13.63), and dropping the primes, the new form of the Lagrangian is

$$\mathcal{L} = \left[\frac{1}{2}(\partial_\mu\eta)(\partial^\mu\eta) - \mu^2\eta^2\right] + \left[-\frac{1}{16\pi}F^{\mu\nu}F_{\mu\nu} + \frac{1}{2}\left(\frac{q}{\hbar c}\frac{\mu}{\lambda}\right)^2 A_\mu A^\mu\right]$$
$$+ \left\{\frac{\mu}{\lambda}\left(\frac{q}{\hbar c}\right)^2 \eta(A_\mu A^\mu) + \frac{1}{2}\left(\frac{q}{\hbar c}\right)^2 \eta^2(A_\mu A^\mu) - \lambda\mu\eta^3 - \frac{1}{4}\lambda^2\eta^4\right\} + \left(\frac{\mu^2}{2\lambda}\right)^2.$$

$$\tag{13.66}$$

The very last term is just a constant and can be ignored, because it does not affect the Euler–Lagrange equations.

So for some reason, nature chooses this gauge, the Goldstone boson does not really exist (or rather, it is there, but the corresponding field is identically zero), and we are left with the Higgs field η, which nobody has yet observed, and the much desired massive gauge field A^μ. Put like that, it does not sound very convincing. But this is just a toy model to illustrate the mechanism.

The GWS theory refers to spinor fields to represent the various fermions that undergo weak and electromagnetic interactions, and there are three massive vector fields and one massless vector field. The former correspond to $W^\pm$ and Z and the latter to the photon. What is more, the masses of the W and Z particles are related, and can be predicted with knowledge of another parameter. Everyone has heard about the tremendous success of this theory, with the W and Z being found in particle accelerator experiments at the predicted masses. So it is a theory that has to be taken seriously.

What exactly are the down sides of this model? Here is a short list:

- In the above, and in GWS theories of electroweak unification, we have to guess the mass and self-interaction of the Higgs field. This amounts to trying out different forms for $\mathcal{U}$ in (13.60) and then testing what they predict in the accelerator. Note, however, that the potential has to be quartic here for the resulting theory to be renormalisable, so there are some constraints.
- The Higgs particle has not been observed yet, and we do not really know how much energy may be required to produce it, because our theory does not tell us its mass.

• It is not obvious why nature chooses this gauge, or put another way, why the Goldstone boson is unphysical.

An up-to-date review of the Higgs particle is available at the website of the Particle Data Group [11].

It is interesting to see how Griffiths defends the last point [10]. He stresses that the Lagrangians in (13.62), (13.64), and (13.66) all describe the same physical system. We have merely selected a convenient gauge in (13.65) and rewritten the fields as fluctuations about a particular ground state. The latter statement refers to the fact that the Feynman diagrams correspond to a perturbative expansion, but the link with real physics is somewhat obscured here by the complexity of quantum field theory. Still, it sounds reasonable enough. Put another way, we do not have to choose a particular gauge here, but if we do not, the theory contains an unphysical ghost particle. According to Griffiths, *we* eliminate the latter by this choice, but one could also say that nature chooses the gauge, for some reason, because we do not really know what makes the Goldstone boson unphysical.

A further gloss here is obtained by talking about degrees of freedom. While the gauge field is massless, it has two degrees of freedom in the form of two possible transverse polarisations, for example, but when it acquires a mass, it also acquires another degree of freedom, in the form of a longitudinal polarisation. This extra freedom is inherited from the Goldstone boson, which subsequently goes out of service. This does not explain, however, but merely describes in a new language.

On several occasions throughout this book, we have been saying that the Higgs field is also needed to explain the inertial masses of the quarks and leptons, whereas so far we have only seen very roughly how it might explain the masses of the vector bosons mediating the weak interactions, via the Higgs mechanism. We cannot possibly spell out all the details of that here, but we can certainly sketch how this is done through the Lagrangian approach to field theory.

The idea is this. In the Lagrangian, we put a term like

$$i\hbar c \overline{\psi} \gamma^\mu \partial_\mu \psi \,,$$

for each quark or lepton, where ψ is a Dirac spinor field, as in (13.54). This corresponds to a massless particle. But we also include a Higgs field ϕ and an interaction term of the form

$$\mathscr{L}_{\text{int}} = -\alpha \overline{\psi} \psi \phi \,,$$

for each spinor field, where α is a coupling constant. When the symmetry is spontaneously broken by the Higgs field selecting one of its ground states, the spinor field Lagrangian acquires a mass term. It is as simple as that.

In a sense, the Lagrangians provide a kind of code for building field theories. We reorganise their terms until we can interpret them physically. But from there to the idea that moving in a Higgs field is like moving through honey, there is a long and arduous trail through the vicissitudes of quantum field theory, which is also rather like moving through a viscous fluid. For the moment, this is the price to pay for a spectacularly successful set of theories that we comfortably call the Standard Model.

13.5 Alternatives

The tone here is critical, as befits any scientific discussion. The reader who consults the review of the Higgs field available at the website of the Particle Data Group [11] will find that the problems mentioned here are taken very seriously indeed, and that there is no shortage of potential solutions to them, all aiming to maintain the Higgs mechanism and its offshoots as a healthy denizen of the particle garden.

But let us remember also the possibility that it may all be wrong. Anyone investigating this line of reasoning must support the burden of replacing all the successes of the Standard Model by an alternative explanation, which is no easy task. The first problem would be to explain the masses of the intermediate vector bosons $W^{\pm}$ and Z. Note here that there are alternatives. One such is the spin gauge theory of Chisholm and Farwell [38–43].

It will be worth sketching the way fermion masses are treated here. Note, however, that this can only be an outline sketch, because the spin gauge theories are all built inside Clifford algebras (also known as geometric algebras), and we cannot reduce such a vast and beautiful mathematical paradigm to the space of a few paragraphs. But the reader can nevertheless get a qualitative glimpse beyond the Higgs horizon.

For every finite dimensional vector space with pseudometric, such as Minkowski spacetime with Lorentz pseudometric, we can construct a finite-dimensional algebra containing the vector space in a way we shall illustrate only for the spacetime algebra. We start with a pseudo-orthonormal basis $\gamma_0, \gamma_1, \gamma_2, \gamma_3$ for an inertial frame, i.e., such that

$$\gamma_\mu \cdot \gamma_\nu = \eta_{\mu\nu} \, ,$$

and generate an algebra formally using these four elements, but under the condition

$$\{\gamma_\mu, \gamma_\nu\} := \gamma_\mu \gamma_\nu + \gamma_\nu \gamma_\mu = 2\eta_{\mu\nu} \, . \tag{13.67}$$

We take the spacetime metric to be $\eta = (1, -1, -1, -1)$. The condition (13.67) thus tells us that any two different elements of the basis anticommute, that γ_0 squares to unity, and that γ_i squares to -1 for $i = 1, 2, 3$. The resulting algebra is a 16D vector space spanned by:

I	$\{\gamma_\mu\}$	$\{\gamma_\mu \wedge \gamma_\nu\}$	$\{\gamma_5\gamma_\mu\}$	γ_5
1 scalar	4 vectors	6 bivectors	4 trivectors	1 pseudoscalar

where I is the unit of the algebra, the pseudoscalar is defined by

$$\gamma_5 := \gamma_0\gamma_1\gamma_2\gamma_3 \, ,$$

and the wedge product is defined by

$$\gamma_\mu \wedge \gamma_\nu := \gamma_\mu \gamma_\nu - \gamma_\nu \gamma_\mu \ .$$

This Clifford algebra is denoted by $C_{1,3}$. Today there are several good references on this subject and every physics undergraduate would be advised to know something about Clifford algebra (or geometric algebra) techniques [44, 45].

We can illustrate one of the key ideas of the spin gauge theories with the following example. We begin with the Lagrangian for a massive Dirac electron represented by the Dirac spinor ψ, and an electromagnetic field represented by A^μ [42]:

$$\mathscr{L} = \frac{1}{2}\left[\overline{\psi}i\gamma^\mu(\partial_\mu - ieA_\mu)\psi + \text{h.c.}\right] - m\overline{\psi}\psi \ . \tag{13.68}$$

The four objects γ^μ are just the linear combinations $\gamma^\mu := \eta^{\mu\nu}\gamma_\nu$ of the vectors γ_ν generating the above Clifford algebra $C_{1,3}$, and the spinors themselves are taken to be elements of some minimal left ideal in the algebra, which thus carries an irreducible representation of the algebra. Everything in these theories belongs to some Clifford algebra, including all the values of the Lagrangian density, which is thus a Clifford-algebra-valued function on spacetime. This allows one to add together things that would otherwise be too disparate to add together. In other words, the beauty of it is precisely that it provides a unified framework for building theories.

The expression $\partial_\mu - ieA_\mu$ builds in what is usually known as the minimal coupling between the EM field represented by A_μ and the spinor matter field ψ. It is in fact a covariant derivative in the language of gauge theories [compare with (13.61)]. The explanation for this terminology is simple: in contrast to the ordinary partial derivative of the given field, the covariant derivative of the field transforms in exactly the same way as the field itself under the local gauge transformations. This explains how it can be used to build a gauge invariant Lagrangian, like the one in (13.62). The notion of covariant derivative is key to understanding any gauge theory.

Now in the algebra $C_{1,3}$, we have

$$\gamma^\mu\gamma_\mu = 4I \ ,$$

where $\gamma^\mu := \eta^{\mu\nu}\gamma_\nu$ as mentioned a moment ago. This is used to factorise the mass term $m\overline{\psi}\psi$ in (13.68), to obtain the new expression

$$\mathscr{L} = \frac{1}{2}\left\{\overline{\psi}i\gamma^\mu\left[I(\partial_\mu - ieA_\mu) + \frac{1}{4}mi\gamma_\mu\right]\psi + \text{h.c.}\right\} \ . \tag{13.69}$$

We now have what Chisholm and Farwell call the extended covariant derivative

$$D_\mu := I(\partial_\mu - ieA_\mu) + \frac{1}{4}mi\gamma_\mu \ . \tag{13.70}$$

Note that $I(\partial_\mu - ieA_\mu)$ is a scalar in the algebra, while γ_μ is a vector, but also in the algebra, so we are allowed to add them together. This is crucial to the present exercise, and explains the benefit of constructing everything within $C_{1,3}$.

But there is one other step we need to take in order to see what the above ploy can do for us. If we move to a curved spacetime, we can choose a basis $\{\gamma_\mu(x)\}$ for the tangent space to the spacetime manifold at each point x, and generate a Clifford algebra at each point of spacetime, replacing the condition (13.67) by

$$\{\gamma_\mu, \gamma_\nu\} := \gamma_\mu \gamma_\nu + \gamma_\nu \gamma_\mu = 2g_{\mu\nu}, \tag{13.71}$$

where $g_{\mu\nu}(x)$ are the components of the curved metric g relative to the chosen basis. We obtain a copy of the spacetime algebra $C_{1,3}$ at each point of spacetime. The field of vectors $\{\gamma_\mu(x)\}$ is called the frame field, and it models gravity by standard general relativity, because it effectively encodes the metric via (13.71).

The spinor field ψ is now defined on some region of the curved spacetime, and A_μ and γ_μ are on a par in our covariant derivative, because they are both spacetime dependent. We have an analogy between the coupling constant e that tags along with $A_\mu(x)$ and the constant $m/4$ that tags along with the field $\gamma_\mu(x)$. The frame field is regarded as a physical field and the mass constant $m/4$ as the coupling of the fermion field to the frame field. In this view, mass is not an intrinsic property of a particle, but rather a kind of friction between the particle and the frame field, making it travel at a speed less than what might otherwise be its natural speed, viz., the speed of light.

That description, taken almost verbatim from [42], makes it sound rather like the Higgs field in its effect, and that reminds us that we were trying to illustrate an alternative to the Higgs mechanism for explaining why truly elementary particles like quarks and leptons might have inertia. Unfortunately, we cannot spell out all the details of these fascinating theories, only summarise some of the results. The key point is that the presence of the frame field in the extended covariant derivative (13.70) gives rise to mass terms for the bosons mediating the weak interactions, without the need for spontaneous symmetry breaking and the Higgs mechanism.

In fact, it does a lot more than that. It gives rise to the Einstein–Hilbert gravitational Lagrangian, so in a sense, this theory unifies electroweak and gravitational effects. (Note also that the theory adds a spin gravity term that is quadratic in the curvature and modifies Einstein's theory at short distances.) Better still, building everything within the Clifford algebra $C_{4,7}$, it is possible to unify electroweak, gravitational, and strong SU(3) interactions, by imposing a mathematical condition on the construction of the Lagrangian, which amounts to a kind of normalisation of the different terms. Chisholm and Farwell even predicted the mass of the top quark in 1991 using this kind of theory [43]. Their calculated value of 152 GeV/c^2 is not so far from the the measured mass found shortly afterwards (see Table 13.1).

The aim here was not really to advocate this theory in itself, although the fact that it can predict elementary particle masses, in contrast to the Standard Model, ought to be a strong argument in its favour. An argument against it is that it is highly mathematical, and the procedure of normalising the Lagrangian is physically unmotivated. Even the clever trick of extending the covariant derivative is mathematically rather than physically motivated. The above physical interpretation of an interaction with the frame field comes afterwards. What this theory probably illustrates is that we can unify all sorts of things if we include enough mathematical complexity. String theory is a case in point.

The approach we really want to advocate in this book is rather the opposite. The self-force explanation of inertia is simply the hypothesis that *all* particles have some kind of structure and that their resistance to acceleration comes from this. It contrasts with the Higgs field or frame field explanations, where inertial mass is not an intrinsic property of a particle. But it contrasts also in being physically rather than mathematically motivated.

Chapter 14
Summary and Conclusion

One of the themes of this book is that our theories of the fundamental forces (mainly the theory of electromagnetism in this account) are telling us things that we may not have heard. It was Maxwell's electromagnetism that told us about the special theory of relativity, although there is a clear tendency today to turn things around and start with relativity in a dry and mathematical way. We should not forget where the theory of relativity came from. The view in this book is that Maxwell's electromagnetism may be able to tell us more, if only we would listen.

Chapter Two

Chapter 2 is a fast track to the really fundamental aspects of classical electromagnetism, the bits that often get relegated to the ends of textbooks that one never quite reaches, or spread out through them as notes so that they lose their coherence. However, it closely follows the basis one would get from an exceptionally good standard textbook, viz., the Feynman lectures. We have Maxwell's equations and their complete solution for any charge distribution, a dimensionally homogeneous relativistic notation for them and related formulas, such as the Lorentz force law, and a discussion of the key idea of retarded times. The energy and momentum of the fields are dealt with explicitly via the energy–momentum tensor and Poynting vector. The chapter then discusses the idealistic case of a point particle, leading to the Lienard–Wiechert retarded potential, and several formulas for the EM fields of a point particle with arbitrary motion. The power radiated by the particle is also examined, leading to the Larmor formula. Although EM radiation is not the key issue in the book, it has its place because there is no classical explanation for EM radiation by point particles. The remarkable case of a point charge moving with constant velocity is also considered.

Lyle, S.N.: *Summary and Conclusion.* Lect. Notes Phys. **796**, 379–396 (2010)
DOI 10.1007/978-3-642-04785-5_14

Chapter Three

This chapter burgeons out from the chapter in Vol. II of the Feynman lectures with the same title, i.e., electromagnetic mass. It is a manifesto for the ideas presented in this book, and at the same time a demonstration that the subject here is not some weird offshoot of standard physics, but a straight application of the latter to a context that may have been largely forgotten simply because it is not very tractable from a mathematical standpoint. The connection is made between the fact that particles are treated as mathematical points and the need for renormalisation, both in classical and quantum physics.

The infinite energy of the Coulomb field of a point particle is the first step, followed by the finite energy when the particle is treated as a charge shell of small radius. The momentum in the fields of such a shell is found when it moves with constant velocity. We encounter the problem of relativistic contraction, and also an apparent discrepancy between the energy and momentum of the EM fields of the shell when it has uniform motion. Indeed, the momentum-derived and energy-derived EM masses are different.

The discussion here centers around the nature of inertial mass. We do not know why things have mass. That is, we do not know how to predict the resistance something will show to being accelerated, which is quantified by its inertial mass. In elementary particle physics, some particles are indeed generally considered to be elementary, in the sense of not being composed of anything smaller, e.g., all the leptons (electron, muon, taon, their corresponding neutrinos, and all the associated antineutrinos), but also all the quarks (three generations, three colour charges for each, and all their associated antiparticles). The masses of all these particles must simply be fed into the Standard Model of particle physics as parameters whose values we attempt to provide experimentally.

Then there are an enormous range of bound state particles, built up from the elementary particles. The inertial mass of such a particle is found by adding up the relativistic masses of the component particles, adjusted for whatever kinetic energy they may have within the system, and including the binding energies in an appropriate way. It is interesting to ask students of physics why one should add in kinetic energies of constituents, or binding energies, divided by c^2. An easy answer would just be that it is an application of $E = mc^2$. Here we have something like a principle from special relativistic dynamics, and as a principle, it requires no further explanation. One of the aims of this book is precisely to explain this feature of the inertial mass of a bound state.

But how does the Standard Model explain the inertia of the truly elementary particles? The answer is the Higgs mechanism. The Standard Model predicts the existence of the Higgs boson, although it has not yet been detected. In popular accounts, moving through the Higgs field is rather like trying to move through honey. The particle gets its inertia from the outside, as it were, rather than from any intrinsic structure. The analogy is not perfect, however, because viscosity also opposes uniform velocities, while the Higgs field presumably does not. One problem with the Higgs particle is that the theory cannot tell us what energy will be required of the particle accelerator that is to generate it, so the only solution is to keep ramping up the energy and hope that it will eventually be found. Since increasing the energy is something that we would have done anyway, out of pure curiosity, there is nothing to be lost by this strategy. But it does raise the question as to whether we should not have alternative theories up our sleeves, just in case the Higgs particle never shows up. In this book, we revive the old bootstrap idea that there may in fact be no elementary particles, and that it is the very structure of each particle that causes it to resist its acceleration.

And so we come to the idea of self-force: any extended charge distribution will exert EM forces on itself when accelerated. In simple cases, one finds that these forces always lie exactly along the direction of acceleration, either opposing or assisting. This means that they can be considered as contributing to the inertia of the system, either negatively or positively. This is an old idea: the bootstrap effect. It is definitely there. Even in modern particle physics, hidden away in the quantum theoretical formalism. One crucial feature of the self-force is that it requires the system to be accelerated, i.e., there is no self-force when the system merely has a uniform velocity.

In Chap. 3, we consider the EM self-force of a charge shell when it is accelerated, as discussed in the Feynman lectures. We show how this force contains a term that can be considered to renormalise the inertial mass of the system, and another that powers its EM radiation, first step toward the Lorentz–Dirac equation, which attempts to adjust the Lorentz force law so that it takes into account the radiation reaction force. Finally, we describe a very simple system that can be used to make EM self-force calculations: a charge dumbbell, consisting of two point charges, or two small charge shells, of like or unlike charge, held a certain distance apart by some unspecified binding effect.

Chapter Four

This is an excursion into general relativity, to show that, in a quite trivial way, EM contributions to inertia must contribute equally to passive gravitational mass, i.e., the measure of how much an object will be attracted gravitationally to another object. There is a discussion of the way the general and experimentally well established equivalence of inertial mass and passive gravitational mass is used in Newtonian theory and becomes a founding experimental observation for general relativity itself, something which makes the above discovery look somewhat circular! On the other hand, if all inertia comes from self-forces, then the strong equivalence principle used to transfer non-gravitational bits of physics to general relativity combines with our theories of the other forces of nature to provide a physical explanation for this famous equivalence of inertial mass and passive gravitational mass.

What we propose is a new law, viz.,

$$\boxed{\sum_{\text{fields}} F_{\text{self}} + F_{\text{supp}} = 0} \tag{14.1}$$

which would replace Newton's second law $\mathbf{F} = m\mathbf{a}$ and its direct extensions to GR with the help of SEP. Newton's second law in its usual form follows from (14.1) by analysing the self-forces into some multiple of the four-acceleration, and the whole problem of the research program suggested in this book is to show that this is always possible, not just for EM forces, but for the other forces too, and then to show that there is no other mechanical mass. So a dynamical law, viz., (14.1), is still necessary here, but from it, at least in the case where the inertia is entirely due to self-force effects, we can deduce results that were merely imposed previously.

For one thing, we understand physically why a supporting force is needed to oppose free fall, namely to balance self-forces. In GR as it is usually presented, the supporting force is needed because the particle has non-zero four-acceleration, but we do not know why a non-zero four-acceleration should require a (supporting) force any more than we know why an acceleration should require a force in Newtonian physics.

Another point is that self-forces make a distinction between uniform velocities and changing velocities. The self-force is zero when the particle has a uniform velocity, and only becomes non-zero when the particle velocity is changing. So we understand from (14.1) why no force F_{supp} is required on the particle to keep it in free fall. And we understand the contrast between Newton's first and second laws, in the same way as the self-force idea explains this contrast in Newtonian physics.

Some authors claim that Einstein's theory actually explains inertia, because the geodesic equation follows from Einstein's equations. This view is refuted here. Even in those cases where geodesic motion does follow for a test particle, which requires drastic assumptions about the particle, it is clear that we feed in everything we generally assume about inertia in the form of the action. It is also argued that the strong equivalence principle is essential for understanding motion in general relativity, even when no external forces act on a system (free fall).

It is noted that the equivalence of active and passive gravitational mass is not explained by self-force considerations and remains a mystery. In Newtonian physics, where each massive object attracts each other massive object in a symmetrical way, this equivalence looks reasonable enough. However, in GR, gravitational forces are replaced by curved spacetime, and active gravitational mass goes into an energy–momentum tensor that acts as a source for curving. The whole issue of this previously natural-looking equivalence is obscured.

There is a brief discussion of Mach's principle and the Brans–Dicke theory. The aim is to see why general relativity does not implement the Machian program, and why the Brans–Dicke extension of general relativity does not explain inertia, because it still basically assumes the geodesic principle.

Of course the view advocated here is that one might make better progress in explaining inertia by paying more attention to the fact that test particles are not likely to be well modelled by mathematical points. We discuss the concrete example of the spinning particle and the effect of curvature on its motion. A similar effect occurs when the particle is a source of some classical force field, typically electromagnetic. The spatially extended particle then exerts forces on itself and in simple cases it can be shown that these forces oppose acceleration in flat spacetime and explain why a force is needed to keep the particle off a geodesic in curved spacetime. If all inertia were due to these self-forces, the geodesic equation, or relevant extension of $\mathbf{F} = m\mathbf{a}$, would then be replaced by an equation of the form $\sum F = 0$, where the F summed over include self-forces.

Treating elementary particles like electrons as spatially extended does not make it easier to model them physically, and the point particle approximation has proven its worth in many ways. The idea in this book is not to reject all the successes of point particle models. And furthermore, we only consider classical theory here, so the wonderfully successful world of quantum theory barely gets a mention. But the origin of inertia is nevertheless worth the detour, and once a classical explanation is found, there is no obvious reason why a quantum version of it should not be constructed.

Chapter Five

Chapter 5 is a key chapter of the book, because we get down to some concrete calculations using a very simple extended charge distribution, viz., a charge dumbbell. The beauty of this is that one can actually evaluate the energy and momentum in the fields of the system, and even the EM self-force, using the Lienard–Wiechert formula exposed in Chap. 2.

First, regarding energy, it is shown how the energy-derived EM mass gets two contributions, one from its smallest structure, viz., the tiny shells of charge at each end, and another from its larger structure, of dimension equal to the separation between the two shells, which is of course basically due to an interaction between the two shells. This suggests that, for a Coulomb type of potential, ever smaller structure will contribute ever greater EM mass.

We calculate the momentum in the fields of the system when it is moving with constant velocity along its axis. Once again, most of the momentum comes from the smaller structure of the charge shells, but there is a contribution due to interference between the fields of the two shells which is inversely proportional to the distance between them. If particles have a hierarchy of structure, a Coulomb type of potential will lead to a hierarchy of mass contributions that is inversely proportional to the spatial dimensions of the level of structure.

We carry out both non-relativistic and relativistic calculations. The latter is interesting for several reasons. To begin with, we have to make assumptions about the shape of the system when it is moving. The shells become ellipsoids and the distance between them is assumed to contract by the usual relativistic factor. Those may seem reasonable assumptions when the system has uniform velocity, but they are assumptions. They depend on what holds each shell together as a shell, and they depend on what holds the two shells in tandem. One assumption that would presumahly imply this is that all the forces within the system are described by Lorentz symmetric theories. The EM forces are, but what about the binding forces that we try to avoid considering explicitly? Of course, we expect them to be Lorentz symmetric, whatever they are. What is important is to see that even the mysterious relativistic contraction comes about because of dynamical things occurring in our systems, a point that is almost taboo in the world of physics, and particularly the philosophy of physics.

Another interesting point is that the momentum-derived EM mass contains a gamma factor. It increases in precisely the way we are told in the special theory of relativity, without explanation, that inertial mass should increase with speed. Of course, explanation is not considered necessary, because this is viewed as a frame-dependent illusion by many. So for that matter is the relativistic contraction. After all, if it is the observer that goes into uniform motion, rather than our system, she will find that the mass of our system has increased, while nothing has happened to the system to explain why that should be. So there is no explanation. And she will find that the system has contracted spatially, while nothing has happened to the system to explain that. Once again, there is no explanation for the relativistic contraction. It is just a matter of consistency, something which Minkowskian geometry deals with very effectively.

The view here is that Minkowskian geometry provides us with one way of understanding things, but that alongside it, there is another way that is not at all fashionable. We may also consider what happens to the system as it is accelerated relative to some given inertial frame. The philosophers do not like this, even if the given inertial frame can be any frame, and even if we realise that each frame involves taking different spatial cross-sections as hyperplanes of simultaneity, so that we are aware of taking a parochial view by our choice. They say that we miss something, even if we keep the Minkowski spacetime view alongside. The present view is that, on the contrary, we miss something if we forget to describe what is happening to things relative to previously chosen inertial frames.

Actually, in the Minkowski view, nothing happens. The whole notion of something happening depends on first choosing an inertial frame. But once we do that, we can talk about what happens, and it is interesting. We can talk about what is happening within our dumbbell system when it accelerates. We can perhaps understand why it ends up contracted, and if its mass were entirely due to the momentum of the EM fields it creates, we would be able to say that it has inertia precisely because a certain kind of EM field has to be created if we want to get it moving. But we have to free ourselves of the block universe dogma of Minkowski spacetime, no matter how admirable it may be, in order to talk in this way.

A worrying point that has led to much debate is the discrepancy between energy-derived and momentum-derived EM masses. This is illustrated by the charge dumbbell. The energy-derived mass is $e^2/4dc^2$, where d is the spatial separation of the two shells, while the momentum-derived mass is twice this, at least when the system moves along its axis. But when we calculate the momentum-derived mass for motion perpendicular to the system axis, we find that the discrepancy disappears. One significant point here is that the momentum-derived mass of a non-spherically symmetric system like this will depend on the direction of motion relative to its shape. The other point is that we need to explain this discrepancy in the one case and its disappearance in the other.

We also discuss neutral systems, when the charge shells at the ends of the dumb-bell have equal and opposite charges. We find something very important here. The momentum-derived EM mass still gets the equal contributions from each shell, but the interference term inversely proportional to the spatial separation d between them is now negative! This level of structure decreases the inertia of the system. But of course, this is exactly what one would expect. The EM binding energy required to put each shell together is positive, because one must force the like charges to concentrate, while the EM binding energy to bring one shell close to the other is negative, since they attract. But why should the latter fact reduce the inertia? Well, we know from relativistic dynamics that the binding energy of a bound system must be included in its inertial mass, because $E = mc^2$.

The present view is that this is hardly an explanation, merely a rule of thumb that happens to work perfectly for some deeper reasons. The point is that the system exerts an EM force on itself when we try to accelerate it. For like charges on the shells, that self-force opposes the acceleration, while for opposite charges, it assists acceleration. But note that this situation reminds us that there must be other forces in the system, whether it carries a net charge or is neutral. There must be binding forces of another kind, and there is every chance that they too will lead to self-forces under accelerations, and contribute in different ways to the inertia of the system. Indeed, this is why there is a discrepancy between energy-derived and momentum-derived EM masses. It turns out that the discrepancy should indeed be there when only the EM effects are taken into account, something discussed at length later. The point is that the total inertia is the result of all the forces at work within the system (and of course maybe other effects too).

There is a brief discussion of the pions. Feynman uses the spherical shell model to explain the mass difference between the charged pions on the one hand and the neutral pion on the other. Here we use a better (although still very primitive) model, in which the pion is a dumbbell with a quark at one end and an antiquark at the other. This reminds us that the neutron is also supposed to be made of quarks, so there is no difficulty understanding how there might be an EM contribution to its inertia, although very hard indeed to work out for such a three-body system. The neutron is more massive than the proton by a small amount. There is no obvious reason why that difference should not be due to EM effects.

Chapters Six to Nine

Self-forces are calculated in Chaps. 6–9 for four different motions of the charge dumbbell:

1. Linear acceleration perpendicular to the system axis.
2. Linear acceleration along the system axis.
3. Circular orbit with velocity perpendicular to the system axis.
4. Circular orbit with velocity along the system axis.

The calculations are difficult and involve approximation. In each case, we use the Lienard–Wiechert fields to calculate the EM force exerted by each charge on the other, now treating the charges as occupying mathematical points, and simply add those two forces together to get a self-force. We then expand this self-force as a power series in the separation d between the charges.

In the second case the acceleration and velocity both lie along the system axis, while in the first, they both lie perpendicular to it. In each case, for like point charges, the leading term in the self-force, going as $1/d$, opposes the acceleration, i.e., it acts along the line of the acceleration, but in the opposite direction. The self-force-derived mass is $e^2/2dc^2$ for case 2 and $e^2/4dc^2$ for case 1, agreeing in each case with the momentum-derived mass, and even agreeing with the energy-derived mass in case 1. It is a significant fact that momentum-derived and self-force-derived masses agree in each case. The discrepancy with the energy-derived mass, when it occurs, is due to not taking other forces into account. But the momentum in the EM fields is put there by a force overcoming the self-force, and this is why $m_{\text{EM}}^{\text{SFDM}} = m_{\text{EM}}^{\text{MDM}}$.

For case 1, it is shown how to renormalise the mass of the system so as to absorb the $O(d^{-1})$ term into any other inertial mass the system may have. We basically have to rewrite the Lorentz force law. Even the relativistic gamma factor works out correctly to allow this reorganisation of the equation of motion. Likewise in case 2. Once this term has been absorbed into the so-called mechanical mass term (assuming there is one), we can allow $d \to 0$, regardless of the fact that the bare mechanical mass would have had to have been infinite itself in order to end up with a finite resulting inertial mass. This is the miraculous renormalisation procedure, still required in quantum field theory.

We also understand from this that the self-force contribution to the inertial mass expressed in the renormalisation equation will actually vary in the way inertial mass is supposed to vary with speed in the special theory of relativity. This was discovered before the advent of the fully fledged relativity theory. With hindsight one might say that this had to happen because Maxwell's theory, from which this contribution was derived, is Lorentz symmetric, and of course it does not matter whether one is aware of that when deriving this result. However, the complexity of the calculations and the arbitrariness of the structure of our toy electron make this result something less than obvious. To put it another way, one might say that we have here an explanation as to why inertial mass should increase as predicted by relativity theory when a system moves faster: it is because the self-forces within the system increase the way they do, at least as far as the self-force contributions to the inertial mass are concerned.

For case 1, we also calculate the term in the self-force going as d^0. This term cannot be absorbed into the inertial mass, and it clearly remains even if we let the system size tend to zero. The point is that this term is a radiation reaction, i.e., a force back on the system due to the fact that it radiates EM energy, like any charge accelerating relative to an inertial frame. But note that there is no explanation for why a point charge should radiate in this way. Put another way, the self-force explanation for the source of this radiated energy when the charge has spatial extent is lost when it is treated as a point. As pointed out by Feynman [2], when a radio antenna is radiating, the forces come from the influence of one part of the antenna current on another. In the case of a single accelerating electron radiating into otherwise empty space, there is only one place the force could come from, namely, the action of one part of the electron on another.

In case 2, for acceleration along the system axis, we also encounter the problem of rigidity. Without a model for the binding force that holds the system together, and without saying how the system is accelerated, we cannot say what the separation between the charges will be. The same problem occurs for any extended charge distribution under acceleration, e.g., the charge shell. The standard assumption is a rigidity assumption. For example, most commentators assume that the charge shell will always look spherical in its instantaneous rest frame. This is discussed at length in Chap. 12. A slightly simpler simplifying assumption is made for the charge dumbbell.

Cases 3 and 4 above are designed to put the self-force idea to a rather severe test. In case 3 the acceleration is along the system axis, while the velocity is normal to it, and in case 4, the acceleration is normal to the system axis, while the velocity lies along it. The four cases thus cover all possibilities, and in each case, the EM self-force is found to oppose the acceleration, thereby contributing to the inertia of the system. Given the complexity, particularly of the last two calculations, this suggests a general result. And indeed, it was shown by Dirac in 1938 [3] that one could always renormalise the mass in the equation of motion of a charged particle. Dirac was in the process of deriving the Lorentz–Dirac equation, which also takes into account the radiation reaction, but leads to anomalies.

But the point is that something about the theory of electromagnetism makes the divergent term in the EM self-force systematically oppose acceleration. And it is this something that makes the theory renormalisable classically. It would be interesting to identify just what it is. Could it be the fact that this is a gauge theory? After all, it was proven in the 1970s that all gauge theories are renormalisable [34, Chap. 12]. It would also be interesting to carry out the EM self-force calculation for a general motion of a small system, although that could prove rather difficult. Note that Dirac employs an a priori quite different approach in his 1938 paper, using the energy–momentum tensor of the fields, but perhaps that method could be adapted. And finally, it would be fascinating to see how the strong force would contribute to self-forces in bound systems. That would be quite a challenge, because we do not have a classical theory of the strong force, only quantum chromodynamics, but one could perhaps make use of the various potential models [33].

It should be mentioned that, in case 3, the self-force-derived mass is the same as for case 1. These are the two cases where the velocity is normal to the system axis. In case 4, the self-force-derived mass is the same as for case 2. These are the cases where the velocity is along the system axis. It is also important to note that the relativistic gamma factors work out in such a way that one can always renormalise the mass in a relativistic context: one requires a factor of γ^3 when the velocity and acceleration are parallel, as in cases 1 and 2, and a factor of γ when they are orthogonal, as in cases 3 and 4.

Finally, regarding these four self-force calculations, it is a very easy matter to see that, if we replace the like charges on our dumbbell by unlike charges, the electromagnetic self-forces switch sign in each case, to actually assist the acceleration. This is no surprise. In physics, it is well known that the negative binding energy of a bound particle contributes negatively to the inertial mass. The mass renormalisations are all negative then.

Chapter Eleven

Chapter 11 aims to reconcile the discrepancies between different ways of getting the EM mass. The solution to this problem is that one must include *all* the forces when working out self-force contributions to inertia. In other words, one should not expect the energy-derived EM mass to equal the self-force-derived EM mass, except in some special cases.

So why such a long chapter? The point is just that it has been the subject of much debate in the literature, and is often rather carelessly taken as an argument against the whole idea of inertia due to self-force. But the reader should be quite clear that self-force does contribute to inertia, and that this is a fully accepted part of standard physics as we take it today, although in a heavily disguised form (the subject of Chap. 13). Of course, there are other reasons than that. One is that alternative formulas for the EM energy–momentum have been put forward, which are rather ad hoc from a physical standpoint, and the present view is that they need to be critically assessed. And another point is simply that one would need to close the gap between m^{SFDM} and m^{EDM} in order to get a coherent self-force theory of inertia. So one would need to think carefully about what is missing.

What is missing are binding forces that stop our dumbbell flying apart. With these included, and with similar binding forces stopping each charge shell from flying apart, the energy–momentum tensor for all the fields present would be conserved everywhere. And it is precisely the non-conservation of the EM energy–momentum tensor that leads to the problem. The reason is that one would like to define a Lorentz covariant four-momentum for the EM fields from the EM energy–momentum tensor. So part of Chap. 11 discusses the way one would normally do this, and shows why that leads precisely to the discrepancy mentioned above, which is basically a deviation from Lorentz covariance.

A widely supported approach today [21, 22, 24] is to redefine the EM four-momentum of the fields directly in a Lorentz covariant way. Whatever the reasoning put forward to justify this, it amounts to merely taking the four-momentum in the fields of the charge shell when it is not moving relative to some inertial frame, in which case the momentum components are zero and there is only an energy component, and then carrying out a Lorentz transformation of that to find out what the four-momentum of the fields should be in some other inertial frame. With this definition, the four-momentum of the EM fields is *never* what the usual formula gives, except when the charge shell is not moving.

So the reader should have no doubt whatever that this is an ad hoc move to get a Lorentz covariant four-momentum. Naturally, this may have its uses, as claimed in particular by Rohrlich. But note that Rohrlich does not intend to apply his theory to spatially extended particles. Perhaps he considers this as a form of renormalisation. But those of us who are seriously interested in spatially extended objects are condemned to deal with all the complexity that picture involves.

Section 11.2 goes into the way one usually derives a Lorentz covariant four-momentum from a conserved energy–momentum tensor, starting with an analogous and easier problem, namely the derivation of a Lorentz invariant charge from a conserved current density. All this is done with some rigorous integration theory using differential forms. The notorious discrepancy at issue in the present discussion is then derived as an integral that really should be there when one attempts to get a four-momentum from a non-conserved energy–momentum tensor.

Section 11.3 considers a challenge by Boyer [23] that is fully supported here. He shows very clearly why Rohrlich's approach loses the natural physical interpretation of what is happening in the case of a spatially extended object like the charge shell. And the present view is that one is compelled to adopt the type of interpretation proposed by Boyer if committed to non-pointlike particles. We discuss in detail Boyer's thought experiment in which the shell forms by collapse from infinity. One reason is that it involves a great deal of interesting theory of the kind that one would have to take seriously in order to understand spatially extended particles. Section 11.3.6 is important because we explain why one should expect a discrepancy for longitudinal motion of the charge dumbbell, but no discrepancy for transverse motion. The explanation relies on mention of the binding forces required to stabilise the system.

Section 11.4 shows exactly why Rohrlich's redefinition of the four-momentum works, and why it should be viewed as a contrived solution, while Sect. 11.5 explains how this redefinition can be considered as derived from an energy–momentum tensor by integration, but reveals a hidden assumption, namely, that the energy–momentum tensor has to be static. This would rule out dynamic models for our spatially extended particle.

Section 11.6 moves on from the collapsing shell idea and starts to consider binding forces in a more realistic scenario. One approach to show how cohesive forces can save the day is to define them directly as the forces that would be required to balance electromagnetic repulsive forces on the sphere, viz.,

$$f_{\text{coh}}^{\mu} := -\partial_{\alpha}\Theta_{\text{em}}^{\alpha\mu} \,,$$

which would be zero if the energy–momentum tensor of the electromagnetic fields were conserved. In whatever frame we examine things, we now find that the cohesive forces impart a 4-momentum P_{coh}^{μ} to the system. It happens to be zero in the rest frame of the charged shell, but it is not Lorentz covariant. On the other hand, the total 4-momentum due to these forces and the electromagnetic fields (the sum of the two objects P_{coh}^{μ} and P_{em}^{μ}) is a 4-vector, and it is equal to

$$P_{\text{coh}}^{\mu} + P_{\text{em}}^{\mu} = m_{\text{e}}v^{\mu} \,,$$

where v^μ is the 4-velocity of the shell in the given frame and m_e is the electromagnetic mass defined from the Coulomb energy in the electric field around the shell in its rest frame.

This is, of course, a minimalist solution to the real physical problem. A more realistic picture would be one in which the electron is made up of parts with electric charge and with some other type of charge, source of a second field that acts to hold the thing together against the Coulomb repulsion of the charged elements. The total energy–momentum tensor for the system would contain contributions from both types of field and it would be conserved. The total 4-momentum of the two fields defined in the usual way from the total energy–momentum tensor would therefore be Lorentz covariant. The inertial mass of the system would contain contributions from both the electromagnetic self-force and the self-force arising from the other kind of field.

This is a model in which the electron has spatial structure in any Lorentz frame. If one assumes the electron to be a point charge, or if one wishes to take the point limit, the cohesive forces do seem to be rather superfluous. Presumably, this is what motivated Rohrlich to go to such lengths to defend what is actually a totally artificial ploy for defining a Lorentz covariant 4-momentum for the electromagnetic fields.

Chapter Twelve

Chapter 12 faces the problem of approximating the shape of a spatially extended object when it is accelerating. In discussions of charge shells, usually assumed spherical in their rest frame, the tradition has been, and still is [7], to make a rigidity assumption, i.e., the sphere is assumed always to look spherical in its instantaneous rest frame. The reasons for calling such an object rigid are examined in detail from an unusual perspective, and the whole idea is criticised.

The main point for the purposes of this book is that the shape of such an object will depend on the balance between all the forces within it. This means that a realistic model for the shape of the spatially extended object under acceleration will require a realistic model for the binding forces holding it together. This would clearly be a very tough problem indeed, so approximation would still be the order of the day. But let us try to make better approximations.

A well known paper by Bell [5] already does something along these lines in the case of electromagnetism. Bell considers the shape of an electron orbit around an atomic nucleus when the latter is accelerated. The model here is pre-quantum and ignores collapse of the orbit due to radiation losses, but it nevertheless provides a clear physical explanation for the relativistic contraction of a moving object, where it is often insinuated that such a contraction is in fact just an illusion. This physical explanation says that objects contract because the EM fields of the atomic nuclei change when the nuclei are moving, in such a way that the electron orbits shrink in the direction of motion.

There are other forces than the EM fields at work in matter, but all these forces are Lorentz symmetric, and this alone should be enough to prove that all spatial relations between components of a spatially extended object will shrink in the same way due to the establishment of a new equilibrium when the whole thing is moving. The shrinking will of course require a certain time to occur, but in theory one could get quite accurate models for the shape of the object by carefully considering the way it is held together.

It is unusual to see such a physical discussion of how the relativistic contraction comes about. The reason is that it is superfluous in the usual principle approach to relativity, and in particular once one postulates that spacetime is Minkowskian. But of course, the latter postulate, which comes with a ready-made metric telling us directly how long or short things are when moving with constant velocity relative to our own inertial frame (assuming we are at rest relative to such a frame ourselves), cannot tell us how long or short they are when they are themselves making the transition between two states of uniform velocity. This does not mean that the theory with the Minkowski postulate cannot help us there, only that we are eventually condemned to look at the details of the physics operating within the object when it is accelerating.

Actually, one of the conclusions of Chap. 12 is that the rigidity assumption is a way of proceeding without looking inside an object to see what is really going on. The conclusion here is that this may not always be a good enough approximation, and anyway, it is clearly much better to try to model the object in more detail. In this connection, there is some criticism of the uncritical way semi-Euclidean frames are used, supposedly to describe the world as modelled by accelerating observers. The link here is this: if a linearly accelerating observer has a rigid ruler in her hand, it will always have just the right length to measure the semi-Euclidean spatial coordinate in the direction of acceleration. Put another way, it satisfies what is known as the ruler hypothesis. In actual fact, the observer would have to accelerate it more carefully than just pushing one end of it. It is shown in Chap. 12 that no object can always behave rigidly. The way it is accelerated is important.

There is in fact no natural or preferred frame of reference for such an observer according to the current theory of relativity, only one that is adapted in some sense to her motion. But Bell's paper shows by physical argument why there *are* such canonical frames of reference for observers with uniform velocity in a flat spacetime. Or rather, why one might expect there to be a set of such frames related by Lorentz transformations.

Once again all that is superfluous if one simply postulates it, which is effectively what Minkowski spacetime does. The postulate that spacetime is Minkowskian with its well known metric $\eta_{\mu\nu}$, or indeed that it is pseudo-Riemannian with a metric $g_{\mu\nu}$, still requires consideration of the physics going on in our metrological equipment at some point in order to connect the mathematical machine with things that are actually measured out there. The general trend of ignoring this point is considered a serious oversight in the present book. As discussed by several authors recently [14, 16,36,37], the metric in relativity theories gets its physical interpretation by detailed physical arguments. At the outset one has a manifold with a rank two covariant tensor field g on it, and this field can be represented by a 4×4 matrix field $g_{\mu\nu}$ on coordinate patches. But what do the coordinates correspond to in the real world of our length and time measurements? How could we know that without looking at the matter making up our measuring tools, and thinking about the theory of that matter (as extended from pre-general-relativistic physics to the curved manifold context by a strong equivalence principle, another essential ingredient), and then eventually showing by this route that $g_{\mu\nu}$ had a chronogeometric significance?

Surely, by just postulating that interpretation of g, one is missing something, and the view here is that it is not just a pedagogical loss, although that would be reason enough for considering it. The fact is that one day we will have to go beyond relativity theory and replace it by better, or more finely adjusted theories. If we work by principle and postulate, there is a serious risk of dogma. Fortunately, physicists tend not to be dogmatic in the detail of their work. However, some philosophers of physics react strongly to the above ideas, and it is important to take such carefully considered views into account, something that goes beyond the scope of this book.

Before leaving this issue, the specific example of Bell's approach described in [5] suggests a very interesting research problem that could help to throw more light on the notorious discrepancy discussed in Chap. 11 between energy-derived and momentum-derived (or self-force-derived) EM masses. One way that two unlike charges can stay together in a spatially extended system without collapse, and without the need for another force (at least in theory, and in a truncated classical model that forgets radiation losses), is simply for one to orbit around the other. It would be interesting, although presumably rather difficult, to model that with a view to obtaining the energy-derived and momentum-derived EM masses. In theory, they must come out the same here, because the EM energy–momentum tensor is conserved.

Chapter Thirteen

Chapter 13 reviews the way inertial mass is treated in elementary particle physics today. The basic idea is this: some particles are really elementary, with no components, while others are bound states of the elementary particles. Today, we do not know the inertial masses of some elementary particles because we can never get them on their own. These are the quarks. On the other hand, we know the masses of the leptons, i.e., the electron, muon, and tau lepton, and we are beginning to know the masses of their associated neutrinos.

But what is the mass of a bound state particle like the proton, or the pi meson? These are made of quarks and antiquarks. Even if we knew the masses of the components, we would still have a problem understanding the inertia of the bound state, because masses do not just add up. For one thing, we conceive of the component quarks as moving around within the particle, with a certain kinetic energy that must be included in their own relativistic mass. And for another, any energy lost or gained by the system by putting the elementary particles together has to be subtracted from or added to the inertial mass of the system (after dividing by c^2).

But why is this? And anyway, why does an elementary particle like a quark have a bigger inertial mass when it is moving? These are questions we no longer really ask today. Or rather, they are questions that are answered very neatly by some principles: the principles of relativity. The great thing is that we know exactly how the inertial mass of an elementary particle changes with its speed, and we know exactly how much mass to add or subtract to account for binding energy within a bound state particle. So in a very real sense, we do not need to ask why. This is instrumentalism.

Section 13.1 traces through an elementary presentation of special relativistic dynamics to see how far we go towards explaining these phenomena. We see how the notion of Lorentz symmetry can guide us to a new dynamical equation, which maintains something of Newton's way, and we identify the hypothesis that underlies it, namely a relation between the three-force as we knew it before relativity and the coordinate time rate of change of the quantity $m_0 \gamma \mathbf{u}$, where m_0 is the inertial mass of the particle when it is not moving, $\mathbf{u}$ is the coordinate three-velocity of the particle, and $\gamma(u)$ is the usual function. Lorentz symmetry may allow us other possibilities, but this is the obvious one. And it gives a result not so different from the $\mathbf{F} = m\mathbf{a}$ that Newton proposed.

But the big difference is that, with the new equation of motion, the faster the particle is moving, the less it will be accelerated by a given force. This prediction from the bold hypothesis of relativistic dynamics is borne out. Particles have more inertia when they get moving. We do not know why, and we do not ask why. The theory works, and that is enough. Of course, why questions sometimes lead to explanations, but those explanations themselves depend on hypotheses, so maybe it is reasonable enough to stop here. After all, what more do we want than to predict? If we have evolved to reason with nature, then it is because good predictions have allowed more of our genes for reasoning and prediction to get through.

Those are the hard facts. But some of us have genes for asking why even when the theory makes good predictions. This is very likely useful, the day the prediction is not so good and we need to improve the theory. So what does the bootstrap idea have to say about this? We have discovered that the momentum in the EM fields of a charged particle increases with the gamma factor. Actually we have rediscovered one of the pointers that encouraged the pioneers to make the above bold hypothesis. They were not guided by Lorentz symmetry, which requires the hindsight of Minkowski spacetime. And as pointed out in Sect. 6.6.1, we know why the momentum in the fields increases in this way: it is because the self-forces within the system increase in the way they do, at least insofar as the self-force contributions to the inertial mass are concerned.

Here we have a parallel with the refusal today to try to explain or understand relativistic contraction as a dynamical phenomenon. The point is that it is superfluous. We get everything by postulating Minkowski spacetime and Lorentz invariant matter theories. The same goes here. After all, is the increased inertia of a moving particle not just an illusion? Because what would we say if it were the observer that got into uniform velocity, so that nothing had happened to the particle? Then there could be no cause for the observed increase in the inertial mass of the particle. So it must be an illusion. On the other hand, if the observer sits still and the particle is accelerated, then its inertial mass will increase, and we would like somehow to attribute that effect to the accelerating agent. This is the philosophical debate. The present view is that it is worth continuing this debate and fighting against the Minkowski strait-jacket, if only because the Minkowski hypothesis will one day have to cede to better principles.

One of the themes of this book is that we do not understand inertia, i.e., that there is something to be understood and that we have not found it yet. And here is the special theory of relativistic dynamics telling us that $m := m_0\gamma$ is the measure of the particle's inertia, and not m_0. There must be a message here for those who seek to understand why particles resist being accelerated, and in different proportions depending on their nature and circumstances. But the usual attitude is one of instrumentalism, as Popper called it [31]. The theory is simply required to be empirically adequate, in the sense of being well-supported by evidence or by the measure of its predictive success.

Having established (13.3) on p. 314 as the best, possibly even the only way to play the Lorentz invariance game, we then simply apply it and note that it works. Here, surely, is a missed opportunity. We did not know why Newton's law had the form $\mathbf{F} = m\mathbf{a}$, and now that we find exactly how we have to adjust this to get it in line with the principles of relativity, we do not ask why the dynamical law still has the form (13.3), even though we have learnt something quite significant about what contributes to inertia.

These are important points. They concern the way we understand the world around us. But the aim of Sect. 13.1 was to lead up to a rule-of-thumb that we usefully apply without measure, namely the adage $E = mc^2$. It comes out of this discussion of dynamics in a rather restricted way, and gets boosted to the boldest hypothesis of them all. It does not matter what the form of the energy is. It could be electrostatic, magnetic, kinetic, elastic, or any other. All energies are the same thing as inertial mass. And reading the equation backwards, we conclude that inertial mass is energy.

Sections 13.2 and 13.3 of Chap. 13 describe how we apply this hypothesis without question today in elementary particle physics. The main aim is to show that these features, viz., increased inertia of moving entities and contribution of binding energy to the inertia of bound states, which stem directly from bootstrap considerations, constitute a considerable part of the way we understand the inertia of bound state systems, even though we no longer even pay lip service to them. There is perhaps more detail in this chapter than necessary. The hope is that the reader will look again at this new but nevertheless standard material of physics with the help of this old light, and maybe discover something.

Section 13.4 races through the theory of the Higgs mechanism. This is the idea that saves us when we ask: what explains the inertia of a truly elementary particle that occupies only a mathematical point in space? Naturally, it cannot be something about the inner workings of the particle, because it cannot have any. So it has to be some kind of interaction it has with something outside. Moving through the Higgs field, whatever that may turn out to be, is in that sense like moving through treacle.

So Sect. 13.4 asks what we know about the masses of those particles that are considered to be truly elementary today. Why do different quark flavours have different masses? Why are there generations of quarks: (u,d), (s,c), and (b,t), getting more and more massive? Why are there generations of leptons: e, μ, and τ, getting more and more massive? Why do the intermediate vector bosons $W^{\pm}$ and Z have the masses they do?

Today, the whole problem of the mediation of the weak force by massive vector bosons is explained by the spontaneous breaking of a local (gauge) symmetry, and this through the Higgs mechanism. The price to pay is one or more as yet unobserved fields whose dynamics is governed by a potential with the given gauge symmetry. This symmetry in the potential allows the field a continuum of possible ground states, and for some reason nature has chosen one of them for it, just as the spins of the microscopic components in a ferromagnet select some common alignment when the conditions are right.

The gauge fields associated with the local symmetry would have behaved as though they were massless if no particular ground state had been chosen, but when one is selected, they behave as though they have a mass. Ultimately, they acquire their mass through an interaction with the Higgs field in its chosen ground state. The same goes for all the other truly elementary particles, viz., the quarks and leptons. They would be massless if the Higgs had not dropped into some particular ground state. But when it does, for whatever reason (one thinks of a cooling of some kind), then they behave as though they have a mass, because they interact in a suitable way with the Higgs field.

In this view, all the truly elementary particles are actually massless. It is just that they behave as though they have mass when the Higgs field, which interacts with all of them, selects one of the continuum of ground states available to it. This is in many ways an elegant idea, underscored by a truly physical mechanism that is actually exemplified in the world. The trouble is that we do not know enough about this handy field, and the problem of the quark and lepton masses is replaced by the problem of their couplings with the Higgs field, because the masses they ultimately appear to have depend directly on those couplings.

In Sect. 13.5, we describe the spin gauge theories in order to illustrate that the well known electroweak unification of Glashow, Weinberg, and Salaam can be achieved without the Higgs mechanism. Otherwise this theory proposes a rather similar idea to explain the masses of the truly elementary particles, viz., an interaction with some other field. Admittedly, the field in question is a known one here, because the frame field in these theories is basically a form of the gravitational field, usually represented by the metric in general relativity. And better, this theory has a predictive value with regard to the masses, but via a rather mathematical principle regarding the construction of the Lagrangian, which on the face of things does not help us to understand physically why quarks and leptons should resist acceleration to the various degrees we observe. So the main reason for mentioning the spin gauge theories as far as this book is concerned is really just to point out that electroweak unification with massive intermediate vector bosons can be achieved without the need for a Higgs field.

Hopefully, some readers will baulk at the term 'truly' elementary. There is certainly something rather naive about the idea that any object could be perfectly represented by a mathematical point, or by a mathematical curve in spacetime. It may of course be a good approximation, making certain problems tractable, and the discussion of spatially extended charge distributions in this book, with all its difficulty and complexity, will no doubt help to justify such an approximation. But the results of calculations like those in Chaps. 6–9 are clearly intriguing, showing that even a very well known theory like Maxwell's may be able to tell us more than we have so far heeded.

The aim here has certainly not been to reject the Standard Model out of hand, nor to advocate a return to pre-quantum methods of calculation, or anything of the sort. It is just intended as a reminder that, if all particles do in fact have substructure, then there is a wealth of explanation available there for the inertia of such entities. Any spatial distribution of sources for some fundamental force will exert a force on itself when it is accelerated, and what would be interesting is to investigate whether such a bootstrap effect would always contain an inertial component to highest order, opposing the acceleration.

Since this question is intimately related with the possibility of mass renormalisation in the classical context, and since all gauge theories are, it appears, renormalisable, then maybe there is a connection between the nature of self-forces due to gauge fields and the fact that massive particles are massive, without the need for ad hoc external fields to slow them down by interaction.

References

1. R.P. Feynman, R.B. Leighton, M. Sands: *The Feynman Lectures on Physics*, Vol. I, Addison-Wesley, Reading, MA (1964)
2. R.P. Feynman, R.B. Leighton, M. Sands: *The Feynman Lectures on Physics*, Vol. II, Addison-Wesley, Reading, MA (1964)
3. P.A.M. Dirac: Classical theories of radiating electrons, Proc. Roy. Soc. London A **167**, 148 (1938)
4. S.N. Lyle: *Uniformly Accelerating Charged Particles. A Threat to the Equivalence Principle*, Springer-Verlag, Berlin, Heidelberg (2008)
5. J.S. Bell: *Speakable and Unspeakable in Quantum Mechanics*, 2nd edn., Cambridge University Press, Cambridge (2004), Chap. 9
6. W. Rindler: *Introduction to Special Relativity*, Oxford University Press, New York (1982)
7. A.D. Yaghjian: *Relativistic Dynamics of a Charged Sphere*, Lecture Notes in Physics 686, Springer-Verlag, New York (2006)
8. S. Parrott: *Relativistic Electrodynamics and Differential Geometry*, Springer, New York (1987)
9. V. Petkov: *Relativity and the Nature of Spacetime*, 2nd edn., Frontiers Series, Springer-Verlag, Berlin, Heidelberg (2009)
10. D. Griffiths: *Introduction to Elementary Particles*, John Wiley, New York (1987); 2nd edn. Wiley-VCH, Weinheim (2008)
11. Particle Data Group: `pdg.lbl.gov/`
12. S.N. Lyle: *Extending Bell's Approach to General Relativity*, unpublished (2008)
13. M. Göckeler, T. Schücker: *Differential Geometry, Gauge Theories, and Gravity*, Cambridge University Press, Cambridge (1987)
14. H.R. Brown: *Physical Relativity: Space-Time Structure from a Dynamical Perspective*, Clarendon Press, Oxford (2005)
15. B.S. DeWitt: *Stanford Lectures on Relativity*, Springer-Verlag, Berlin, Heidelberg (to be published)
16. J. Butterfield: Reconsidering relativistic causality, Intl. Studies Phil. Sci. **21**, issue 3 (2007)
17. F. de Felice, C.J.S. Clarke: *Relativity on Curved Manifolds*, Cambridge University Press, Cambridge (1995)
18. M. Friedman: *Foundations of Space–Time Theories*, Princeton University Press, NJ (1983)
19. J.V. Narlikar: *Introduction to Cosmology*, Cambridge University Press, Cambridge (1993)
20. T.Y. Cao: *Conceptual Developments of Twentieth Century Field Theories*, Cambridge University Press, Cambridge (1997) Sect. 4.3
21. F. Rohrlich: Self-energy and stability of the classical electron, Am. J. Phys. **28**, 639 (1960)
22. F. Rohrlich: Comment on the preceding paper by T.H. Boyer, Phys. Rev. D **25** (12), 3251–3255 (1982)

Lyle, S.N.: *References*. Lect. Notes Phys. **796**, 397–398 (2010)
DOI 10.1007/978-3-642-04785-5 © Springer-Verlag Berlin Heidelberg 2010

23. T. Boyer: Classical model of the electron and the definition of electromagnetic field momentum, Phys. Rev. D **25** (12), 3246–3250 (1982)

24. P. Moylan: An elementary account of the factor of 4/3 in the electromagnetic mass, Am. J. Phys. **63**, No. 9, 818–829 (1995)

25. S.W. Hawking and G.F.R. Ellis: *The Large Scale Structure of Spacetime*, Cambridge University Press, Cambridge (1973)

26. J.D. Jackson: *Electrodynamics*, 2nd edn., Wiley, New York (1975)

27. Full details of any missing calculations can be obtained from the author

28. M. Blau: *Lecture Notes on General Relativity*, available on the Web

29. E. Pierce: The lock and key paradox and the limits of rigidity in special relativity, Am. J. Phys. **75** (7), 610–614, 2007

30. M. Longair: *Theoretical Concepts in Physics*, Cambridge University Press, Cambridge (1984)

31. K.R. Popper: *Realism and the Aim of Science*, Hutchinson, London (1983)

32. W. Greiner, B. Müller: *Quantum Mechanics. Symmetries*, Springer-Verlag, Berlin, Heidelberg (1989)

33. W. Greiner, S. Schramm, E. Stein: *Quantum Chromodynamics*, 3rd edn., Springer-Verlag, Berlin, Heidelberg (2007)

34. J.C. Collins: *Renormalization*, Cambridge University Press, Cambridge (1984)

35. D. Dürr, S. Teufel: *Bohmian Mechanics*, Springer-Verlag, Berlin Heidelberg (2009)

36. H.R. Brown, O. Pooley: Minkowski spacetime. A glorious non-entity. In: *The Ontology of Spacetime*, ed. by D. Dieks, Elsevier (2006). Also available at `philsci-archive.pitt.edu/archive/00001661`

37. H.R. Brown, O. Pooley: The origin of the spacetime metric. Bell's 'Lorentzian pedagogy' and its significance in general relativity. In: *Physics Meets Philosophy at the Planck Scale*, ed. by C. Callender, N. Huggett, Cambridge University Press (2001). Also available at `gr-qc/9908048`

38. J.S.R. Chisholm, R.S. Farwell: Electroweak spin gauge theories and the frame field, J. Phys. A Math. Gen. **20**, 6561–6580 (1987)

39. J.S.R. Chisholm, R.S. Farwell: Unified spin gauge theory of electroweak and gravitational interactions, J. Phys. A Math. Gen. **22**, 1059–1071 (1989)

40. J.S.R. Chisholm, R.S. Farwell: Unified spin gauge theories of the four fundamental forces, Proc. of the IMA Conf.: The Interface of Mathematics and Physics, Oxford University Press, Oxford (1990)

41. J.S.R. Chisholm, R.S. Farwell: Tetrahedral structure of idempotents of the Clifford algebra $C_{3,1}$. In: *Clifford Algebras and their Applications in Mathematical Physics*, ed. by A. Micali et al., Kluwer Academic Publishers (1992) pp. 27–32

42. J.S.R. Chisholm, R.S. Farwell: Unified spin gauge theories of the four fundamental forces. In: *Clifford Algebras and their Applications in Mathematical Physics*, ed. by A. Micali et al., Kluwer Academic Publishers (1992) pp. 363–370

43. J.S.R. Chisholm, R.S. Farwell: A fermion–boson mass relation and the top mass, J. Phys. G Nucl. Part. Phys. **18**, L117–L122 (1992)

44. C. Doran, A. Lasenby: *Geometric Algebra for Physicists*, Cambridge University Press, Cambridge (2003)

45. I.M. Benn, R.W. Tucker: *An Introduction to Spinors and Geometry with Applications in Physics*, Adam Hilger, Bristol, Philadelphia (1987)

Index